Fractals and Disordered Systems

Springer

Berlin
Heidelberg
New York
Barcelona
Budapest
Hong Kong
London
Milan
Paris
Santa Clara
Singapore
Tokyo

Armin Bunde   Shlomo Havlin (Eds.)

Fractals and Disordered Systems

Second Revised and Enlarged Edition
With 165 Figures and 10 Color Plates

Springer

Professor Dr. Armin Bunde

Institut für Theoretische Physik
Universität Giessen
Heinrich-Buff-Ring 16
D-35392 Giessen
Germany

Professor Dr. Shlomo Havlin

Department of Physics
Bar-Ilan University
Ramat Gan
Israel

ISBN 3-540-56219-2 2nd ed. Springer-Verlag Berlin Heidelberg New York

ISBN 3-540-54070-9 1st ed. Springer-Verlag Berlin Heidelberg New York

Library of Congress Cataloging-in-Publication Data. Fractals and disordered systems /
Armin Bunde, Shlomo Havlin (eds.). - 2nd ed. p. cm. Includes bibliographical references
and index. ISBN 3-540-56219-2 (alk. paper). 1. Fractals. 2. Chaotic behavior in systems.
3. Order-disorder models. I. Bunde, Armin, 1947- . II. Havlin, Shlomo.
QA614.86.F72 1996 530.4'13-dc20 95-41926

Camera-ready copy from the authors/editors using a Springer TeX macro package
Printing: Appl, Wemding
Binding: aprinta, Wemding
SPIN: 10074548   56/3144-543210 - Printed on acid-free paper

Preface to the Second Edition

The growing interest in fractals and disordered systems made a second edition of this book necessary. We took advantage of this situation to make, together with the other authors, major revisions in order to include new trends and achievements in this fast-developing field. Many sections were revised completely and several sections and subsections were added. For example, H. Eugene Stanley added, in Chap. 1, new sections on DLA (glove and skeleton), multiscaling and self-affine surfaces. In Chaps. 2 and 3, we revised completely the sections dealing with the important chemical distance concept and added sections on branched polymers and elasticity. In Chap. 5, Hans Herrmann added a section on hydraulic fracture, which is of high technological importance. Bernhard Sapoval revised Chap. 6 significantly by considering transfer across irregular surfaces in a more general approach. In Chap. 9, Dietrich Stauffer added a section on recent biologically motivated developments of cellular automata. Benoit Mandelbrot added ten pages to Chap. 10, where he presented new ideas and results on multifractal and Minkowski measures.

Also, all the tables and references have been updated. We wish to thank Dr. Hans Kölsch and Petra Treiber from Springer-Verlag for their very helpful and constructive cooperation.

Giessen   *Armin Bunde*
Ramat-Gan   *Shlomo Havlin*
September 1995

Preface to the First Edition

Disordered structures and random processes that are self-similar on certain length and time scales are very common in nature. They can be found on the largest and the smallest scales: in galaxies and landscapes, in earthquakes and fractures, in aggregates and colloids, in rough surfaces and interfaces, in glasses and polymers, in proteins and other large molecules. Owing to the wide occurrence of self-similarity in nature, the scientific community interested in this phenomenon is very broad, ranging from astronomers and geoscientists to material scientists and life scientists.

Among the major achievements in recent years that have strongly influenced our understanding of structural disorder and its formation by random processes are the fractal concepts pioneered by B. B. Mandelbrot. Fractal geometry is a mathematical language used to describe complex shapes and is particularly suitable for computers because of its iterative nature.

The field of fractals and disordered systems is developing very rapidly, and there exists a large gap between the knowledge presented in advanced textbooks and the research front. The aim of this book is to fill this gap by introducing the reader to the basic concepts and modern techniques in disordered systems, and to lead him to the forefront of current research.

The book consists of ten chapters, written with a uniform notation, with cross references in each chapter to related subjects in other chapters. In each chapter emphasis has been placed on connections between theory and experiment. A special chapter (Chap. 8) discusses experimental studies of fractal systems.

In the first chapter, H. E. Stanley introduces relevant fractal concepts, from self-similarity to multifractality. Several fractal prototypes such as random walks and diffusion limited aggregation (DLA) are discussed in detail in order to demonstrate the various concepts and their utility. The possibility that DLA is a prototype fractal describing many different phenomena in biology, chemistry and physics, is stressed.

The next two chapters, written by A. Bunde and S. Havlin, present a detailed description of the static (Chap. 2) and dynamical (Chap. 3) properties of percolation systems. Percolation is a common model for disordered systems, and its relation to the fractal concept is emphasized. Many theoretical and numerical approaches such as scaling theories, series expansions, renormalization, and exact enumeration of random walks are discussed.

In Chap. 4, A. Aharony applies the ideas of fractals and multifractals to several models of random growth processes; examples are the Eden model, invasion percolation and general aggregation processes. The relation between DLA and electrodeposition processes, viscous fingering, dielectric breakdown, and the Laplacian growth processes is discussed.

Chapter 5 by H. J. Herrmann discusses microscopic models for the formation and the complex structure of fractures, which represent a subject of great technological importance. The microscopic modeling of fractures is quite a new topic. Previous work was mainly devoted to phenomenological properties of fractures.

In Chap. 6, B. Sapoval studies the physical and chemical properties of irregular surfaces such as electrodes, membranes and catalysts, interest in which has been revived by the concept of fractal geometry. Many natural or industrial processes take place through surfaces or across the interfaces between two media. For example tree roots exchange water and inorganic salts with the earth through the surface of the roots; the larger the interface area, the more effective the process. Porous fractals are objects that have this property.

This is followed in Chap. 7 by several models for rough surfaces and interfaces presented by J. F. Gouyet, M. Rosso and B. Sapoval. The authors begin with a general characterization of rough surfaces, followed by deposition models and diffusion fronts and ending with fluid–fluid interfaces, membranes, and tethered surfaces.

In Chap. 8, J. K. Kjems presents a broad review on the experimental techniques used to study fractal features in disordered systems. The chapter deals with the questions of how fractal realizations can be made, how to measure the fractal dimension, and how to measure physical properties of fractals.

In Chap. 9, D. Stauffer provides systematic approaches as well as computational techniques for the study of cellular automata where the fractal concept is relevant, including the Wolfram characterization and, among others, the Kaufman model for genetics.

In the last chapter, B. B. Mandelbrot and C. J. G. Evertsz focus on exactly self-similar left-sided multifractals, which seem to be relevant to fully developed turbulence and DLA. The authors present explicit examples of multiplicative multifractals having the property that the multifractal scaling relation fails to hold, either for small enough negative moments, or for high enough positive moments.

We wish to thank the authors for their cooperation and many of our colleagues, in particular M. A. Denecke, R. Nossal, H. E. Roman, and H. Taitelbaum for many useful discussions. We kindly acknowledge the help of J. D. Chen, F. Grey, A. Hansen, C. Kolb, R. A. Masland, P. Meakin, C. Roessler, S. Roux, S. Schwarzer, B. L. Trus, D. A. Weitz, and S. Wolfram, who produced the beautiful color figures appearing in the book. We hope that the book can be used as a first textbook for graduate students, for teachers at universities for preparing regular courses or seminars, and for researchers who encounter fractals and disordered systems.

Hamburg   *Armin Bunde*
Washington   *Shlomo Havlin*
June 1991

Contents

1 Fractals and Multifractals: The Interplay of Physics and Geometry

By H. Eugene Stanley (With 30 Figures)

2   Percolation I

By Armin Bunde and Shlomo Havlin (With 24 Figures)

3   Percolation II

By Shlomo Havlin and Armin Bunde (With 20 Figures)

4   Fractal Growth

By Amnon Aharony (With 4 Figures)

5   Fractures

By Hans J. Herrmann (With 18 Figures)

6   Transport Across Irregular Interfaces: Fractal Electrodes, Membranes and Catalysts

By Bernard Sapoval (With 8 Figures)

7   Fractal Surfaces and Interfaces

By Jean-François Gouyet, Michel Rosso and Bernard Sapoval
(With 27 Figures)

8   Fractals and Experiments

By Jørgen K. Kjems (With 18 Figures)

9 Cellular Automata

By Dietrich Stauffer (With 6 Figures)

10 Exactly Self-similar Left-sided Multifractals

By Benoit B. Mandelbrot and Carl J.G. Evertsz
with new Appendices B and C
by Rudolf H. Riedi and Benoit B. Mandelbrot (With 10 Figures)

List of Contributors

Amnon Aharony

School of Physics and Astronomy,
Raymond and Beverly Sackler Faculty of Exact Sciences,
Tel Aviv University, Tel Aviv 69978, Israel
and
Institute of Physics, University of Oslo, Norway

Armin Bunde

Institut für Theoretische Physik, Justus-Liebig-Universität,
Heinrich-Buff-Ring 16, D-35392 Giessen, Germany

Carl J.G. Evertsz

Fachbereich Mathematik, Universität Bremen,
D-28334 Bremen, Germany

Jean-François Gouyet

Laboratoire de Physique de la Matiere Condensée, Ecole Polytechnique,
F-91128 Palaiseau, France

Shlomo Havlin

Department of Physics, Bar-Ilan University,
Ramat Gan, Israel,
and
Center of Polymer Studies and Department of Physics, Boston University,
Boston, MA 02215, USA

Hans J. Herrmann

L.H.M.P, E.S.P.C.I.,
F-75231 Paris, France
and
Institut für Computeranwendungen I, Universität Stuttgart,
D-70569 Stuttgart, Germany

Jørgen K. Kjems

Risø National Laboratory,
DK-4000 Roskilde, Denmark

Benoit B. Mandelbrot

Physics Department, IBM T.J. Watson Research Center,
Yorktown Heights, NY 10598, USA
and
Mathematics Department, Yale University, Box 2155 Yale Station,
New Haven, CT 06520, USA

Rudolf H. Riedi

Mathematics Department, Yale University, Box 2155 Yale Station,
New Haven, CT 06520, USA

Michel Rosso

Laboratoire de Physique de la Matière Condensée, Ecole Polytechnique,
F-91128 Palaiseau, France

Bernard Sapoval

Laboratoire de Physique de la Matière Condensée, Ecole Polytechnique,
F-91128 Palaiseau, France

H. Eugene Stanley

Center of Polymer Studies and Department of Physics, Boston University,
Boston, MA 02215, USA

Dietrich Stauffer

Institut für Theoretische Physik, Universität Köln,
D-50923 Köln, Germany

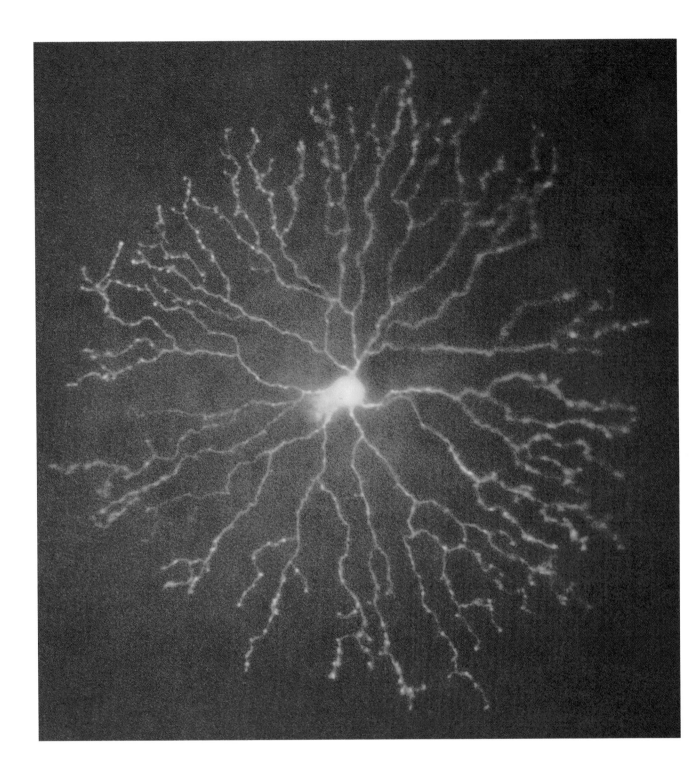

1 Fractals and Multifractals: The Interplay of Physics and Geometry

H. Eugene Stanley

1.1 Introduction

In recent years, a wide range of complex structures of interest to scientists, engineers, and physicans have been quantitatively characterized using the idea of a *fractal* dimension: a dimension that corresponds in a unique fashion to the geometrical shape under study, and often is not an integer [1.1–15]. The key to this progress is the recognition that many random structures obey a symmetry as striking as that obeyed by regular structures. This "scale symmetry" has the implication that objects look the same on many different scales of observation.

Nonspecialists are also familiar with fractal objects. Everyone has seen fractal objects – probably at an early stage in life. Perhaps we once photographed scenery from the back of a train and noticed that the photograph looked the same at all stages of enlargement. Perhaps we noticed that entire cities (especially the Metro of Paris) have tenuous structures that look similar at different size scales [1.16–17]. Perhaps we saw that snow crystals all have the same pattern, each part of a branch being similar to itself. In fact, to "see" something at all – fractal or nonfractal – requires that the nerve cells in the eye's retina send a signal, and these retinal nerve cells are themselves fractal objects (Fig. 1.0).

So long as we associate a unit mass with each pixel, a single scaling exponent d_f completely characterizes the structure of the object. In recent years, however, very interesting phenomena have been studied which appear to require not one but an infinite number of exponents for their description. Such phenomena, termed *multifractal phenomena*, have recently become an extremely active area of investigation for many researchers from various fields of endeavor ranging from physics and chemistry on the one hand to fluid dynamics and meteorology

---

◄ **Fig. 1.0.** Neuron from the retina – a quasi-two-dimensional fractal object. Courtesy of R.H. Masland

on the other. The purpose of this opening chapter is to provide the nonspecialist with a brief introduction to fractal and multifractal phenomena.

Although there are many different types of fractal and multifractal phenomena, we shall concentrate on a few examples which we hope will prove useful to the reader.

1.2 Nonrandom Fractals

Fractals fall naturally into two categories, *nonrandom* and *random*. Fractals in physics belong to the second category, but it is instructive to first discuss a much-studied example of a nonrandom fractal – the Sierpiński gasket. We simply iterate a *growth rule*, much as a child might assemble a castle from building blocks. Our basic unit is a triangular-shaped tile shown in Fig. 1.1a, which we take to be of unit *mass* ($M = 1$) and of unit edge *length* ($L = 1$).

The Sierpiński gasket is defined operationally as an "aggregation process" obtained by a simple iterative process. In stage one, we join three tiles together to create the structure shown in Fig. 1.1b, an object of mass $M = 3$ and edge $L = 2$. The effect of stage one is to produce a unit with a lower density. If we define the density to be

$$\varrho(L) \equiv M(L)/L^2, \tag{1.1}$$

then the density decreases from unity to 3/4 as a result of stage one.

Now simply iterate – i.e., repeat this rule over and over *ad infinitum*. Thus in stage two, join together – as in Fig. 1.1c – three of the $\varrho = 3/4$ structures

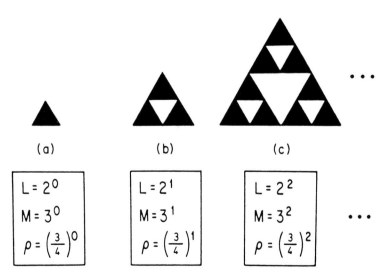

Fig. 1.1a-c. First few stages in the aggregation rule which is iterated to form a Sierpiński gasket fractal. After [1.10]

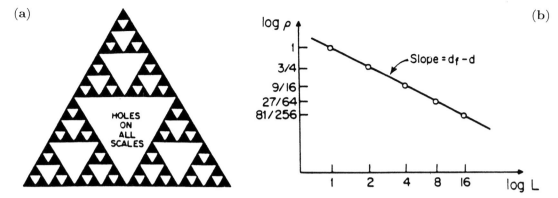

Fig. 1.2. (a) Sierpiński gasket fractal after four stages of iteration. (b) A log–log plot of ϱ, the fraction of space covered by black tiles, as a function of L, the linear size of the object. After [1.18]

constructed in stage one, thereby building an object with $\varrho = (3/4)^2$. In stage three, join three objects identical to those constructed in stage two. Continue until you run out of tiles (if you are a physicist) or until the structure is infinite (if you are a mathematician!). The result after stage four – with 81 black tiles and $27 + 36 + 48 + 64$ white tiles (Fig. 1.2a) may be seen in floor mosaics of the church in Anagni, Italy, which was built in the year 1104. Thus although this fractal is named after the 20th century Polish mathematician W. Sierpiński, it was universally known some eight centuries earlier to every churchgoer of this village!

The citizens of Anagni did not have double-logarithmic graph paper in the 12th century. If they had possessed such a marvelous invention, then they might have plotted the dependence of ϱ on L. They would have found Fig. 1.2b, which displays two striking features:

- $\varrho(L)$ decreases monotonically with L, without limit, so that by iterating sufficiently we can achieve an object of *as low a density as we wish*, and
- $\varrho(L)$ decreases with L in a *predictable* fashion – a simple power law.

Power laws have the generic form $y = \mathcal{A} x^\alpha$ and, as such, have two parameters, the "amplitude" $\mathcal{A}$ and the exponent α. The amplitude is not of intrinsic interest, since it depends on the choice we make for the definitions of M and L. The exponent, on the other hand, depends on the process itself – i.e., on the "rule" that we follow when we iterate. Different rules give different exponents. In the present example, $\varrho(L) = L^\alpha$ so the amplitude is unity. The exponent is given by the slope of Fig. 1.2b,

$$\alpha = \text{slope} = \frac{\log 1 - \log(3/4)}{\log 1 - \log 2} = \frac{\log 3}{\log 2} - 2. \qquad (1.2)$$

Finally we are ready to define the fractal dimension d_f, through the equation

$$M(L) \equiv \mathcal{A}\, L^{d_f}. \qquad (1.3)$$

If we substitute (1.3) into (1.1), we find

$$\varrho(L) = \mathcal{A}\, L^{d_f - 2}. \tag{1.4}$$

Comparing (1.2) and (1.4), we conclude that the Sierpiński gasket is indeed a fractal object with fractal dimension

$$d_f = \log 3 / \log 2 = 1.58\ldots\quad. \tag{1.5}$$

Classical (Euclidean) geometry deals with regular forms having a dimension the same as that of the embedding space. For example, a line has $d = 1$, and a square $d = 2$. We say that the Sierpiński gasket has a dimension intermediate between that of a line and an area – a kind of "fractional" dimension – and hence the term *fractal*.

1.3 Random Fractals: The Unbiased Random Walk

Real systems in nature do not resemble the floor of the Anagni church – in fact, *nonrandom* fractals are not found in nature. Nature exhibits numerous examples of objects which by themselves are not fractals but which have the remarkable feature that, if we form a *statistical average* of some property such as the density, we find a quantity that decreases linearly with the length scale when plotted on double-logarithmic paper. Such objects are termed *random fractals*, to distinguish them from the nonrandom *geometric fractals* discussed in the previous section.

Consider the following prototypical problem in statistical mechanics. At time $t = 0$ an ant[1] is parachuted to an arbitrary vertex of an infinite one-dimensional lattice with lattice constant unity: we say $x_{t=0} = 0$. The ant carries an *unbiased* two-sided coin, and a clock. The dynamics of the ant is governed by the following rule. At each "tick" of the clock, it tosses the coin. If the coin is heads, the ant steps to the neighboring vertex on the east [$x_{t=1} = +1$]. If the coin is tails, it steps to the nearest vertex on the west [$x_{t=1} = -1$].

There are *laws of nature* that govern the position of this drunken ant. For example, as time progresses, the average of the *square* of the displacement of the ant increases monotonically. The explicit form of this increase is contained in the following "law" concerning the *mean square displacement*:

$$\langle x^2 \rangle_t = t. \tag{1.6}$$

---

[1] The use of the term "ant" to describe a random walker is used almost universally in the theoretical physics literature – perhaps the earliest reference to this colorful animal is a 1976 paper of de Gennes that succeeded in formulating several general physics problems in terms of the motion of a "drunken" ant with appropriate rules for motion.

Equation (1.6) may be proved by induction, by demonstrating that (1.6) implies $\langle x^2 \rangle_{t+1} = t + 1$.

Additional information is contained in the expectation values of higher powers of x, such as $\langle x^3 \rangle_t$, $\langle x^4 \rangle_t$, and so forth. We can immediately see that $\langle x^k \rangle_t = 0$ for all *odd* integers k, while $\langle x^k \rangle_t$ is nonzero for *even* integers. Consider, e.g., $\langle x^4 \rangle_t$. We may easily verify that

$$\langle x^4 \rangle_t = 3t^2 - 2t = 3t^2 \left(1 - \frac{2/3}{t} \right). \tag{1.7}$$

1.4 The Concept of a Characteristic Length

Let us compare (1.6) and (1.7). What is *the* displacement of the randomly walking ant? On the one hand, we might consider identifying this displacement with a length $\mathcal{L}_2$ defined by

$$\mathcal{L}_2 \equiv \sqrt{\langle x^2 \rangle} = \sqrt{t}. \tag{1.8}$$

On the other hand, it is just as reasonable to identify this displacement with the length $\mathcal{L}_4$ defined by

$$\mathcal{L}_4 \equiv \sqrt[4]{\langle x^4 \rangle} = \sqrt[4]{3} \sqrt{t} \left(1 - \frac{2/3}{t} \right)^{1/4}. \tag{1.9}$$

The remarkable and significant point is that both lengths display the same *asymptotic* dependence on the time. We call the leading exponent the *scaling exponent*, while the nonleading exponents are termed *corrections-to-scaling*.

The reader may verify that the same scaling exponent is found if we consider any length $\mathcal{L}_k$ (provided k is even),

$$\mathcal{L}_k \equiv \sqrt[k]{\langle x^k \rangle} = \mathcal{A}_k \sqrt{t} \, [1 + \mathcal{B}_k t^{-1} + \mathcal{C}_k t^{-2} + \cdots + \mathcal{O}(t^{-k/2+1})]^{1/k}. \tag{1.10}$$

The subscripts on the amplitudes indicate that these depend on k. Equation (1.10) exemplifies a robust feature of random systems: *regardless of the definition of the characteristic length, the same scaling exponent describes the asymptotic behavior.* We say that all lengths scale as the square root of the time, meaning that whatever length $\mathcal{L}_k$ we choose to examine, $\mathcal{L}_k$ will double whenever the time has increased by a factor of four. This scaling property is not affected by the fact that the amplitude $\mathcal{A}_k$ in (1.10) depends on k, since we do not inquire about the absolute value of the length $\mathcal{L}_k$ but only how $\mathcal{L}_k$ *changes* when t changes.

1.5 Functional Equations and Fractal Dimension

We have seen that several different definitions of a characteristic length ξ all scale as $\sqrt{t}$. Equivalently, if $t(\xi)$ is the characteristic time for the ant to "trace out" a domain of linear dimension ξ, then

$$t \sim \xi^2. \tag{1.11}$$

More formally, for all positive values of the parameter λ such that the product $\lambda\xi$ is large, $t(\xi)$ obeys the asymptotic equation

$$t(\lambda^{1/2}\xi) \sim \lambda t(\xi). \tag{1.12}$$

Equation (1.12) is called a *functional equation* since it provides a constraint on the form of the function $t(\xi)$. In fact, (1.11) is the *solution* of the functional equation (1.12) in the sense that any function $t(\xi)$ satisfying (1.12) also satisfies (1.11) – we say that power laws are the solution to the functional equation (1.12). To see this, we note that if (1.12) holds for all values of the parameter λ, then it holds in particular when $\lambda = 1/\xi$. With this substitution, (1.12) reduces to (1.11).

It is also straightforward to verify that any function $t(\xi)$ obeying (1.11) *also* obeys (1.12). Thus (1.12) implies (1.11) *and conversely*! This connection between power law behavior and a symmetry operation, called *scaling symmetry*, is at the root of the wide range of applicability of fractal concepts in physics.

Writing (1.11) in the form

$$t \sim \xi^{d_f} \tag{1.13a}$$

exhibits the fact that the scaling exponent d_f explicitly reflects the asymptotic dependence of a characteristic "mass" (the number of points in the trail of the ant) on a characteristic "length" ξ. For the random walk $d_f = 2$, and in general d_f is the *fractal dimension* of the random walk.

If we express (1.12) in the equivalent form

$$t(\lambda\xi) \sim \lambda^{d_f} t(\xi), \tag{1.13b}$$

then we see that d_f plays the role of a scaling exponent governing the *rate* at which we must scale the time if we want to trace out a walk of greater spatial extent. For example, if we want a walk whose trail has twice the size, we must wait a time 2^{d_f}.

1.6 An Archetype: Diffusion Limited Aggregation

It is a fact that fractals abound. In fact, almost any object for which randomness is the basic factor determining the structure will turn out to be fractal over some range of length scales – for much the same reason that the unbiased random walk is fractal: there is simply nothing in the microscopic rules that can set a length scale so the resulting macroscopic form is "scale-free"; scale-free objects obey power laws and lead to functional equations of the form of (1.12).

Today, there are roughly of the order of 10^3 recognized fractal systems in nature, though a decade ago when Mandelbrot's classic *Fractal Geometry of Nature* [1.1] was written, many of these systems were not known to be fractal. Diffusion limited aggregation (DLA) alone has about 50 realizations in physical systems and much current interest on fractals in nature focuses on DLA [1.19,20]. DLA structures arise naturally when studying many phenomena of current interest to physicists and chemists, ranging from electrochemical deposition [1.21] and dendritic solidification [1.22-25] to various "breakdown phenomena" such as dielectric breakdown [1.26], viscous fingering [1.27], chemical dissolution [1.28], and the rapid crystallization of lava [1.29].

Like many models in statistical mechanics, the rule defining DLA is simple. At time 1, we place in the center of a computer screen a white pixel, and release a random walker from a large circle surrounding the white pixel. The four perimeter sites have an equal *a priori* probability p_i of being stepped on by the random walker (Fig. 1.3a), so we write

$$p_i = \frac{1}{4} \qquad (i = 1, \dots, 4). \qquad (1.14a)$$

The rule is that the random walker sticks irreversibly – thereby forming a cluster of mass $M = 2$.

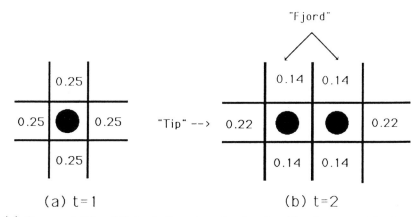

Fig. 1.3. (a) Square lattice DLA at time $t = 1$, showing the four growth sites, each with growth probability $p_i = 1/4$. (b) DLA at time $t = 2$, with six growth sites, and their corresponding growth probabilities p_i. After [1.20]

There are $N_p = 6$ possible sites, henceforth called *growth sites* (Fig. 1.3b), but now the probabilities are *not* all identical: each of the growth sites of the two tips has a growth probability $p_{\max} \cong 0.22$, while each of the four growth sites on the sides has a growth probability $p_{\min} \cong 0.14$. Since a site on the tip is 50% more likely to grow than a site on the sides, the next site is more likely to be added to the tip – it is like capitalism in that the rich get richer. One of the main features of our approach to DLA is that instead of focusing on the tips who are "getting richer", we focus on the fjords who are "getting poorer".

Just because the third particle is *more likely* to stick at the tip does not mean that the next particle *will* stick on the tip. Indeed, the most that we can say about the cluster is to specify the *growth site probability distribution* – i.e., the set of numbers,

$$\{p_i\}, \qquad i = 1, \ldots, N_p, \qquad (1.14b)$$

where p_i is the probability that perimeter site ("growth site") i is the next to grow, and N_p is the total number of perimeter sites. The recognition that the set of $\{p_i\}$ gives us essentially the *maximum* amount of information we can have about the system is connected to the fact that tremendous attention has been paid to these p_i – and to the analogs of the p_i in various closely related systems.

Fig. 1.4. Large DLA cluster on a triangular lattice, with noise reduction and lattice anisotropy. Cluster sites are color-coded according to the time of arrival. After [1.23]

If the DLA growth rule is simply iterated, then we obtain a large cluster characterized by a range of growth probabilities that spans several orders of magnitude – from the tips to the fjords. Figure 4.0 shows such a large cluster, where each pixel is colored according to the time it was added to the aggregate. ¿From the fact that the "last to arrive" particles (green pixels) are never found to be adjacent to the "first to arrive" particles (white pixels), we conclude that the p_i for the growth sites on the tips must be vastly larger than the p_i for the growth sites in the fjords.

There are reasons for the ubiquity of structures relevant to the DLA algorithm. Firstly, aggregation phenomena based on random walkers correspond to a Laplace equation ($\nabla^2 \Pi(r, t) = 0$) for the probability $\Pi(r, t)$ that a walker is at position r and time t. Secondly there is a range of phenomena that at first sight seem to have *nothing to do with random walkers*. These include viscous fingering phenomena, for which the pressure P at every point satisfies a Laplace equation (Table 1.1). Similarly, dielectric breakdown phenomena, chemical dissolution, electrodeposition, and a host of other displacement phenomena may be members of a suitably defined *DLA universality class*. If anisotropy is added, then the DLA universality class grows to include even dendritic crystal growth [1.24] and snowflake growth [1.25] (Figs. 1.4 and 1.5).

Recently, several phenomena of *biological* interest have attracted the attention of DLA aficionados. These include the growth of bacterial colonies [1.30], the retinal vasculature [1.31], and neuronal outgrowth [1.32,33] (Figs. 1.0 and

Table 1.1. A "Rosetta stone" connecting the physics underlying (a) an electrical problem (dielectric breakdown), (b) a fluid mechanics problem (viscous fingering), and (c) a diffusion problem (dendritic solidification). After [1.20]

(a) Electrical	(b) Fluid mechanics	(c) Dendritic solidification
Electrostatic potential:	**Pressure:**	**Concentration:**
$\phi(r, t)$	$P(r, t)$	$c(r, t)$
Electric field:	**Velocity:**	**Growth rate:**
$E \propto -\nabla\phi(r, t)$	$v \propto -\nabla P(r, t)$	$v \propto -\nabla c(r, t)$
Conservation:		
$\nabla \cdot E = 0$	$\nabla \cdot v = 0$	$\nabla \cdot v = 0$
Laplace equation:		
$\nabla^2 \phi = 0$	$\nabla^2 P = 0$	$\nabla^2 c = 0$

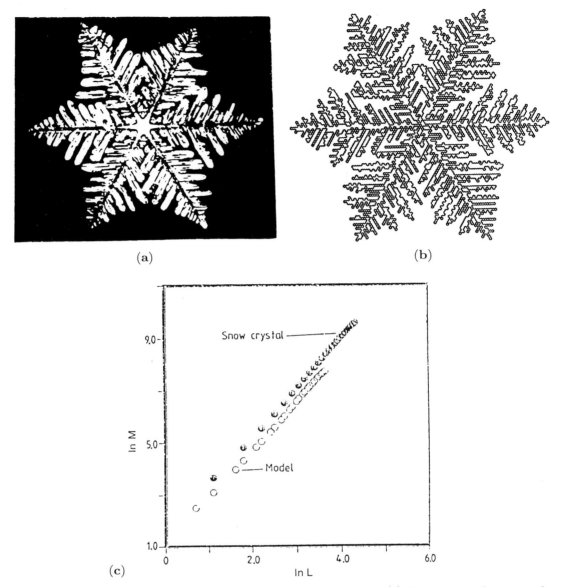

(a)   (b)

(c)

Fig. 1.5. (a) A typical snow crystal. (b) A DLA simulation. (c) Comparison between the fractal dimensions of (a) and (b) obtained by plotting the number of pixels inside an $L \times L$ sandbox logarithmically against L. The same slope, $d_f = 1.85 \pm 0.06$, is found for both. The experimental data extend to larger values of L, since the digitizer used to analyze the experimental photograph has 20 000 pixels while the cluster has only 4000 sites. After [1.25]

1.6). The last example is particularly intriguing: if evolution indeed chose DLA as the morphology for the nerve cell, then can we understand "why" this choice was made? What evolutionary advantage does a DLA morphology convey? Is it significant that the Paris Metro evolved with the same morphology – even a fractal dimension – or is this fact just a numerical coincidence? Can we use the answer to these questions to better design the next generation of computers? These are important issues that we cannot hope to quickly resolve, but already

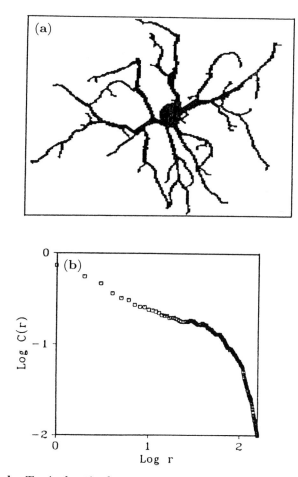

Fig. 1.6a,b. Typical retinal neuron and its fractal analysis. After [1.32]

we appreciate that a fractal object is the most efficient way to obtain a great deal of intercell "connectivity" with a minimum of "cell volume," so the next question is "which" fractal did evolution select, and why?

1.7 DLA: Fractal Properties

It is awe-inspiring that remarkably complex objects in nature can be quantitatively characterized by a single number, d_f. It is equally awe-inspiring that such complex objects can be described by various models with extremely simple rules. It is also an intriguing fact that even though no two natural fractal objects that we are likely to ever see are identical, nonetheless every DLA has a generic 'form' that even a child can recognize. The analogous statement holds for many random structures in nature. For example no two snowflakes are the same yet every snowflake has a generic form that every child recognizes.

Perhaps most awesome to a student of theoretical physics is the fact that geometrical models – with no Boltzmann factors – suffice to capture features of real statistical mechanical systems. What does this mean? If we understand the essential physics of an extremely robust model, such as the Ising model, then we say that we understand the essential physics of the complex materials that fall into the universality class described by the Ising model. In fact, by understanding the pure Ising model, we can even understand most of the features of *variants* of the Ising model (such as the n-vector model) that may be appropriate for describing even more complex materials. Similarly, we feel that if we can understand DLA, then we are well on our way to understanding *variants* of DLA. And just as the Ising model is a paradigm for all systems composed of interacting subunits, so also DLA may be a paradigm for all kinetic growth models.

One useful definition of the fractal dimension d_f is by the "window box scaling" operation:

(1) First place an imaginary window box of edge L around an arbitrarily-chosen occupied DLA site ("local origin").
(2) Then count the number of occupied pixels $M(L)$ within that window box.
(3) Next choose many different local origins to obtain good statistics.
(4) Finally, make a log–log plot of $M(L)$ vs L, and interpret the fractal dimension d_f as the "asymptotic" ($L \to \infty$) slope of this plot.

Conventionally, we write

$$M(L) \sim L^{d_f}, \tag{1.15}$$

where the tilde denotes "asymptotically equal to".

The difficulty of extrapolating from finite L to infinite L has motivated ever more clever algorithms for generating ever larger DLA clusters. Most of the world records are held by Meakin and his collaborators [1.34]:

$$M_{\mathrm{max}} = \begin{cases} 12 \times 10^6 & \text{[square lattice DLA]}, \\ 10^6 & \text{[off-lattice DLA]}. \end{cases} \tag{1.16}$$

The corresponding estimates for d_f are [1.34]

$$d_f = \begin{cases} 1.55 & \text{[square lattice DLA]}, \\ 1.715 \pm 0.004 & \text{[off-lattice DLA]}. \end{cases} \tag{1.17}$$

The result for square lattice DLA is based on theoretical arguments [1.34], for the simulations themselves are not conclusive in that the estimate of d_f simply decreases slowly with increasing cluster mass. Equation (1.17) suggests that "anisotropic DLA" (DLA grown on a lattice) is in a different universality class.

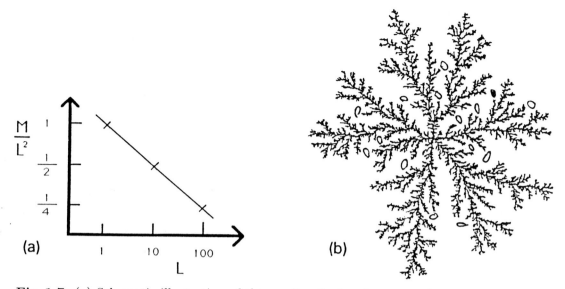

Fig. 1.7. (a) Schematic illustration of the results of a hands-on experiment to actually see that DLA is indeed a fractal since the density decreases linearly with the size L of the observation window (or inverse wave vector q^{-1}), and that the fractal dimension is given by roughly $d_f - d \simeq \log_{10} \frac{1}{2} \simeq -0.301$. (b) Off-lattice DLA cluster of 10^5 sites indicating some of the channels that serve to delineate voids [courtesy of P. Meakin]. After [1.20]

We can actually "see with our eyes" that $d_f \simeq 1.7$ by means of a simple *hands-on* demonstration. We begin with a large DLA cluster (Fig. 1.7). Suppose we take a sequence of boxes with $L = 1, 10, 100$ (in units of the pixel size), and estimate the fraction of the box that is occupied by the DLA. This fraction is called the density,

$$\varrho(L) \equiv M(L)/L^d, \tag{1.18}$$

where $d = 2$ here. Combining (1.15) and (1.18), we find

$$\varrho(L) \sim L^{d_f - d}. \tag{1.19}$$

Now (1.19) is equivalent to the functional equation

$$\varrho(\lambda L) = \lambda^{d_f - d} \varrho(L). \tag{1.20}$$

Carrying out this operation on Fig. 1.7 with $\lambda = 10$ will reveal

$$\varrho(L) \cong \begin{cases} 1 & L = 1, \\ 1/2 & L = 10, \\ 1/4 & L = 100. \end{cases} \tag{1.21}$$

Here the result of (1.21),

$$\varrho(10L) \cong \frac{1}{2}\varrho(L),$$

convinces one that $10^{d_f - 2} \cong 1/2$. So

$$d_f - 2 \cong \log_{10} \frac{1}{2} = -0.301, \qquad (1.22a)$$

leading to

$$d_f \cong 1.70. \qquad (1.22b)$$

1.8 DLA: Multifractal Properties

1.8.1 General Considerations

Until relatively recently, most of the theoretical attention paid to DLA has focused on its fractal dimension. We now have estimates of d_f that are accurate to roughly 1%. However, we lack any way to *interpret* this estimate! This is in contrast to other useful models in statistical physics. For example, for both the $d = 2$ Ising model [1.35] and $d = 2$ percolation [1.36], we can calculate the various exponents *and* interpret them in terms of "scaling powers."

For DLA, what we *can* interpret is the distribution function $\mathcal{D}(p_i)$, which describes the histogram of the number of perimeter sites with growth probability p_i. The key idea is to focus on how this distribution function $\mathcal{D}(p_i)$ *changes* as the cluster mass M increases. The reason why this approach is fruitful is that the $\{p_i\}$ contain the maximum information we can possibly extract about the dynamics of the growth of DLA. Indeed, specifying the $\{p_i\}$ is analogous to specifying the four "growth" probabilities $p_i = 1/4 \; [i = 1 \cdots 4]$ for a random walker on a square lattice.

The set of numbers $\{p_i\}$ may be used to construct a histogram $\mathcal{D}(\ln p_i)$. This distribution function can be described by its *moments*,

$$Z_\beta \equiv \sum_{\ln p} \mathcal{D}(\ln p) \mathrm{e}^{-\beta(-\ln p)}, \qquad (1.23)$$

which is a more complex way of writing

$$Z_\beta = \sum_i p_i^\beta. \qquad (1.24)$$

It is also customary to define a dimensionless "free energy" $F(\beta)$ by the relation

$$F(\beta) = -\frac{\log Z_\beta}{\log L}, \qquad (1.25)$$

which can be written in the suggestive form

$$Z_\beta = L^{-F(\beta)}. \qquad (1.26)$$

Table 1.2. Rosetta stone connecting our notation and other notations in use. The function $f(\alpha)$ is defined in the Appendix. Adapted from [1.37]

$$\beta \longleftrightarrow q \qquad\qquad F(\beta) \longleftrightarrow \tau(q)$$
$$E \longleftrightarrow \alpha \qquad\qquad S(E) \longleftrightarrow f(\alpha)$$

The form (1.23) and the notation used suggest that we think of β as an *inverse temperature*, $-\ln p/\ln L$ as an *energy*, and Z_β as a *partition function*. The notation we have used is suggestive of thermodynamics (cf. Appendix 1.A). Indeed, the function $F(\beta)$ has many of the properties of a free energy function – for example, it is a convex function of its argument and can even display a singularity or "phase transition". However, for most critical phenomena problems, exponents describing moments of distribution functions are linear in their arguments, while for DLA $F(\beta)$ is not linear – we call such behavior *multifractal*. Multifractal behavior is characteristic of random multiplicative processes, such as arise when we multiply together a string of random numbers.

In the literature there exist other symbols, and a brief dictionary is presented in Table 1.2.

Multifractal approaches in statistical physics have a rich history [1.38–46], and were first introduced for describing DLA in 1985 by Meakin and collaborators [1.44–46]. The key idea is to focus on the set of growth probabilities $\{p_i\}$ and how their distribution function $\mathcal{D}(p_i)$ changes as the cluster mass M increases.

1.8.2 "Phase Transition" in 2d DLA

What can we do with this *thermodynamic formalism*? One approach that we have found to be particularly revealing is the analog for DLA of the successive approximation ("series expansion") approach pioneered by Cyril Domb and his collaborators. In fact, Lee *et al.* [1.37] have recently extended the renormalization ideas of Nagatani [1.47] to actually obtain exact results for a $L \times L$ cell for a sequence of values of L up to and including $L = 5$. This work is described elsewhere [1.37], so we focus on one key result – the apparent singularity in the quantity

$$C(\beta) \equiv -\frac{\partial^2 F(\beta)}{\partial \beta^2}. \tag{1.27}$$

Figure 1.8, which shows $C(\beta)$ for a sequence of L values, is reminiscent of the famous *finite-size-scaling* plot of $C_H(\beta)$ for the $L \times L$ Ising model [1.48]. We interpret the maximum in $C(\beta)$ as heralding the existence of a singularity in $C(\beta)$ – a "phase transition"– at some critical value β_c (Fig. 1.9). Mandelbrot terms such a phase transition *left-sided multifractality* since the function $f(\alpha)$ of the Appendix is not defined for its "right side" [1.49].

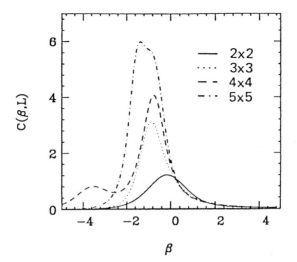

Fig. 1.8. Analog of Fisher/Ferdinand plot for DLA. Shown is the dependence on β of $\partial^2 F/\partial\beta^2$, where $F \equiv -\log Z/\log L$ and $Z_\beta \equiv \sum_i (p_i)^\beta$. After [1.37]

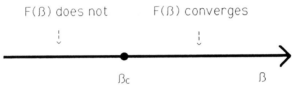

Fig. 1.9. Schematic illustration of the phase diagram for DLA as a function of the "control parameter" β. After [1.20]

1.8.3 The Void-Channel Model of 2d DLA Growth

Despite remarkable recent progress [1.50], the "final theory" of DLA is not universally agreed on. There have been attempts to measure and understand the distribution of p_i, and its lower cutoff $p_{\min}$, as well as to understand the the discovery of a "phase transition" in the behavior of the moments of the distribution, and the connection between multifractality and multiscaling.

The mass dependence of $p_{\min}$ has proved particularly useful. There are suggestions of an exponential [1.51,52],

$$p_{\min} \sim \exp\left(-AM^x\right), \tag{1.28}$$

a power-law [1.53,54]

$$p_{\min}(L) \sim M^{-\gamma}, \tag{1.29}$$

and an "intermediate" behavior [1.55–58]

$$p_{\min}(L) \sim M^{-\log M} \qquad [\log p_{\min} \sim -(\log M)^2]. \tag{1.30}$$

Each of these forms can be related to a possible fjord structure:

(a) The exponential form corresponds to narrow necks.

(b) The power-law form corresponds to wedge-type fjords.

(c) The "intermediate" behavior can be explained in terms of a structural model of DLA, which has self-similar *voids* connected by *necks*.

Our own group's numerical results [1.55–56] support possibility (c), and may provide a clue for the underlying puzzle of understanding DLA structure. Specifically, we have proposed a "void-neck" model of DLA [1.55–58] in order to explain the result (1.30). The void-neck model states that each fjord is characterized by a hierarchy of voids separated from each other by narrow "necks" or "gateways." The key features of the model are: (i) the distribution of voids must be *self-similar*, (ii) the voids are separated by necks ("channels," or "gateways"): a random walker can pass from one void to the next only by passing through a gateway.

What is the evidence supporting the void-neck model of 2d DLA growth dynamics?

(1) First, we note that if necks "dominate", then (1.28) would have to be satisfied. The numerics rule this out.

(2) Second, we note that if self-similar voids dominate, then (1.29) would have to be satisfied. Again, the numerics rule this out.

(3) Photos of large DLA clusters reveal the presence of such voids and channels. Moreover, when the DLA mass is doubled, we find that outer branches "grow together" to form new channels (enclosing larger and larger voids).

(4) The void-channel model can be *solved* [1.55,56] under the approximation that the voids are strictly self-similar and the gates are narrow. The solution demonstrates that (1.30) holds.

(5) The void-channel model is consistent with recent work [1.59] suggesting that DLA structures can be partitioned into two zones:

   (a) an inner *finished zone,* typically with $r \leq R_g$ (where R_g is the radius of gyration), for which the growth is essentially "finished" in the sense that it is overwhelmingly improbable that future growth will take place;

   (b) an outer *unfinished zone* (typically $r \geq R_g$) in which the growth is unfinished.

   Thus future growth will almost certainly take place in the region $r > R_g$. Now $2R_g \approx \frac{1}{2}L$, where L is the spanning diameter. Hence only about $1/4$ of the total "projected area" of DLA is finished, the rest of the DLA being *unfinished.* We suggest that the finished region will be created from the unfinished region by tips in the unfinished region growing into juxtaposition (thereby forming voids).

(6) The complete distribution function for the $\{p_i\}$ has been calculated [1.56], and the results are consistent with the result (1.30).

Indeed, two tips will grow closer and closer until their growth probabilities become so small that no further narrowing will occur. This observed phenomenon can perhaps be better understood if one notes that the growth probabilities $\{p_i\}$ of a given DLA cluster are identical to normalized values of the electric field $\{E_i\}$ on the surface of a charged conductor whose shape is identical to the given DLA cluster. Thus as two arms of the DLA "conductor" grow closer to each other, the electric field at their surface must become smaller (since $E_i \propto \nabla\phi_i$, where $\phi \equiv$ constant on the surface of the conductor). That E_i is smaller for two arms that are close together can be graphically demonstrated by stretching a drumhead with a pair of open scissors.

(1) If the opening is big, the tips of the scissors are well-separated and the field on the surface is big (we see that the gradient of the altitude of the drumhead is large between the tips of the scissors).

(2) On the other hand, if the scissor tips are close together, the field is small (we see that the gradient of the altitude of the drumhead is small between the scissor tips).

1.8.4 Multifractal Scaling of 3d DLA

The multifractal spectrum of the growth probability of 3d off-lattice DLA has also been studied [1.60–61]. The results indicate that, in contrast to 2d DLA, there appears to be *no* phase transition in the multifractal spectrum. Why? In both 2d and 3d, "necks" are created by side branches in DLA that grow closer and closer until their growth probabilities become so small that no further narrowing occurs. However, in 3d, even if there are points where tips from different branches of the aggregate come close or meet, there is no significant screening of growth due to this configuration, because no volume is cut off from the exterior and particles can enter the cluster from a direction perpendicular to the loop. Simply stated, one cannot cut off a volume with branches in the same way one can cut off an area. Thus we interpret the apparent absence of a phase transition for 3d as the effect of the topological differences between 2 and 3 dimensions. We further note that as d increases, d_f becomes closer to $d - 1$; the higher d is, the less dense the clusters are, since $\rho(R) \sim R^{d_f - d}$. Thus it is tempting to conjecture that $d = 2$ is a "lower critical dimension" in the sense that there is a phase transition for $d = 2$ but power-law scaling for all $d > 2$.

Schwarzer *et al.* [1.60] have calculated the $\{p_i\}$ for 50 off-lattice 3d DLA clusters, and compared our analysis to the 2d case which is believed to undergo a phase transition. They find the 3d case is quite different:

(i) the local slopes $\tau(q, M) \equiv \partial \ln Z / \partial \ln M$ do not diverge for $q < 0$ as they do in 2d (here Z denotes the qth moment of the distribution $\{p_i\}$),

(ii) the Legendre transform function $f(\alpha)$ has no systematic mass dependence, as it has in 2d, and

(iii) for 3d p_{min} has a power-law singularity in M [Eq. (1.29)], in contrast to
      the 2d case, where p_{min} vanishes much faster [Eq. (1.30)].

In sum, for 3d, even when tips from different branches are close, there is
no significant screening of growth, since particles can enter from directions
perpendicular to the loop, suggesting a power-law dependence of p_{min} on the
mass M of the cluster. Thus the apparent absence of a phase transition in 3d
DLA can be interpreted as due to the topological difference between 2 and 3
dimensions.

1.9 Scaling Properties of the Perimeter of 2d DLA: The "Glove" Algorithm

Schwarzer *et al.* [1.61] introduced a "glove" algorithm and used it to carry out
a systematic study of various properties of the 2d DLA perimeters (see also
the discussion of the "accessible perimeter" in Sect. 1.15.1). They developed an
algorithm – the "glove" method – which can be applied to study topological
properties of any fractal (or self-affine) surface. In particular, the glove method
can be used to determine:

(i) the total perimeter of a fractal, the set of all nearest-neighbor sites of the
    fractal, and a generalization thereof to neighboring sites of higher order;

(ii) the accessible perimeter of a fractal, which is the set of the perimeter sites
     that can be reached from the exterior of the object, and a generalization
     thereof to neighbor sites of higher order – this quantity has been studied
     experimentally on porous media and fresh fractures, and theoretically on
     percolation clusters;

(iii) the "lagoon"-size distribution, where "lagoons" are generalizations of the
      notion of voids to the case of *loopless* fractals and describe the regions of a
      fractal inaccessible to probes with a given particle size. The glove algorithm
      also enables one to identify unambiguously "necks" in a fractal structure.

1.9.1 Determination of the ℓ Perimeter

We begin by representing the investigated object in discretized form on a lattice
and label its sites with the index 0 (black sites in Fig. 1.10). In the first step,
we find all the nearest-neighbor sites of the object and label them $\ell = 1$, as
shown in Fig. 1.10a. Those sites that are nearest neighbors of sites with $\ell = 1$
and not already labeled are identified as $\ell = 2$ sites. We repeat the procedure
and obtain ℓ values for all sites surrounding the object (see Fig. 1.11). The
number ℓ associated with every lattice site is called the topological distance of
the site to the object. We will use the term "ℓ perimeter" to refer to the set of
sites with label ℓ.

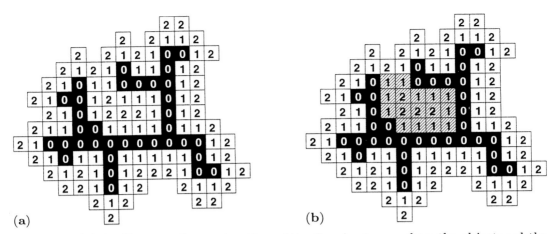

Fig. 1.10. (a) To illustrate the construction of the ℓ perimeter we show the object and the ℓ perimeters with $1 \leq \ell \leq 3$. Sites belonging to the investigated object are displayed in black and bear the label 0. Other numbers denote the order of the ℓ perimeter. (b) Lagoons are constructed by placing flexible gloves, one lattice unit thick, on the object. The gloves cannot penetrate through narrow openings. In our example, the 2-glove has "sealed" all openings, and we identify the enclosed space as lagoons (*grey sites*). After [1.61]

1.9.2 The ℓ Gloves

Next we describe the procedure to determine the "ℓ gloves" of the object. In the first step, instead of labeling all the neighbor sites, we place a flexible "glove," one lattice unit thick, on the object. In general, since gloves cannot penetrate through the narrowest openings (less or equal to $\sqrt{2}$ lattice constants wide), they cannot cover the object completely. The second glove is placed on the union of object and ℓ glove. We iterate the covering process to obtain ℓ gloves up to any desired order. Figure 1.10b illustrates the glove algorithm by showing the gloves of order $\ell = 1, 2, 3$ for a small DLA cluster.

1.9.3 Necks and Lagoons

Significantly, the glove algorithm can be used to extend to the case of loopless fractals the notion of voids as empty spaces in multiply connected fractals. Imagine, e.g., a circle with a small opening of width $w = 2\ell_o$, a simple example of a loopless object. Cover the surface with gloves, one after the other. When the number of gloves reaches ℓ_o, the glove cannot penetrate into the opening. We denote by "lagoon" the set of points left in the interior – now inaccessible from the exterior. The number of enclosed sites is the lagoon size s. The sites where glove ℓ_o touches itself identify a "neck." Note that there exists a one-to-one correspondence of lagoons and necks, so that each lagoon has a unique neck width w given by $w = 2\ell_o$. For the object in Fig. 1.10b, all the necks of lagoons have width $w = 4$. Schwarzer *et al.* [1.61] found that the lagoon size distribution in 2d DLA is consistent with a self-similar structure of the aggregate, but that

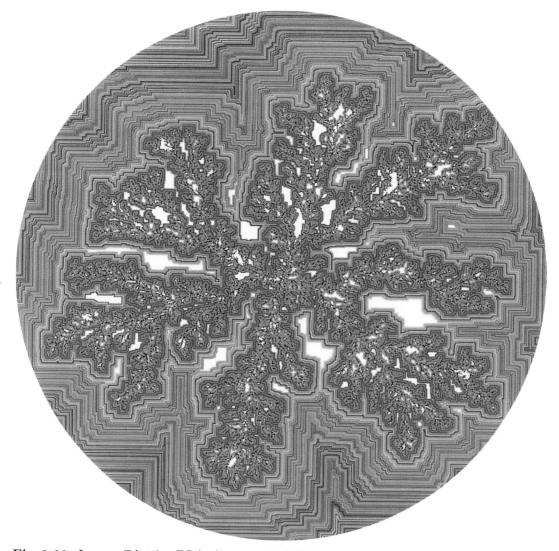

Fig. 1.11. Large off-lattice DLA cluster with 50 000 sites, covered by a series of successive gloves ($\ell = 1, 2, 3 \ldots$). After [1.61]

even for large lagoons the most probable width of the necks that separate the lagoons from the exterior is very small.

1.10 Multiscaling

However complex the above picture of DLA may seem already, there are even richer scaling features in this growth paradigm: DLA also exhibits *multiscaling* [1.59]. To see what this means, let us first introduce the annular mass $\rho_A(x, M)$, where $\rho_A(x, M)dx$ is the number of sites in the annulus $[x, x + dx]$ in a cluster of mass M. We define $x_i \equiv r_i/R$ as the distance of a cluster site i from the

Fig. 1.12. A DLA configuration with the shaded region corresponding to a shell characterized by a value $x \equiv r/R$. After [1.20]

seed, normalized by the radius of gyration R of the cluster (see Figs. 1.12–13). The conventional mass density $\rho(r)$ is related to $\rho_A(x, M)$ by $\rho_A(x, M)dx = 2\pi r \rho(r, M)dr$.

Let us next introduce the set of fractal dimensions $D(x)$ characterizing the mass distribution within each annulus x. In conventional scaling, $D(x) = d_f$ is independent of x. In DLA, however, calculations suggest that $D(x)$ *depends on* x. Such calculations also support the following *multiscaling Ansatz* [1.59],

$$\rho_A(x, M) \sim r^{D(x)} C(x), \qquad (1.31)$$

where $C(x)$ is a "cut-off function" to ensure that $\rho_A = 0$ outside the cluster. The connection to the usual fractal dimension is that d_f, the fractal dimension of the entire cluster, is equal to the maximum of $D(x)$, since asymptotically only sites in the annuli with $D(x) = d_f$ contribute to the total cluster mass.

To gain insight into the conditions under which multiscaling arises, consider the joint distribution function $N(\alpha, x, M)$, where $N(\alpha, x, M)d\alpha dx$ is the number of growth sites in the annulus $[x, x + dx]$ and with α values in the interval $[\alpha, \alpha + d\alpha]$ [the parameter α, formally defined in (1.61) in the Appendix, is determined from the moments of the growth probabilities]. The distribution $N(\alpha, x, M)$ contains all the information of $n(\alpha, M)$, since $n(\alpha, M)$ can be recovered by integration of $N(\alpha, x, M)$ with respect to x:

$$n(\alpha, M) \equiv \frac{\int dx N(\alpha, x, M)}{\int dx d\alpha N(\alpha, x, M)}. \qquad (1.32)$$

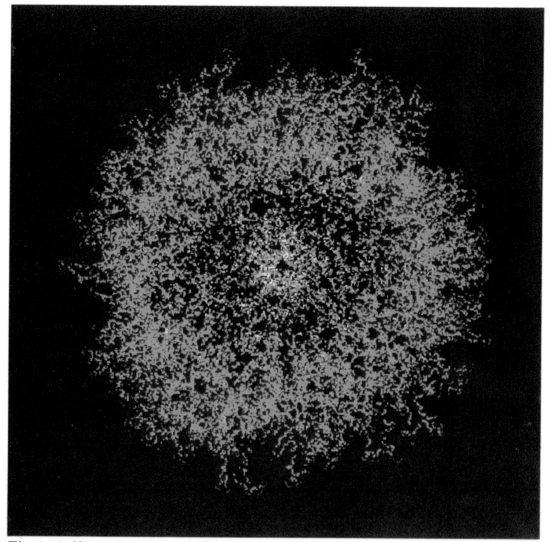

Fig. 1.13. Nineteen superposed 2d off-lattice DLA clusters. Three subsets of the growth sites are colored according to their α values (the parameter α, formally defined in the Appendix, is related to the fractal dimension of the subset). Red denotes the range $0.4 < \alpha < 0.8$, green $0.8 < \alpha < 1.2$, white $3.8 < \alpha < 4.2$. Note that large α values correspond to sites with small growth probabilities. We see that growth sites with specific values of α are located in approximate annuli characterized by different average positions $\bar{x}(\alpha, M)$ (where $x \equiv r/R$) and different widths $\xi(\alpha, M)$. After [1.20]

A reasonable approximation to $N(\alpha, x, M)$ is a Gaussian in x, namely

$$N(\alpha, x, M) = \frac{M^{f(\alpha)}}{\sqrt{2\pi\xi^2(\alpha, M)}} \exp\left[-\frac{(x - \bar{x}(\alpha, M))^2}{2\xi^2(\alpha, M)}\right], \qquad (1.33)$$

with mean square width $\xi^2(\alpha, M)$ and center $\bar{x}(\alpha, M)$; the function $f(\alpha)$ is discussed in the Appendix. The function $\bar{x}(\alpha, M)$ appears to converge for large M to a limit $\bar{x}(\alpha)$ [1.59].

Since the growth sites of DLA are associated with cluster sites, we expect that $\rho_A(x, M)$ is proportional to the the density profile of growth sites, which follows from (1.33) by integration over α, i.e.,

$$\rho_A(x, M) \sim \int d\alpha N(\alpha, x, M). \tag{1.34}$$

There are three distinct possibilities for the functional dependence of the width $\xi(\alpha, M)$. We discuss these and their implications next.

Case (i). "Constant Width": $\xi(\alpha, M) = A(\alpha)$. A constant width corresponds to both the average location $\bar{x}(\alpha)$ of the α sites and the width of the growth zone not being mass dependent. This implies that both length scales are proportional to the cluster radius R. Substituting (1.33) into (1.34) and performing a steepest descent of the resulting integral, leads to the cluster density profile

$$\rho_A(x, M) \sim r^{d_f} C(x). \tag{1.35}$$

Since the exponent d_f is the fractal dimension of the cluster, in case (i) conventional scaling arises.

Case (ii). "Strong Localization": $\xi(\alpha, M) = A(\alpha) M^{-y}$. In this case a similar analysis to that of case (i) reveals that the annular density displays multiscaling,

$$\rho_A(x, M) \sim r^{f(\alpha^*(x)) + y d_f} C(x), \tag{1.36}$$

where $\alpha^*(x)$ is the inverse function of $\bar{x}(\alpha)$ and f is from (1.33). The form in (1.36) is not altogether surprising. Because in this case the width of the distribution $N(\alpha, x, M)$ for fixed α tends to zero as $M \to \infty$; in this limit almost all the sites with a specific α value are located at distance $\bar{x}(\alpha)$ from the cluster seed. That is why we refer to case (ii) as "strong localization." Conversely, a specific location x singles out an α value $\alpha^*(x)$.

Case (iii). "Weak Localization": $\xi(\alpha, M) = A(\alpha)/(\ln M)^{1/2}$. The logarithmic factor in $\xi(x, M)$ in case (iii) changes the exponential term of (1.33) into a power law, which gives rise to qualitatively new scaling phenomena. As in case (ii), the exponent $D(x)$ is x dependent so that multiscaling arises [1.59]. In case (ii) only the typical α values, namely those for which the average location $\bar{x}(\alpha) = x$, enter $D(x)$. But in case (iii) we observe contributions from other α values. Thus, we refer to case (iii) as a case of "weak localization." There is still some localization since the width is still vanishing with increasing mass.

1.11 The DLA Skeleton

DLA has a loopless tree structure [1.62,63]. This observation is apparent in simulations of off-lattice aggregates [1.64,65] where an incoming particle is added to the cluster when its distance to the cluster is below a specific sticking distance. Its "parent" particle is the one that it was closest to at the moment of incorporation into the cluster. The child–parent relationship allows us to uniquely assign a generation number or "chemical distance from the seed" to every cluster particle. To this end, we assign to the seed particle of the DLA cluster the number $\ell = 0$. The children of the seed are assigned the chemical distance $\ell = 1$; children of $\ell = 1$ particles have chemical distance $\ell = 2$. The chemical distance turns out to be a very useful quantity in studies of branching properties [1.66–69] and Schwarzer *et al.* use it to define the DLA skeleton, a concept not entirely unrelated to the backbone discussed in Sect. 1.14.3.

In Fig. 1.14, we display the skeletons of a growing DLA cluster at mass $M = 5\,000,\ 50\,000,\ 500\,000$. In this figure, ℓ_c is chosen to be half the "chemical radius" $\Lambda/2$ of the cluster. Note that the termination points of the skeleton are located almost on a circle, although we use the chemical and not the Euclidean distance from the seed to define the skeleton. This indicates that DLA grows radially outward without forming loops.

One physical interpretation of the skeleton in DLA can be obtained if we consider the aggregate as a conductor situated between the grounded seed and a circular electrode (a sphere or hypersphere in 3d or 4d, respectively) of "radius" ℓ_c. Then the skeleton is the collection of paths that contribute to the current through the aggregate.

Schwarzer *et al.* [1.62] have determined the skeleton of comparatively large ($\approx 10^6$ sites) off-lattice DLA clusters in 2d, 3d, and 4d. They find that, asymptotically, in 2d the skeleton of DLA suggests a fixed number of $N_b \approx 7.5 \pm 1.5$ main branches and a self-similar structure. In 3d and 4d, and possibly in all spatial dimensions $d > 4$, the DLA skeleton is a ramified object, which displays branching over the whole range of ℓ values for which it is defined. For all dimensions Schwarzer *et al.* find strong finite-size effects corresponding to a slow change in the structure of DLA as the cluster size increases. The presence of strong corrections to scaling of logarithmic character in 2d DLA is in agreement with findings of deviations from self-similar behavior.

1.12 Applications of DLA to Fluid Mechanics

Many recent advances have occurred in our understanding of the role of fluctuations in fluid mechanics and dendritic solidification. In this and the following section, we shall argue that if one understands completely the DLA model or the closely related dielectric breakdown model (DBM), then one understands

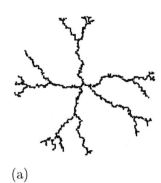

(a)

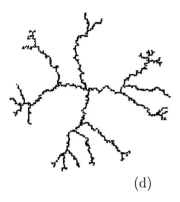

(d)

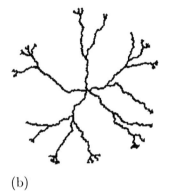

(b)

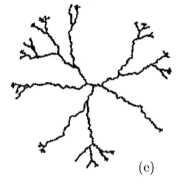

(e)

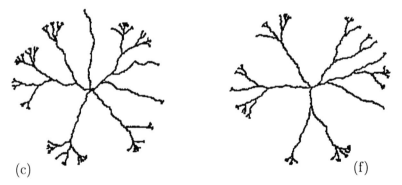

(c)

(f)

Fig. 1.14a–f. Skeleton for two growing DLA clusters: (a–c) first cluster, (d–f) second cluster. Growth of the cluster has been interrupted at cluster masses $M = 5\,000$ (a,d), $50\,000$ (b,e) or $500\,000$ (c,f). The determination of the skeleton is based on a value of $\ell_c/\Lambda = 0.5$. The skeleton is then rescaled so that independent of cluster mass the same size results. After [1.62]

the role of fluctuations in a range of fluid mechanical systems, as well as in dendritic solidification. The detailed descriptions of such systems require suitably chosen variants, such as DBM with anisotropy and noise reduction.

We first argue that we can approach these experimental subjects of classic difficulty with the same spirit that has been used in recent years to approach problems associated with phase transitions and critical phenomena. This approach is to carefully choose a microscopic model system that captures the essential physics underlying the phenomena at hand, and then study this model until we understand "how the model works." Then we reconsider the phenomena at hand, to see if an understanding of the model leads to an understanding of the phenomena. Sometimes the original model is not enough, and a variant is needed, and we shall see that this is the case here also. Fortunately, however, we shall see that the same underlying physics is common to the model and its variants.

1.12.1 Archetype 1: The Ising Model and Its Variants

We begin, then, with the classic Ising model [1.70]. Over 1000 papers have been published on this model, but only since 1977 have we known that if one understands the Ising model thoroughly, one understands the essential physics of many materials, since they are simply *variants* of the Ising model. For example, a large number of systems are related to special cases of the n-vector model, which in turn is an Ising model in which the spin variable $\mathbf{s}$ has not one component but rather n separate components s_j: $\mathbf{s} \equiv (s_1, s_2, \ldots, s_n)$.

The Ising model solves the puzzle of how it is that nearest-neighbor interactions of *microscopic* length scale 1Å "propagate" their effect cooperatively to give rise to a correlation length ξ_T of *macroscopic* length scale near the critical point (Fig. 1.15). In fact, ξ_T increases without limit as the coupling $K \equiv J/kT$ increases to a critical value $K_c \equiv J/kT_c$,

$$\xi_T \sim \mathcal{A}\left(\frac{K - K_c}{K_c}\right)^{-\nu_T}. \tag{1.37a}$$

The "amplitude" $\mathcal{A}$ has a numerical value of the order of the lattice constant $a_\circ$. A snapshot of an Ising system shows that there are fluctuations on all length scales from $a_\circ$ (≈ 1Å) to ξ_T (which can be from $10^2 - 10^4$Å in a typical experiment).

1.12.2 Archetype 2: Random Percolation and Its Variants

In percolation (see Chap. 2), one randomly occupies a fraction p of the sites of a d-dimensional lattice (the case $d = 1$ is shown schematically in Fig. 1.15b). Again, phenomena occurring on the local 1Å scale of a lattice constant are "amplified" near the percolation threshold $p = p_c$ to a macroscopic length ξ_p.

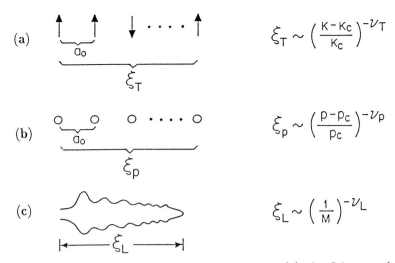

Fig. 1.15. Schematic illustration of the analogy between (a) the Ising model, which has fluctuations in spin orientation *on all length scales* from the microscopic scale of the lattice constant a_o up to the macroscopic scale of the thermal correlation length ξ_T, (b) percolation, which has fluctuations in characteristic size of clusters *on all length scales* from a_o up to the diameter of the largest cluster – the pair connectedness length ξ_p, and (c) the DLA/DBM problem, whose clusters have fluctuations *on all length scales* from the microscopic length $d_o = \gamma/L$ (γ is the surface tension and L the latent heat) up to the diameter of the cluster ξ_L. Also shown, on the right side, is the analogy between the scaling behavior of the three length scales ξ_T, ξ_p, and ξ_L. After [1.20]

Here p plays the role of the coupling constant K of the Ising model. When p is small, the characteristic length scale is comparable to 1Å. However when p approaches p_c, there occur phenomena on all scales ranging from a_o to ξ_p, where ξ_p increases without limit as $p \to p_c$:

$$\xi_p \sim \mathcal{A}\left(\frac{p - p_c}{p_c}\right)^{-\nu_p}. \tag{1.37b}$$

Again, the amplitude $\mathcal{A}$ is roughly a lattice constant ($\approx$1Å).

Each phenomenon of thermal critical phenomena has a corresponding analog in percolation, so that the percolation problem is sometimes called a geometric or *connectivity* critical phenomenon. Any connectivity problem can be understood by starting with pure random percolation and then adding interactions. Thus, for example, we understand why the critical exponents describing the divergence to infinity of various geometrical quantities (such as ξ_p) are the same regardless of whether the elements interact or are noninteracting [1.71]. Similarly, the same connectivity exponents are found regardless of whether the elements are constrained to the sites of a lattice or are free to be anywhere in a continuum [1.71–73].

1.12.3 Archetype 3: The Laplace Equation and Its Variants

Just as variations in the Ising and percolation problems were found to be sufficient to describe a rich range of thermal and geometric critical phenomena, so, we have found, are variants of the original Laplace equation useful in describing puzzling patterns in fluid mechanics, dendritic growth, and various breakdown phenomena.

In the Ising model, we place a spin on each pixel (site) of a lattice. In percolation we allow each pixel to be occupied or empty. In fluid mechanics, we assign a number – call it ϕ – to each pixel. We might think of ϕ as being the pressure or chemical potential at this region of space.

The spins in an Ising model interact with their neighbors. Hence the state of one Ising pixel depends on the state of all the other pixels in the system – up to a length scale given by the thermal correlation length ξ_T. The "global" correlation between distant pixels in an Ising simulation arises from the fact that neighboring pixels at i and j have a "local" exchange interaction J_{ij}. Similarly, the correlation in connectivity between distant pixels in the percolation problem arises from the "propagation" of local connectivity between neighboring pixels. In fluid mechanics, the pressure on each pixel is correlated with the pressure at every other pixel because the pressure obeys the Laplace equation.

One can calculate an equilibrium Ising configuration by "passing through the system with a computer" and flipping each spin with a probability related to the Boltzmann factor. Similarly, one can calculate the pressure at each pixel by "passing through the system" and re-adjusting the pressure on each pixel in accord with the Laplace equation. If we were to arbitrarily flip the configuration of a single pixel in the Ising problem (from $+1$ to -1), we would significantly influence the configuration of the system out to a length scale of the order of ξ_T. Similarly, if we were to arbitrarily impose a given pressure on a single point of a system obeying the Laplace equation, we would drastically change the resulting pattern out to a length scale that we shall call ξ_L.

Does ξ_L obey a "scaling form" analogous to (1.37a) and (1.37b) obeyed by the functions ξ_T and ξ_p for the Ising model and percolation? We believe that the answer to this question is "yes," although our ideas on this subject remain somewhat tentative and subject to revision.

The best way to see the fluctuations inherent in structures grown according to the Laplace equation is to first introduce some specific models. There are two models that were once thought to be fully equivalent, although it is now recognized that the actual patterns produced by each have a different "susceptibility to lattice anisotropy" [1.74]. The first of these models is diffusion limited aggregation (DLA). As discussed above, in DLA one releases a random walker from a large circle surrounding a seed particle placed at the origin. When the random walker touches a *perimeter* site of the seed, it "sticks" (i.e., the perimeter site becomes a cluster site), and we have a cluster of mass $= 2$ and a second random walker is then released. In contrast, for the dielectric breakdown model (DBM), the random walker does not stick to the cluster when it steps on a perimeter site. Rather, the walker survives until it steps on a cluster site, whereupon it sticks to the perimeter site it visited on the previous time step. Both the DLA and DBM models are believed to belong to the same universality class, and indeed both models can be obtained by solving the Laplace equation, the only difference being whether the potential is held constant on the cluster (DBM) or on the cluster plus its perimeter sites (DLA).

In thermal critical phenomena (or percolation) the length L introduced when we have a finite system size scales the same as the correlation lengths ξ_T (or ξ_p). Hence for DLA we expect that there will be fluctuations on length scales up to ξ_L, where ξ_L itself increases with the cluster mass according to

$$\xi_L \sim \mathcal{A}\left(\frac{1}{M}\right)^{-\nu_L} \qquad \left[d_f \equiv \frac{1}{\nu_L} = \text{fractal dimension}\right]. \qquad (1.37c)$$

Here the amplitude $\mathcal{A}$ is again on the order of 1Å. Note that (1.37c) is analogous to (1.37a) and (1.37b) if we think of $M \to \infty$ as being analogous to $K \to K_c$. Note also that $\nu_L = 1/d_f$ plays the role of the critical exponents ν_T and ν_p of (1.37a) and (1.37b). Suppose one tests this idea, qualitatively, by examining the largest DLA clusters in detail. One finds that indeed there are fluctuations in mass on length scales less than, say, the width W of the side branches. If one makes a log–log plot of W against mass M, one finds the same slope $1/d_f$ that one finds when one plots the diameter against M [1.23].

1.13 Applications of DLA to Dendritic Growth

1.13.1 Fluid Models of Dendritic Growth

By analogy with the Ising model and its variants, we can modify DLA/DBM to describe other fluid-mechanical phenomena. One of the most intriguing of these concerns a variation of the viscous fingering phenomenon in which anisotropy is present. Ben Jacob et al. [1.22] imposed this anisotropy by scratching a lattice of lines on their Hele-Shaw cell. They found patterns that strongly resemble snow crystals! If viscous fingers are described by DLA, then can the Ben Jacob patterns be described by DLA with imposed anisotropy?

Nittmann and Stanley [1.25] attempted to answer this question – specifically, they attempted to reproduce the Ben Jacob patterns with suitably modified DLA. A scratch in a Hele-Shaw cell means that the plate spacing b is increased along certain directions, and the permeability coefficient k relating growth velocity to ∇P is proportional to b^2 ($k \propto b^2$). Hence Nittmann and Stanley calculated DLA patterns for the case in which a periodic variation was imposed on the k. It is significant that their simulations reproduce snow-crystal-type patterns, just like the experiments. These simulations relied for their efficacy on the presence of noise reduction.

1.13.2 Noise Reduction

The original DLA and DBM models are prototypes of completely chaotic systems. No discernible pattern emerges. If there is a weak anisotropy, we expect that the resulting pattern reflects this anisotropy. For example, if the simulations are carried out on a lattice, then the presence of the lattice imposes a weak anisotropy (e.g., on a square lattice, it is more likely that particles attach to the westernmost tip if they approach from the west than if they aopproach from the north or south). This weak anisotropy is not visually apparent unless large clusters are grown. However, the largest DLA clusters made [1.8], with a mass of about 4 million sites, clearly display the anisotropy (Fig. 1.16).

Unfortunately, no one can afford the computer resources to make such "mega-DLA" clusters each time a new phenomenon is to be modeled. Noise reduction is a computational trick that seems to have the property that it speeds up the attainment of this asymptotic limit. In the absence of noise reduction, a perimeter site becomes a cluster site whenever it is chosen (e.g., whenever a random walker lands on that site). "Noise reduction" means that we associate a counter with each perimeter site; each time that site is chosen, the counter increments by one. The perimeter site becomes a cluster site only after the counter reaches a predetermined threshold value termed s [1.25,75,76]. When $s = 1$, we recover the original noisy DLA. Growth is dominated by the

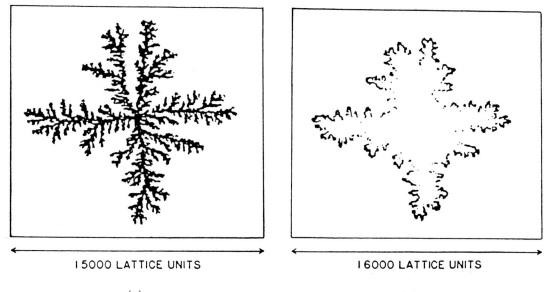

| 5000 LATTICE UNITS | 6000 LATTICE UNITS

(a)   (b)

Fig. 1.16a,b. A huge DLA cluster with a mass of 4 million sites grown on a square lattice. Shown is only the last 5% of the growth. In reality, there is structure on all scales less than the width W of the four arms. Moreover, W scales with the cluster mass as $W \sim (1/M)^{-1/d_f}$, just in the same way as the quantity ξ_L defined in (1.37c) does. The spontaneous appearance of side branches is reminiscent of experimental dendritic growth patterns. Meakin (private communication)

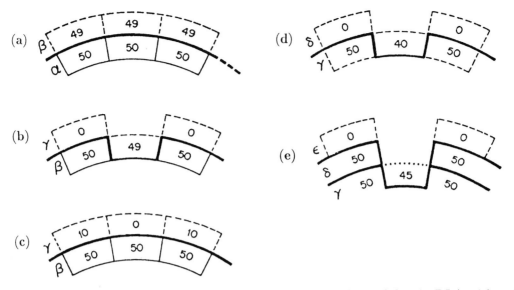

Fig. 1.17a-e. Schematic illustrations of the genesis of a fjord instability in DLA with noise reduction. After [1.23]

stochastic randomness in the arrival of random walkers. If s is very large, then growth is determined by the actual probability distribution.

For example, suppose we start with a large disc as a seed particle (instead of a single site). The growth probability at all points on the disc surface will be equal, if we assume a continuum.

By the D'Arcy growth law this disc should evolve in time into a larger disc. On the other hand, for ordinary DLA ($s = 1$), as soon as a random walker touches a single perimeter site on the disc, this site will become part of the cluster and the disc will lose its circular symmetry. The growth probabilities will all be re-calculated, and the perimeter sites close by the one that just grew will have higher growth probabilities. Thus the disc with a single site added to it will be more likely to grow in the direction of that single site. At a later time we will almost certainly not find a cluster with circular symmetry.

Figure 1.17 illustrates the difference between an outward ("positive") and an inward ("negative") interface fluctuation. A positive fluctuation tends to be damped out rather quickly, as mass quickly attaches to the side of the extra site that is added. On the other hand, a negative fluctuation grows, in the sense that mass accumulates on both sides of the tiny notch. The notch itself has a lower and lower probability of being filled in, as it becomes the end of a longer and longer fjord. This is the underlying mechanism for the tip-splitting phenomenon when no interfacial tension is present. Figure 1.17a shows the advancing front (row α) of a cluster with $s = 50$. The heavy line separates the cluster sites (all of which were chosen 50 times) from the perimeter sites (all of which have counters registering less than 50). In Figure 1.17a, no fluctuations in the counters of these three sites have occurred yet, and all three perimeter counters register 49. Figure 1.17b shows a negative fluctuation, in which the central perimeter site is chosen slightly less frequently than the two on either side; the latter now register 50, and so they become cluster sites in row β. The perimeter site left in the notch between these two new cluster sites grows much less quickly because it is shielded by the two new cluster sites. For the sake of concreteness, let us assume it is chosen 10 times less frequently. Hence by the time the notch site is chosen one more time, the two perimeter sites at the tips have been chosen 10 times (Fig. 1.17c). The interface is once again smooth (row γ), as it was before, except that the counters on the three perimeter sites differ. After 40 new counts per counter, the situation in Fig. 1.17d arises. Now we have a notch whose counter lags behind by 10, instead of by 1 as in Fig. 1.17b. Thus the original fluctuation has been amplified, due to the tremendous shielding of a single notch. Note that no new fluctuations were assumed: the original fluctuation of 1 in the counter number is amplified to 10 solely by electrostatic screening. This amplification of a negative "notch fluctuation" has the effect that the tiny notch soon becomes the end of a long fjord. To see this, note that Fig. 1.17e shows the same situation after 50 more counts have been added to each of the two tip counters, and hence (by the 10 : 1 rule) five new counts have been added to the notch counter. The tip counters therefore become part of the cluster, but the notch counter has not yet reached 50 and remains a perimeter site. The notch has become an incipient fjord of length 2, and the potential at the end of this fjord is now exceedingly low. Indeed it is quite possible that the counter will never pass from 45 to 50 in the lifetime of the cluster. In our simulations we can see tiny notch fluctuations become the ends of long fjords, and all of the above remarks on the time-dependent dynamics of tip splitting are confirmed quantitatively.

Clearly if s is very large, then the initial growth will preserve an almost circular structure. This is because, before the first site is added to the circular seed, all the perimeter sites will acquire large numbers in their counters ($s - 1$, $s - 2$, etc.). After the first site is added, these additional perimeter sites will be very close to the threshold for growth, whereas the new perimeter sites that were born when the first cluster site is added will all have counters initialized at zero.

At first sight, there is little economy in computational speed, since one needs "s times as many" random walkers to reach a given cluster size. Thus to grow a cluster with merely 4000 sites with $s = 1000$ requires almost as much time as needed to generate a mega-DLA with 4 000 000 sites and $s = 1$. Fortunately, there is a way around this problem. Instead of using random walkers to solve the Laplace equation (to sample the growth probabilities p_i on each perimeter site), we can directly solve the Laplace equation numerically. This is the approach used when the dielectric breakdown model was first proposed (Fig. 1.18). Whether one calculates the growth probabilities by sending in random walkers or by solving the Laplace equation is immaterial: the difference between DLA and DBM is in the boundary conditions, not the method of calculation.

The advantage of the Laplace equation approach when s is large is obvious: one need re-solve the Laplace equation only after a site is actually added to the

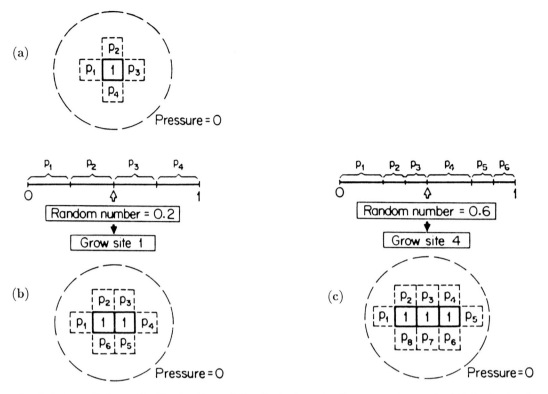

Fig. 1.18a-c. Schematic illustration of the first steps in the generation of a DLA cluster by solving directly the Laplace equation on a square lattice. (a) $t = 0$, (b) $t = 1$, (c) $t = 2$. After [1.77]

cluster. In between adding sites, one simply chooses random numbers weighted by the growth probabilities of each perimeter site. This is a relatively rapid procedure for the computer, compared with its counterpart of sending random walkers.

1.13.3 Dendritic Solid Patterns: "Snow Crystals"

Of course, real dendritic growth patterns (such as snow crystals) do not occur in an environment with periodic fluctuations in $k(x, y)$. Rather, the *global* asymmetry of the pattern arises from the *local* asymmetry of the constituent water molecules. Can this local asymmetry give rise to global asymmetry? Buka *et al.* [1.22] replaced the Ben Jacob experiment (isotropic fluid, anisotropic cell) by the reverse: isotropic cell but anisotropic fluid!

To accomplish this, they used a nematic liquid crystal for the high-viscosity fluid. Thus the analog of the water molecules in a snow crystal are the rod-shaped anisotropic molecules of a nematic. This experiment shows that the underlying anisotropy can as well be in the fluid as in the environment.

Snow-crystal formation is thought to involve mainly the aggregation of tiny ice particles and droplets of supercooled water [1.78–80]. To the extent that snow crystals grow by adding water molecules previously in the vapor or liquid phase, the growth rate is thought to be limited by the diffusion away from the growing snow crystal of the latent heat released by these phase changes. Under conditions of small Peclet number, the diffusion equation describing the space and time dependence of the temperature field $T(\mathbf{r}, t)$ reduces to the Laplace equation. Thus a reasonable starting point is DLA, independent of whether we wish to focus on particle aggregation, heat diffusion, or both.

While the various deterministic models of snow crystals produce patterns that are much too "symmetric," the DLA approach suffers from the opposite problem: DLA patterns are too "noisy." That DLA is too noisy has long been recognized as a defect of this otherwise physically appealing model. Recently, an approach has been proposed [1.25] that retains the "good" features of DLA and at the same time produces patterns that resemble real (random) snow crystals. We introduce controlled amounts of noise reduction. We do not explicitly introduce anisotropy – the only anisotropy present is the six-fold anisotropy arising from the underlying triangular lattice.

The patterns obtained [1.25] have the same general features for all values of s greater than about $s = 100$. The fjords between the six main branches contain much empty space; some snow-crystals have such wide "bays" but some do not. A better model would seem to require some tunable parameter that enables the complete range of snow-crystal morphologies to be generated. One such parameter, η, has the desired effect of reducing the difference in the ratio of the growth probabilities between the tips and fjords. Specifically, we relate by the rule $p_i \propto (\nabla \phi)^\eta$ the growth probability p_i (the probability that perimeter site i is the next to grow) to the potential ϕ.

We used η to tune the balance between tip growth and fjord growth and found growth patterns that resemble better the wide range of snow crystal morphologies that have been experimentally observed [1.25]. Moreover, our values for d_f agreed remarkably well with values we obtained by digitizing photographs of experimentally observed snow-crystals (Fig. 1.5).

1.13.4 Dendritic Solid Patterns: Growth of NH_4Br

Dendritic crystal growth has been a field of immense recent progress, both experimentally and theoretically. In particular, Dougherty *et al.* [1.81] have recently made a detailed analysis of stroboscopic photographs, taken at 20 s intervals, of dendritic crystals of NH_4Br (Fig. 1.19a). They have found three surprising results: (i) the side branches are nonperiodic at any distance from the tip, with random variations in both phase and amplitude, (ii) side branches on opposite sides of the dendrite are essentially uncorrelated, and (iii) the rms side-branch amplitude is an exponential function of the distance from the tip, with no apparent onset threshold distance.

How can we understand these new experimental facts? Many existing models reflect the essential physical laws underlying the growth phenomena, but

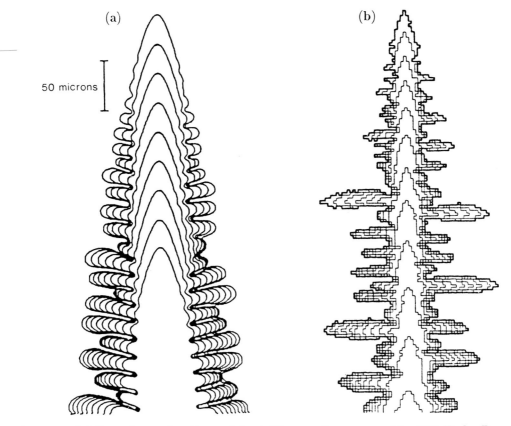

Fig. 1.19. (a) Experimental pattern of dendritic growth, measured for NH_4Br by Dougherty *et al.* [1.81], (b) DLA simulation with noise-reduction parameter $s = 200$. After [1.24]

fail to find a tractable mechanism to incorporate the effects of noise on the growth. Growth of a dendrite from solution is controlled by the diffusion of solute towards the growing dendrite. In the limit of a small Peclet number, the diffusion equation reduces to the Laplace equation (as mentioned above). The Laplace equation for a moving interface (the growing dendrite) brings to mind the diffusion limited aggregation model (DLA). Growth patterns produced by the various DLA simulation algorithms do *not* resemble dendritic growth patterns: DLA patterns are much too chaotic in appearance. We shall discuss here a related model [1.24] whose asymptotic structure does resemble the patterns found experimentally – both in broad qualitative features and in quantitative detail. The picture that emerges is one of Laplacian growth, where noise arises from the fact that there are concentration fluctuations in the vicinity of the growing dendrite (roughly $\pm 10^5$ NH_4Br molecules per cubic micrometer).

Our starting point is the observation that minute amounts of anisotropy become magnified as the mass of a cluster increases. In fact, even the weak anisotropy of the underlying lattice structure can become so amplified that clusters of 4 000 000 particles take on a cross-like appearance (Fig. 1.16). A real dendrite has a mass of roughly 10^{16} particles; it is impossible to generate clusters of this size on a computer, since even clusters of size 10^6 require hundreds of hours on the fastest available computers. Fortunately, noise reduction speeds the convergence of the pattern toward its asymptotic infinite mass limit.

A typical result [1.24] for a mass of 4000 particles is shown in Fig. 1.19b. After each 333 particles are added, a contour has been drawn. If noise reduction is used to tune the effect of noise, and cubic anisotropy is introduced through the use of an underlying square lattice, the resulting patterns obtained strongly resemble the experimental patterns of Dougherty *et al.* [1.81]. Side branching arises from the fact that an approximately flat interface in the DLA problem grows "trees" (which resemble "bumps" in the presence of noise reduction); these compete for the incoming flux of random walkers. If one "tree" gets ahead, it has a further advantage for the next random walker and so gets ahead still more. Thus some side branches grow while others do not. The patterns we obtain are reasonably independent of details of the simulation in that similar patterns are obtained when we vary the surface tension parameter σ over a modest range; we can also alter the boundary conditions of the model with some latitude and even allow for nonlinearity in the growth process ($\eta \neq 1$).

1.14 Other Fractal Dimensions

A single fractal dimension d_f is not sufficient to describe all fractal objects. For example, percolation and DLA both have $d_f = 2.5$ when $d = 3$, yet any child can immediately see that these two fractals look completely different from each other. We shall see that when we introduce a new fractal dimension, $d_{\min}$, to describe the tortuosity of the fractal, we find that $d_{\min}$ has quite different values for DLA and for percolation.

In fact, it is necessary to introduce roughly ten distinct fractal dimensions [1.82]. Although this may sound overwhelming, these arise in as natural a fashion as the ten or so distinct critical point exponents that most students know and use regularly. Just as the critical point exponents were found to be not all independent of one another, so also the ten fractal dimensions are not all independent quantities. Rather, they are related by relations not altogether unlike the scaling laws relating critical exponents. Here we shall review only a few of these ten; a more complete discussion appears in [1.82].

1.14.1 The Fractal Dimension d_w of a Random Walk

Suppose we wish to know the mean number of sites visited when a random walker stumbles around randomly on a fractal substrate. De Gennes [1.83] has termed this problem the "ant in a labyrinth", but a "drunk in a gulag" might seem more descriptive of the actual picture (Fig. 1.20). A random walker (drunken ant or drunken prisoner) has been parachuted onto a randomly chosen

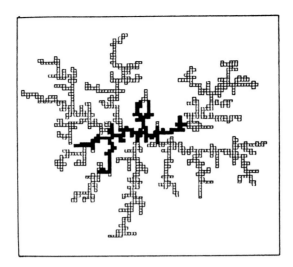

Fig. 1.20. A random walk of 2500 steps on a DLA fractal substrate with 1000 sites. The sites visited by this walk have been indicated by solid squares. This is a visualization of the de Gennes "ant in a labyrinth" problem. After [1.84] and [1.85]

site, which is then colored black. At each successive time step, the ant moves to a neighboring fractal site, and that becomes black also. At each time step t, the ant (*la fourmi*) calculates her range $r \equiv \sqrt{\langle r^2 \rangle}$, the rms displacement from the local origin where the parachute landed. The fractal dimension d_w determines how t scales with r:

$$r \sim t^{1/d_w}. \tag{1.38a}$$

We anticipate that d_w is considerably *larger* than 2 since many of the neighbors at each step are unavailable to the ant so she is obliged to return in the direction of her parachute point. Hence r increases much less rapidly with t than for an unconstrained random walk (Fig. 1.21).

Now let us ask how the number of visited sites (one "mass") scales with the time (a different "mass"). The ant compactly fills a region of the fractal, so the first mass scales with fractal dimension d_f. The second mass scales with fractal dimension d_w. Hence the number of visited sites scales with time as the *ratio* of the two fractal dimensions:

$$\langle s \rangle \sim t^{d_f/d_w}. \tag{1.38b}$$

The ratio of two fractal dimensions is termed an *intrinsic dimension*. The quantity d_f/d_w was first introduced by Alexander and Orbach [1.86] (AO) in connection with studying the density of states for phonon-like excitations on a fractal substrate. They defined the fracton dimension to be

$$d_s \equiv 2d_f/d_w. \tag{1.39a}$$

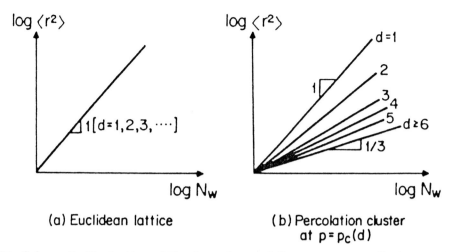

Fig. 1.21. Schematic illustration of the dependence of the mean square displacement on the number of steps of a random walk, plotted double logarithmically, for (a) a Euclidean lattice, and (b) a percolation cluster at the percolation threshold. For d above 6, the slope sticks at the $d = 6$ value of $1/3$ – so that the fractal dimension of the walk sticks at $d_w = 6$. After [1.82]

AO tabulated d_s for percolation fractals on a d-dimensional Euclidean lattice and noted that while d_f and d_w change dramatically with d (below $d_c = 6$), d_s does not. Hence they were led to the "AO conjecture" that for percolation

$$d_s = 4/3 \qquad [2 \le d \le 6]. \tag{1.39b}$$

The term *super-universal* was coined to express the possibility that an exponent could be independent not only of lattice details (as in ordinary universality) but also of d itself. Shortly after the AO conjecture appeared, it motivated several careful calculations [1.87,88]. Additional studies tested whether the AO conjecture might hold for other random phenomena, such as random walks on DLA clusters [1.84] or diffusive annihilation between random walkers who are constrained to move on a percolation fractal substrate [1.89]. Leyvraz and Stanley [1.90] found conditions under which the AO conjecture should hold rigorously, and these conditions were demonstrated to hold for $d \ge 6$.

As researchers became intrigued by the AO conjecture, it was put to ever more severe numerical tests (see, e.g., [1.91,92]), and eventually it was concluded that d_s is 1%–2% smaller than 4/3. It is a pity the AO idea does not work, since if it did we would have a relation between two fractal dimensions: d_f, which describes "statics", and d_w, which describes "dynamics." (For more on diffusion, see [1.92] and Chap. 3.)

1.14.2 The Fractal Dimension $d_{\min} \equiv 1/\tilde{\nu}$ of the Minimum Path

You are a soldier in a mine-infested battlefield. General X wants you to carry a message to his boss, General Y, who is 10 km distant (Fig. 1.22). You do

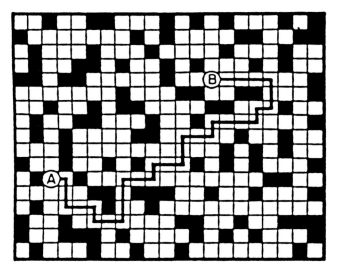

Fig. 1.22. Schematic illustration of the minimum path followed by a soldier in carrying an important message from General X (at position A) to General Y (at position B) in a battlefield that has been randomly infested by mines in sufficient concentration that the pair connectedness length ξ is larger than the Pythagorean distance L between A and B. This figure is adapted from Ritzenberg and Cohen [1.93] who used it to model the spread of electrical excitations in the heart tissue

this, and note that it takes one hour. General Y then asks you to carry the same message to *his* boss, General Z, who is 20 km away. It is critical that the message be delivered within two hours. You explain to the ignorant and impatient General Y that it will, statistically, take *more* than two hours to travel twice the distance, since when land mines are placed at random (by the presumably drunken enemy troops), they form clusters of all possible sizes. The minimum path between two generals is a convoluted path since it must avoid these clusters. If the "Pythagorean distance" r is short, there is little probability of encountering a huge cluster. Since General X and General Z are separated by $2R$, it is possible that you will encounter *much* larger mine clusters and hence your path length will *more* than double. More precisely, you tell your boss that the minimum path ℓ obeys the functional equation [1.94–97]

$$\ell(\lambda r) = \lambda^{d_{\min}} \ell(r), \tag{1.40a}$$

where $d_{\min} > 1$ (roughly 9/8 and 4/3, respectively, for $d = 2, 3$ percolation fractals [1.98]). The solution is found on setting $\lambda = 1/r$:

$$\ell \sim r^{d_{\min}}. \tag{1.40b}$$

Can one relate the exponent $d_{\min}$ to the other exponents? To do this, it is first important to develop some feeling for $d_{\min}$. To this end, we have calculated $d_{\min}$ for various fractal substrates, including percolation [1.98], lattice animals [1.99], DLA [1.100], and cluster–cluster aggregation [1.100]. We find that in all cases except DLA (Fig. 1.23), $d_{\min}$ increases monotonically with d up to the

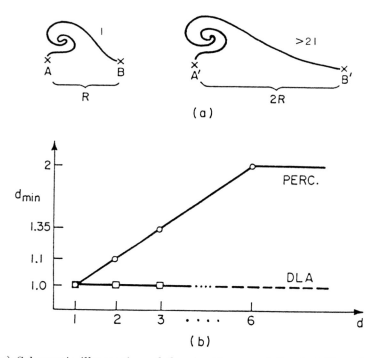

Fig. 1.23. (a) Schematic illustration of the significance of the fact that $d_{\min}$ is above 1: If the distance between points A and B is doubled, then the shortest path length increases by a factor $2^{d_{\min}}$, which is larger than 2. (b) Numerical values of $d_{\min}$ for percolation (*top curve*) and DLA (*bottom curve*). The fact that $d_{\min}$ sticks at 2 above the upper marginal dimensions d_c means that the shortest path between two is a random walk. The fact that $d_{\min}$ does not seem to approach 2 for DLA means that there is no d_c for DLA. This figure is based on data in [1.100]

critical dimension d_c. At and above d_c,

$$d_{\min} = 2 \qquad [d \geq d_c]; \tag{1.40c}$$

this means that the minimum path between two points in a fractal above d_c has the *same statistics as a random-walk* or Gaussian chain. Thus we can understand d_c in a different fashion: all fractals have different values of d_f, d_w, etc. above d_c but all have $d_{\min} = 2$. As d is decreased, this continues to hold until, at $d = d_c$, the shortest path between two points is "straighter" than a random-walk, and $d_{\min} < 2$. That $d_{\min} = 1$ for DLA [1.99] is consistent with the idea that DLA has no upper critical dimension.

The utility of d_f in providing a quantitative characterization of a fractal form was discussed above – so also was the utility of d_w in characterizing diffusion ("random-walks") on a fractal substrate. What is the utility of $d_{\min}$? Clearly $d_{\min}$ applies to physical phenomena that propagate *efficiently* from site to site, not re-visiting previous sites. In a rough sense, we can say that d_w describes random-walk propagation where re-visiting fractal sites is possible while $d_{\min}$ describes a sort of self-avoiding walk where re-visits are not. Our "runner" finds the minimum path from General X to General Y by considering all possible self-avoiding walks from X to Y and choosing the shortest.

Physically, this is the same as imagining that every cluster site of the fractal had a tree on it. If the system is near the percolation threshold, then the trees form a self-similar substrate for length scales up to the connectedness length $\xi \sim (p-p_c)^{-\nu}$. Now at time $t = 1$ let us ignite the trees on the site occupied by General X. At $t = 2$, we ignite the trees on the neighboring sites of X, and so forth until after some time delay the site occupied by General Y is ignited. This time delay is the minimum path ℓ between X and Y, and we have described a very elaborate procedure of determining ℓ which consists of igniting successive "chemical shells" around General X.

We can now introduce a second intrinsic fractal dimension, the "chemical dimension" or "spreading dimension"[2] d_ℓ. Suppose we ask for the total mass of burning trees at time t. From the definition of the fractal dimension d_f, we have

$$M(t) \sim r^{d_f}, \tag{1.41}$$

where here $r = r(t)$ is the radius of gyration. We asked for the dependence upon t of $M(t)$, where $t = \ell$ is the minimum path length between the newest shell of burning and the origin of the fire. Combining (1.40b) and (1.41), it follows that [1.94–97]

$$M(\ell) \sim \ell^{d_\ell} \qquad [\ell \ll r^{d_{\min}}], \tag{1.42a}$$

with

$$d_\ell \equiv d_f/d_{\min}. \tag{1.42b}$$

What is the velocity v with which the fire front is propagated? Below p_c the fire is localized to finite clusters of burning trees, so $v = 0$. Above p_c it can spread without limit, and v rises rapidly as p increases above p_c with an exponent ψ defined through

$$v \sim (p - p_c)^\psi. \tag{1.43}$$

We have used the symbol ψ to suggest that v plays the role of an order parameter in this problem. Clearly the velocity is given by

$$v = dr/dt = dr/d\ell = (d\ell/dr)^{-1} = r^{1-d_{\min}} = (p - p_c)^{\nu(d_{\min}-1)}. \tag{1.44}$$

Combining (1.43) and (1.44), one finds [1.93,102,103]

$$\tilde{\psi} \equiv \psi/\nu = d_{\min} - 1. \tag{1.45}$$

---

[2]   The term spreading dimension is not to be confused with the phenomenon of "damage spreading" in which one monitors the difference between two systems which initially have the same configuration except for a single spin or group of spins that are forced to be different – see [1.101].

Fig. 1.24. A large machine-generated finite cluster just below the percolation threshold of the square lattice bond problem. The bonds of this cluster are of three sorts: "yellow" dangling ends (light), singly connected "red" bonds (heavy), and multiply connected "blue" bonds (dotted) [1.104]. The red and blue bonds form the backbone [1.105], which (a) carries stress if the two ends of the cluster are pulled, (b) carries current if a battery is applied across the cluster, (c) carries fluid if the bottom of the cluster is oil and the top is my car (assuming the cluster represents a model of a randomly porous material), (d) propagates spin order, if we view this as a large cluster of Ising spins just below p_c (see, e.g., [1.106]). After [1.104]

For $d = 2$, $d_{min} - 1 \approx 1/8$, very close to zero [1.97,98]. Hence the increase in velocity just above p_c is remarkably steep: a fire which fails to propagate at all just below p_c will propagate extremely rapidly just above.

1.14.3 Fractal Geometry of the Critical Path: "Volatile Fractals"

How does oil flow through randomly porous material just at that point where it can "break through" and reach the surface? How does electricity flow through a random resistor network where the fraction of intact resistors is close to the percolation threshold p_c? How does one describe the shape of the incipient infinite cluster that appears near p_c when one considers the polyfunctional condensation of monomers? These are but three of a host of questions that one can ask that are of some practical interest and require for their answer a clear specification of cluster structure (Fig. 1.24).

It has long been recognized [1.105] that the dangling ends shown in Fig. 1.24, do not contribute to transport. The structure after the dangling ends have been

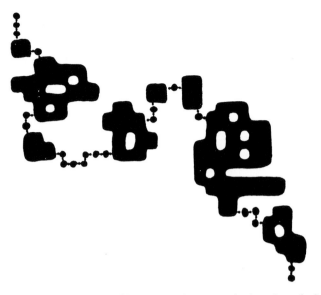

Fig. 1.25. Actual simulation of a backbone in site percolation just below p_c for a square lattice. The decomposition into blobs of all sizes from 1 to ∞ is apparent. After [1.107]

decapitated is termed the backbone (Figs. 1.25 and 1.26). Is the backbone self-similar up to length scales r of the order of the connectedness length ξ? If so, what is its fractal dimension d_B? It appears that the backbone is indeed a fractal object, with

$$N_B \sim r^{d_B} \qquad [1 \ll r \ll \xi]. \tag{1.46}$$

The fractal dimension d_B sticks at the value 2 for $d > d_c = 6$, and then decreases "slowly" as d decreases below 6. For $d = 2$, d_B has decreased to about 13/8 [1.107,108]. One of the problems associated with calculating d_B is obtaining a fast algorithm (Herrmann's burning algorithm [1.109] is quite efficient).

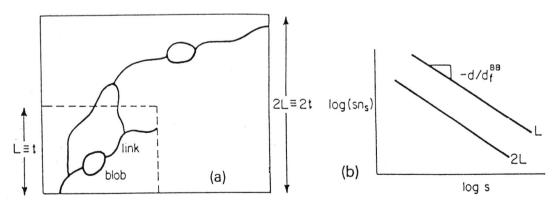

Fig. 1.26. (a) Schematic illustration of how small blobs become part of larger blobs when the box size L is doubled; (b) schematic illustration of how the cluster size distribution $n_s(L)$ is uniformly depressed when L increases. After [1.107]

Although we now have estimates of various fractal dimensions that are accurate to roughly 1%, we lack any way to *interpret* them. This is in contrast to both the $d = 2$ Ising model and $d = 2$ percolation, where we can calculate the various exponents and interpret them in terms of scaling powers. Thus [1.35]

$$y_h = \frac{15}{8}, \qquad y_T = 1 \qquad \text{[Ising model]}, \tag{1.47a}$$

and [1.36]

$$y_h = d_f = \frac{91}{48}, \qquad y_T = d_{\text{red}} = \frac{3}{4} \qquad \text{[percolation]}, \tag{1.47b}$$

where d_{red} is the fractal dimension of the singly connected "red" bonds of the incipient infinite cluster [1.104]. The rationale for the terms red and blue is that the red bonds carry all the current from one bus bar to the other, and so are "hotter" than the blue bonds which "share" the current with the other blue bonds.

It follows that both scaling fields in percolation are equal to geometric properties of the incipient cluster! What about d_B? To what is *this* fractal dimension related? This is a problem worthy of attention.

1.15 Surfaces and Interfaces

1.15.1 Self-Similar Structures

Next we turn to the subtle and fascinating subject of disordered surfaces (see also Chaps. 8 and 9). But what do we mean by "the" surface of a fractal object? In fact, we shall see that there are many different surfaces, depending on the physical process in question (Fig. 1.27). Even nonfractal objects can have fractal surfaces, and indeed the subject of fractal surfaces has attracted a great deal of recent interest [1.110].

a) **Total perimeter.** It is a rigorous result for d-dimensional percolation [1.111] that the total number of perimeter sites N_p scales in the same fashion as the total number of cluster sites:

$$N_p \sim N_f \sim L^{d_f}. \tag{1.48}$$

b) **External perimeter (hull).** The total number of *external* surface sites, or hull, scales with the radius of gyration r as

$$N_{\text{hull}} \sim r^{d_h}. \tag{1.49a}$$

For $d = 2$ percolation, it had been conjectured that [1.112]

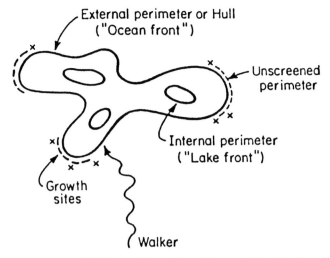

Fig. 1.27. Schematic illustration of different fractal surfaces arising in the description of a percolation cluster; (a) the total perimeter has a fractal dimension d_f, equal to that of the total bulk mass of the cluster [1.111]. (b) The external "ocean front" perimeter (or hull) has a fractal dimension d_h [1.112,113]; since $d_f > d_h$, it follows that the internal "lake front" perimeter must have the same fractal dimension d_f of the total perimeter. (c) The unscreened perimeter, where an incoming random-walker is more likely to hit, has fractal dimension d_u [1.114], and (d) the growth sites, having fractal dimension d_g, depend for their definition on the rules by which the cluster is grown [1.115]. After [1.85]

$$d_h = 1 + d_{\mathrm{red}} = 1 + \frac{3}{4}, \qquad (1.49b)$$

where $d_{\mathrm{red}} = 1/\nu = 3/4$ is the fractal dimension of the singly connected "red" bonds [1.104]. The result (1.49b) has been proved rigorously [1.113].

For $d = 2$ percolation $d_f = 91/48$; the fact that $d_h < d_f$ means that the ratio of *external* perimeter sites to *total* perimeter sites approaches zero at the percolation threshold. As the clusters get larger the internal perimeter sites ("lake front" sites) completely swamp the external perimeter sites ("ocean front" sites).

c) Accessible perimeter. For $d = 2$ percolation, Grossman and Aharony [1.116] have noticed that there are many regions of the hull where a loop is almost formed. They introduced a new concept, the *accessible perimeter*, and an associated fractal dimension:

$$N_{\mathrm{ap}} \sim r^{d_{\mathrm{ap}}}. \qquad (1.50a)$$

Their calculations indicated that

$$d_{\mathrm{ap}} \approx \frac{4}{3}. \qquad (1.50b)$$

This value is possibly related to the fractal dimension $d_{\theta'}$ describing the configuration of a kinetic growth walk (a SAW model that never intersects itself), or to the fractal dimension d_θ describing the conformation of a $d = 2$ interacting SAW near its θ temperature [1.117].

1.15.2 Self-Affine Structures

(a) Self-Affinity. The scale transformation we described for self-similar fractals is isotropic, which means that dilation increases the size of the system *uniformly* in every spatial direction. Fractal objects that must be rescaled using an *anisotropic* transformation are called *self-affine* fractals.

Scaling – For quantifying disorderly surfaces, we are interested in a special subclass of anisotropic fractals, described by single-valued functions called *self-affine functions*. The analog of the scaling relation for a self-affine function can be formulated as

$$h(x) \simeq b^{-\alpha} h(bx), \qquad (1.51a)$$

where α (in some texts denoted by H or χ) is called the Hölder or self-affine exponent and gives a quantitative measurement of the 'roughness' of the function $h(x)$. A self-affine function must be rescaled in a different way horizontally and vertically: if we 'blow up' the function with a factor b horizontally ($x \to bx$), it must be 'blown up' with a factor b^α vertically ($h \to b^\alpha h$) in order that the resulting object overlaps the object obtained in the previous generation. For the special case $\alpha = 1$, the transformation is isotropic and the system is self-similar.

An important consequence of (1.51a) concerns the scaling of the height difference $\Delta(\ell) \equiv |h(x_1) - h(x_2)|$ between two points separated by a distance $\ell \equiv |x_1 - x_2|$. For self-affine systems $\Delta(\ell)$ obeys (1.51a). The solution of the 'functional equation' (1.51a) is a power law

$$\Delta \sim \ell^\alpha. \qquad (1.51b)$$

(b) An example. Paper is an inhomogeneous material, a prototype of the porous inhomogeneous rock that holds oil! One difference between fluid flow in the paper towel and in oil-bearing rock arises from the length scales at which these phenomena take place. This difference is an advantage: we can use a 20 cm paper towel system (Fig. 1.28) to help develop our understanding of the 20 km oilfield problem. For example, we can characterize the wet–dry interface using scaling laws, whose form is predicted by simple models that capture the essential mechanisms contributing to the morphology. This 'benchtop exercise' is an example of some of the current experiments being carried out on idealized systems which are yielding new insights into practical interface problems.

(c) Pinning by directed percolation. In this section, we discuss the directed percolation depinning (DPD) model, which has a diverging nonlinear term. The scaling exponents can be obtained exactly by mapping the depinning problem onto a variant of percolation termed *directed percolation*. The connection between surface growth with quenched disorder and directed percolation was originally proposed by two independent studies [1.119,120]. A number of seemingly different models also belong to the same universality class [1.121–128].

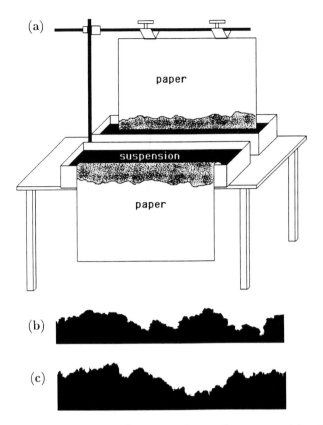

Fig. 1.28. (a) Schematic illustration of an experimental setup probing interface motion in random media. Parameters such as type of paper, temperature, humidity, direction of growth and concentration of coffee can be varied systematically. These changes affect the area, the speed of wetting, and the global width of the rough surface, but they do not affect the scaling properties of the surface. (b) Digitized *experimental* interface; the horizontal size of the paper was 20 cm. (c) Typical result of a discrete *model* mimicking interface motion in disordered media. After [1.118]

The model is defined as follows: on a square lattice of edge L (with periodic boundary conditions), we *block* a fraction p of the cells to correspond to the quenched disorder. Blocked cells will try to stop the growth, while the interface is free to advance on unblocked cells. At $t = 0$, the 'interface' is the bold horizontal line shown in Fig. 1.29a. At $t = 1$ we randomly choose a cell (labeled X in Fig. 1.29b) from among the unblocked cells that are nearest neighbors to the interface. We 'wet' cell X and *any cells that are below it in the same column.* This process is then iterated. For example, Fig. 1.29c shows that at $t = 2$ we choose a second unblocked cell, cell Y, to wet, while Fig. 1.29d shows that at $t = 3$ we wet cell Z *and also cell Z′ below it.*

For p below a critical threshold $p_c = p_c(L)$ the interface propagates without stopping, while for p above p_c the interface is pinned by the blocked cells. Figure 1.30 displays a typical interface after the growth has stopped; this occurs just as the surface meets a spanning path of blocked cells connected through nearest and next-nearest neighbors. The definition of the model precludes this path

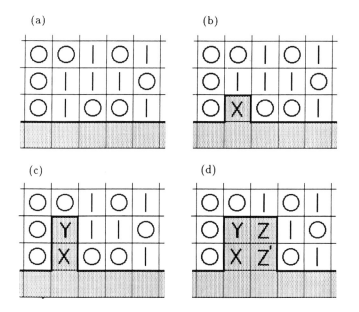

Fig. 1.29a-d. The DPD (directed percolation depinning) model for interface growth with erosion of overhangs. 'Wet' cells are shaded, while dry cells are randomly blocked with probability p (indicated by O) or unblocked with probability $(1-p)$ (indicated by |). The interface between wet and dry cells is shown by a heavy line. (a) $t = 0$, (b) $t = 1$, (c) $t = 2$ and (d) $t = 3$. After [1.120]

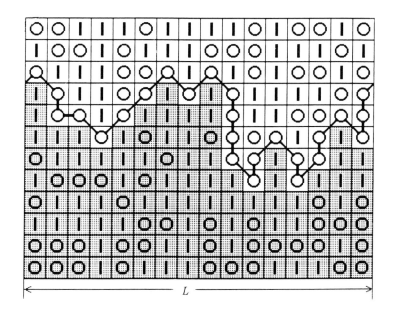

Fig. 1.30. Shown as a bold line is a *spanning* path formed by connected nearest-neighbor and next-nearest-neighbor blocked cells which pin the interface. Note that various *nonspanning* clusters of blocked cells (found inside the wet region) are not sufficient to pin the interface. After [1.120]

having overhangs, and thus it is equivalent to a spanning path on a square lattice that goes from left to right, is able to turn up and down, but can never turn left. Such a directed path is, in fact, a path on a directed percolation cluster. Indeed, one can show that the pinned interface in the model corresponds to the hull of a directed percolation cluster.

When the probability of blocked cells p is above p_c, the growth is halted by a spanning path of a directed percolation cluster. Such a directed path is characterized by a correlation length parallel to the interface, $\xi_\parallel$, and one perpendicular to it, $\xi_\perp$. Their meaning is the following: blocking a fraction p of the sites on a lattice, the *blocked* sites form directed paths whose average length is $\xi_\parallel$ and whose width is $\xi_\perp$. The two correlation lengths diverge in the vicinity of p_c,

$$\xi_\parallel \sim |p - p_c|^{-\nu_\parallel}, \qquad \xi_\perp \sim |p - p_c|^{-\nu_\perp}. \tag{1.52}$$

From calculations on directed percolation, it is known that $\nu_\parallel \simeq 1.733$ and $\nu_\perp \simeq 1.097$ (see Chap. 2).

For $p \ll p_c$, $\xi_\parallel \ll L$. The directed paths *locally* pin the interface, but the interface eventually moves with a nonzero average velocity by advancing in between the blocked paths. Complete pinning appears when $\xi_\parallel$ becomes equal to the system size L. The width, w, of such a path is of the order of $\xi_\perp$. Hence using (1.52) we obtain the scaling of the width

$$w \sim \xi_\perp \sim |p - p_c|^{-\nu_\perp} \sim \xi_\parallel^{\nu_\perp/\nu_\parallel} \sim L^{\nu_\perp/\nu_\parallel} = L^\alpha, \tag{1.53}$$

from which we obtain for the roughness exponent

$$\alpha = \frac{\nu_\perp}{\nu_\parallel} \simeq 0.633 \pm 0.001. \tag{1.54}$$

This result is confirmed by numerical studies on the scaling of the pinned interface. The critical probability p_c must coincide with the critical probability of the underlying directed percolation problem, giving $p_c \simeq 0.47$.

1.A Appendix: Analogies Between Thermodynamics and Multifractal Scaling

The purpose of this appendix is to offer some appreciation of the analogies of multifractals with thermodynamics [1.129,130] and to introduce the function $f(\alpha)$, which plays a role analogous to the entropy.

We begin by considering again the sum in (1.24), which we write in the form (see the notation dictionary of Table 1.2)

$$Z(q) = \sum_p e^{G(p)} \tag{1.55}$$

where

$$G(p) = \log n(p) + q \log p. \tag{1.56}$$

The sum in (1.55) is dominated by some value $p = p^*$, where $p*$ is the value of p that maximizes $G(p)$. Thus

$$Z(q) \sim e^{G(p*)} = n(p*)(p*)^q. \tag{1.57}$$

For fixed q, $p*$ and $n(p*)$ both depend on the system size L, leading one to define the new q-dependent exponents α and f by

$$p* \sim L^{-\alpha}; \qquad n(p*) \sim L^f. \tag{1.58}$$

Substituting (1.58) into (1.57) we find

$$Z(q) \sim L^{f - \alpha q}. \tag{1.59}$$

Comparing (1.59) with (1.58), we find the desired result

$$\tau(q) = q\alpha(q) - f(q). \tag{1.60}$$

From (1.56) it follows that

$$\frac{d}{dq}\tau(q) = \alpha(q). \tag{1.61}$$

Hence we can interpret $f(\alpha)$ as the negative of the Legendre transform of the function $\tau(q)$:

$$f(\alpha) = -(\tau(q) - q\alpha) \qquad [\alpha \equiv d\tau/dq]. \tag{1.62}$$

The function $Z(q)$ is formally analogous to the partition function $Z(\beta)$ in thermodynamics, so that $\tau(\beta)$ is like the free energy. The Legendre transform $f(\alpha)$ is thus the analog of the entropy, with α being the analog of the energy E. Indeed, the characteristic shape of plots of $f(\alpha)$ against α are reminiscent of plots of the dependence on E of the entropy for a thermodynamic system.

Acknowledgements.   I wish to thank the many visitors to Boston who have given so generously of their time and of their ideas. Most important, I wish to thank my present and former students from whom I have learned so much.

The specific results discussed in this chapter benefited directly from work of many individuals, among whom are P. Alstrøm, M. Araujo, A.-L. Barabási, S. V. Buldyrev, A. Bunde, F. Caserta, A. Coniglio, G. Daccord, P. Devillard, Z. Djordjevic, F. Family, A.D. Fowler, A. Geiger, S. Glotzer, A.B. Harris, S. Havlin, H.J. Herrmann, G. Huber, N. Jan, J. Kertész, H. Larralde, J. Lee, B.B. Mandelbrot, A. Margolina, P. Meakin, S. Milošević, T. Nagatani, H. Nakanishi, J. Nittmann, C.-K. Peng, P.H. Poole, S. Prakash, P.J. Reynolds, S. Sastry,

S. Schwarzer, F. Sciortino, R.L. Blumberg Selinger, D. Stauffer, B. Stošić, P. Trunfio, T. Vicsek, G.H. Weiss, and T.A. Witten.

Last but not least, I wish to thank ONR, NSF, BP, and NATO for financial support, without which none of this research could have taken place.

References

1.1   H.E. Stanley, N. Ostrowsky, eds.: *On Growth and Form: Fractal and Non-Fractal Patterns in Physics* (Martinus Nijhoff, Dordrecht 1985);
H.E. Stanley, N. Ostrowsky, eds.: *Random Fluctuations and Pattern Growth: Experiments and Theory* (Kluwer, Dordrecht 1988);
H.E. Stanley, N. Ostrowsky, eds.: *Correlations and Connectivity: Geometric Aspects of Physics, Chemistry and Biology* (Kluwer, Dordrecht 1990)

1.2   L. Pietronero, E. Tosatti, eds.: *Fractals in Physics* (Elsevier, Amsterdam 1986)

1.3   N. Boccara, M. Daoud, eds.: *Physics of Finely Divided Matter* (Springer, Berlin, Heidelberg 1986)

1.4   R. Pynn, A. Skjeltorp, eds.: *Scaling Phenomena in Disordered Systems* (Plenum, New York 1986);
R. Pynn, T. Riste, eds.: *Time Dependent Effects in Disordered Materials* (Plenum, New York 1987)

1.5   J. Feder: *Fractals* (Plenum, New York 1988)

1.6   T. Vicsek: *Fractal Growth Phenomena, Second Edition* (World Scientific, Singapore 1993)

1.7   D. Avnir, ed.: *The Fractal Approach to Heterogeneous Chemistry* (Wiley, Chichester 1989)

1.8   J.P. Hulin, A.M. Cazabat, E. Guyon, F. Carmona, eds.: *Hydrodynamics of Dispersed Media*, Random Materials and Processes, Vol. 1 (North-Holland, Amsterdam 1990);
H.J. Herrmann, S. Roux, eds.: *Statistical Models for the Fracture of Disordered Media*, Random Materials and Processes, Vol. 2 (North-Holland, Amsterdam 1990);
P.J. Reynolds, ed.: *On Clusters and Clustering*, Random Materials and Processes, Vol. 3 (North-Holland, Amsterdam 1993);
D. Bideau, A. Hansen, eds.: *Disorder and Granular Media*, Random Materials and Processes, Vol. 4 (North-Holland, Amsterdam 1993);
G.H. Weiss, ed.: *Aspects and Applications of the Random Walk*, Random Materials and Processes, Vol. 5 (North-Holland, Amsterdam 1994)

1.9   L. Pietronero, ed.: *Fractals: Physical Origin and Properties* (Plenum, London 1990)

1.10   H. Takayasu: *Fractals in the Physical Sciences* (Manchester Univ. Press, Manchester 1990)

1.11   D. Stauffer, H.E. Stanley: *From Newton to Mandelbrot: A Primer in Theoretical Physics, Second Edition* (Springer, Berlin, Heidelberg 1995)

1.12   E. Guyon, H.E. Stanley: *Les Formes Fractales* (Palais de la Découverte, Paris 1991)

1.13   T. Vicsek, M. Shlesinger, M. Matsushita, eds.: *Fractals in Natural Sciences* (World Scientific, Singapore, 1994)

1.14   J.M. Garcia-Ruiz, E. Louis, P. Meakin, L. Sander, eds.: *Growth Patterns in Physical Sciences and Biology* [Proc. 1991 NATO Advanced Research Workshop, Granada, Spain, October 1991], (Plenum, New York, 1993);
D. L. Turcotte: *Fractals and Chaos in Geology and Geophysics* (Cambridge University Press, Cambridge, 1992);
J.B. Bassingthwaighte, L.S. Liebovitch, B.J. West: *Fractal Physiology* (Oxford University Press, New York, 1994);
R. C. Hilborn: *Chaos and Nonlinear Dynamics* (Oxford University Press, New York, 1994)

1.15   A.-L. Barabási, H.E. Stanley: *Fractal Concepts in Surface Growth* (Cambridge University Press, Cambridge, 1995)

1.16 L. Benguigui, M. Daoud: Geog. Analy. **23**, 362 (1991);
      L. Benguigui: Physica A **191**, 75 (1992)
1.17 H.A. Makse, S. Havlin, H.E. Stanley: "Modeling Morphology of Cities and Towns",
      Nature (in press);
      M. Batty, P. Longley: *Fractal Cities* (Academic Press, San Diego, 1994)
1.18 H.E. Stanley: J. Stat. Phys. **36**, 843 (1984)
1.19 T. A. Witten, L.M. Sander: Phys. Rev. Lett. **47**, 1400 (1981);
      T.A. Witten, L.M. Sander: Phys. Rev. B **27**, 5686 (1983);
      L.M. Sander: Nature **332**, 789 (1986)
1.20 H. E. Stanley: Phil. Mag. B **56**, 665 (1987);
      H.E. Stanley, A. Bunde, S. Havlin, J. Lee, E. Roman, S. Schwarzer: Physica A **168**, 23
      (1990);
      H.E. Stanley, A. Coniglio, S. Havlin, J. Lee, S. Schwarzer, M. Wolf, Physica A **205**,
      254 (1994)
1.21 R.M. Brady, R.C. Ball: Nature **309**, 225 (1984);
      M. Matsushita, M. Sano, Y. Hayakawa, H. Honjo, Y. Sawada: Phys. Rev. Lett. **53**, 286
      (1984)
1.22 E. Ben-Jacob, R. Godbey, N.D. Goldenfeld, J. Koplik, H. Levine, T. Mueller, L.M.
      Sander: Phys. Rev. Lett. **55**, 1315 (1985);
      E. Ben-Jacob, P. Garik: Nature **343**, 523 (1990);
      J.S. Langer: Science **243**, 1150 (1989);
      A. Buka, J. Kertész, T. Vicsek: Nature **323**, 424 (1986)
1.23 J. Nittmann, H.E. Stanley: Nature **321**, 663 (1986)
1.24 J. Nittmann, H.E. Stanley: J. Phys. A **20**, L981 (1987)
1.25 J. Nittmann, H.E. Stanley: J. Phys. A **20**, L1185 (1987);
      F. Family, D. Platt, T. Vicsek: J. Phys. A **20**, L1177 (1987)
1.26 L. Niemeyer, L. Pietronero, H.J. Wiesmann: Phys. Rev. Lett. **52**, 1033 (1984);
      L. Pietronero, H.J. Wiesmann: J. Stat. Phys. **36**, 909 (1984)
1.27 H. S. Hele-Shaw: Nature **58**, 34 (1898);
      P. G. Saffman, G. I. Taylor: Proc. R. Soc. London A **245**, 312 (1958);
      J. Nittmann, G. Daccord, H.E. Stanley: Nature **314**, 141 (1985);
      J.D. Chen, D. Wilkinson: Phys. Rev. Lett. **55**, 1892 (1985);
      K. J. Måløy, J. Feder, T. Jøssang: Phys. Rev. Lett. **55**, 2688 (1985);
      H. Van Damme, F. Obrecht, P. Levitz, L. Gatineau, C. Laroche: Nature **320**, 731
      (1986);
      G. Daccord, J. Nittmann, H.E. Stanley: Phys. Rev. Lett. **56**, 336 (1986);
      U. Oxaal, M. Murat, F. Boger, A. Aharony, J. Feder, T. Jøssang: Nature **329**, 32
      (1987);
      R.B. Selinger, J. Nittmann, H.E. Stanley: Phys. Rev. A **40**, 2590 (1989);
      K.R. Bhaskar, B.S. Turner P. Garik, J.D. Bradley, R. Bansil, H.E. Stanley, J.T. LaM-
      ont: Nature **360**, 458 (1992);
      R. Bansil, H.E. Stanley, T.J. LaMont: Ann. Rev. Physiol. **57**, 635 (1995)
1.28 G. Daccord: Phys. Rev. Lett. **58**, 479(1987);
      G. Daccord, R. Lenormand: Nature **325**, 41 (1987);
      T. Nagatani, J. Lee, H.E. Stanley: Phys. Rev. Lett. **66**, 616 (1991);
      T. Nagatani, J. Lee, H.E. Stanley: Phys. Rev. A **45**, 2471 (1992)
1.29 A.D. Fowler, H.E. Stanley, G. Daccord: Nature **341**, 134 (1989)
1.30 M. Matsushita, H. Fujikawa: Physica A **168**, 498 (1990)
1.31 F. Family, B.R. Masters, D.E. Platt: Physica D **38**, 98 (1989)
1.32 H.E. Stanley: Bull. Am. Phys. Soc. **34**, 716 (1989);
      F. Caserta, H.E. Stanley, W. Eldred, G. Daccord, R.E. Hausman, J. Nittmann: Phys.
      Rev. Lett. **64**, 95 (1990);
      F. Caserta, R.E. Hausman, W.D. Eldred, H.E. Stanley, C. Kimmel, Neurosci. Letters
      136, 198 (1992);
      F. Caserta, W.D. Eldred, E. Fernandez, R.E. Hausman, L.R. Stanford, S.V. Buldyrev,
      S. Schwarzer, H.E. Stanley: J. Neurosci. Methods **56**, 133 (1995)
1.33 D. Kleinfeld, F. Raccuia-Behling, G.E. Blonder: Phys. Rev. Lett. **65**, 3064 (1990)
1.34 P. Meakin: CRC Critical Rev. in Solid State and Materials Sciences **13**, 143 (1987);
      S. Tolman, P. Meakin: Phys. Rev. A **40**, 428 (1989);

P. Meakin, R.C. Ball, P. Ramanlal, L.M. Sander: Phys. Rev. A **35**, 5233 (1987);
J-P Eckmann, P. Meakin, I. Procaccia, R. Zeitak: Phys. Rev. A **39**, 3185 (1990)

1.35   H.E. Stanley: *Introduction to Phase Transitions and Critical Phenomena* (Oxford University Press, London 1971)

1.36   D. Stauffer: *Introduction to Percolation Theory* (Taylor & Francis, London 1985)

1.37   M. Jensen, P. Cvitanović, T. Bohr: Europhys. Lett. **6**, 445 (1988);
J. Lee, H.E. Stanley: Phys. Rev. Lett. **61**, 2945 (1988);
J. Lee, P. Alstrøm, H.E. Stanley: Phys. Rev. A **39**, 6545 (1989);
T.C. Halsey, in: [1.9]

1.38   B.B. Mandelbrot: J. Fluid Mech. **62**, 331 (1974)

1.39   P. Grassberger: Physics Lett. **97A**, 227 (1983);
P. Grassberger, I. Procaccia: Physica D **13**, 34 (1984);
P. Grassberger, Phys. Lett. **107A**, 101 (1985)

1.40   H.G.E. Hentschel, I. Procaccia: Physica D **8**, 435 (1983)

1.41   U. Frisch, G. Parisi: In *Turbulence and Predictability in Geophysical Fluid Dynamics and Climate Dynamics*, ed. by M. Ghil, R. Benzi, G. Parisi (North-Holland, Amsterdam 1985)

1.42   G. Paladin, A. Vulpiani: Phys. Rep. **156**, 147 (1987);
R. Benzi, G. Paladin, G. Parisi, A. Vulpiani: J. Phys. A **17**, 3521 (1984);
R. Benzi, G. Paladin, G. Parisi, A. Vulpiani: J. Phys. A **18**, 2157 (1985);
R. Badii, A. Politi: J. Stat. Phys. **40**, 725 (1985);
A. Crisanti, G. Paladin, A. Vulpiani: Phys. Lett. **126A**, 120 (1987)

1.43   L. de Arcangelis, S. Redner, A. Coniglio: Phys. Rev. B **31**, 4725 (1985);
R. Rammal, C. Tannous, P. Breton, A.M.S. Tremblay: Phys. Rev. Lett. **54**, 1718 (1985);
L. de Arcangelis, S. Redner, A. Coniglio: Phys. Rev. B **34**, 4656 (1986)

1.44   A. Coniglio, in: *Statistical Physics* [Proc. STATPHYS-16], ed. by H.E. Stanley (North-Holland, Amsterdam 1986)

1.45   P. Meakin, H.E. Stanley, A. Coniglio, T.A. Witten: Phys. Rev. A **32**, 2364 (1985);
T.C. Halsey, P. Meakin, I. Procaccia: Phys. Rev. Lett. **56**, 854 (1986);
T.C. Halsey, M.H. Jensen, L.P. Kadanoff, I. Procaccia, B. Schraiman: Phys. Rev. A **33**, 1141 (1986);
P. Meakin, A. Coniglio, H.E. Stanley, T.A. Witten: Phys. Rev. A **34**, 3325 (1986)

1.46   C. Amitrano, A. Coniglio, F. di Liberto: Phys. Rev. Lett. **57**, 1016 (1987);
J. Nittmann, H.E. Stanley, E. Touboul, G. Daccord: Phys. Rev. Lett. **58**, 619 (1987);
Y. Hayakawa, S. Sato, M. Matsushita: Phys. Rev. A **36** 1963 (1987);
C. Amitrano, P. Meakin, H.E. Stanley: Phys. Rev. A **40**, 1713 (1989)

1.47   T. Nagatani: Phys. Rev. A **36**, 5812 (1987);
T. Nagatani: J. Phys. A **20**, L381 (1987);
J. Lee, A. Coniglio, H.E. Stanley: Phys. Rev. A **41**, 4589 (1990);
T. Nagatani, H.E. Stanley: Phys. Rev. A **41**, 3263 (1990);
T. Nagatani, H.E. Stanley: Phys. Rev. A **42**, 3512 (1990);
T. Nagatani, H.E. Stanley: Phys. Rev. A **42**, 4838 (1990);
T. Nagatani, H.E. Stanley: Phys. Rev. A **43**, 2963 (1991)

1.48   A.E. Ferdinand, M.E. Fisher: Phys. Rev. **185**, 832 (1969)

1.49   B.B. Mandelbrot: Physica A **168**, 95 (1990)

1.50   L. Pietronero, A. Erzan, C. J. G. Evertsz: Phys. Rev. Lett. **61**, 861 (1988);
L. Pietronero, A. Vespignani, S. Zapperi: Phys. Rev. Lett. **72**, 1690 (1994);
A. Erzan, L. Pietronero, A. Vespignani: Rev. Mod. Phys. (in press).

1.51   R. Blumenfeld, A. Aharony: Phys. Rev. Lett. **62**, 2977 (1989);
J. Lee, P. Alstrøm, H.E. Stanley: Phys. Rev. Lett. **62**, 3013 (1989)

1.52   P. Trunfio, P. Alstrøm: Phys. Rev. B **41**, 896 (1990)

1.53   A.B. Harris, M. Cohen: Phys. Rev. A **41**, 971 (1990);
A.L. Barabási, T. Vicsek: J. Phys. A **23**, L729 (1990)

1.54   B.B. Mandelbrot, T. Vicsek: J. Phys. A **20**, L377 (1989)

1.55   S. Schwarzer, J. Lee, A. Bunde, S. Havlin, H.E. Roman, H.E. Stanley: Phys. Rev. Lett. **65**, 603 (1990);
J. Lee, S. Havlin, H. E. Stanley: Phys. Rev. A **45**, 1035 (1992)

1.56 J. Lee, S. Havlin, H.E. Stanley, J. Kiefer: Phys. Rev. A **42**, 4832 (1990)
S. Schwarzer, J. Lee, S. Havlin, H. E. Stanley, P. Meakin: Phys. Rev. A **43**, 1134 (1991)
1.57 S. Havlin, B.L. Trus, A. Bunde, H.E. Roman: Phys. Rev. Lett. **63**, 1189 (1989);
J. Lee, H.E. Stanley: Phys. Rev. Lett. **63**, 1190 (1989)
1.58 S. Havlin, B.L. Trus: J. Phys. A **21**, L731 (1988)
1.59 A. Coniglio, M. Zannetti: Physica A **163** 325 (1990);
C. Amitrano, A. Coniglio, P. Meakin, M. Zannetti: Phys. Rev. B **44**, 4974 (1991);
P. Ossadnik: Physica A **195**, 319 (1993);
J. Lee, S. Schwarzer, A. Coniglio, H.E. Stanley: Phys. Rev. E **48**, 1305 (1993)
1.60 S. Schwarzer, M. Wolf, S. Havlin, P. Meakin, H.E. Stanley: Phys. Rev. A **46**, R3016 (1992);
S. Schwarzer, S. Havlin, H.E. Stanley: Physica A **191**, 117 (1992)
1.61 S. Schwarzer, S. Havlin, H.E. Stanley: Phys. Rev. E **49**, 1181 (1994)
1.62 S. Schwarzer, S. Havlin, P. Ossadnik, H.E. Stanley, "Number of Branches in Diffusion-Limited Aggregates: The Skeleton," Phys. Rev. E **52**, xx (1995)
1.63 P. Meakin, I. Majid, S. Havlin, H.E. Stanley: J. Phys. A **17**, L975 (1984)
1.64 S. Tolman, P. Meakin: Phys. Rev. A **40**, 428 (1989)
1.65 P. Ossadnik: Physica A **176**, 454 (1991)
1.66 P. Alstrøm, P. Trunfio, H.E. Stanley: in [1.1], pp. 340–342
1.67 E.L. Hinrichsen, K.J. Måløy, J. Feder, T. Jøssang: J. Phys. A **22**, L271 (1989)
1.68 P. Ossadnik: Phys. Rev. A **45**, 1058 (1992)
1.69 S. Havlin, R. Nossal, B. Trus, G. Weiss: J. Phys. A **17**, L957 (1984)
1.70 W. Lenz: Phys. Z. **21**, 613 (1920);
E. Ising: Ann. Physik **31**, 253 (1925)
1.71 A. Geiger, H.E. Stanley: Phys. Rev. Lett. **49**, 1895 (1982)
1.72 E.T. Gawlinski, H.E. Stanley: J. Phys. A **14**, L291 (1981)
1.73 A. Geiger, H.E. Stanley: Phys. Rev. Lett. **49**, 1749 (1982)
1.74 R. Ball: Physica **140A**, 62 (1986)
1.75 C. Tang: Phys. Rev. A. **31** 1977 (1985)
1.76 J. Kertész, T. Vicsek: J. Phys. A **19**, L257 (1986)
1.77 H.E. Stanley: Physica A **163**, 334 (1990)
1.78 U. Nakaya: *Snow Crystals* (Harvard University Press, Cambridge 1954)
1.79 E.R. LaChapelle: *Field Guide to Snow Crystals* (University of Washington Press, Seattle 1969)
1.80 W.A. Bentley, W.J. Humphreys: *Snow Crystals* (Dover, New York 1962)
1.81 A. Dougherty, P.D. Kaplan, J. P. Gollub: Phys. Rev. Lett. **58**, 1652 (1987)
1.82 H.E. Stanley, in [1.1]
1.83 P.G. de Gennes: La Recherche **7**, 919 (1976)
1.84 P. Meakin, H.E. Stanley: Phys. Rev. Lett. **51**, 1457 (1983)
1.85 H.E. Stanley, in: *Scaling Phenomena in Disordered Systems*, ed. by R. Pynn, A. Skjeltorp (Plenum, New York 1986)
1.86 S. Alexander, R. Orbach: J. de Phys. **43**, L625 (1982);
R. Rammal, G. Toulouse: J. de Phys. **44**, L13 (1983)
1.87 R.B. Pandey, D. Stauffer: Phys. Rev. Lett. **51**, 527 (1983)
1.88 S. Havlin, D. Ben-Avraham, J. Phys. A **16**, L483 (1983)
1.89 P. Meakin, H.E. Stanley: J. Phys. A **17**, L173 (1984);
K. Kang, S. Redner: Phys. Rev. Lett. **52**, 955 (1984)
1.90 F. Leyvraz, H.E. Stanley: Phys. Rev. Lett. **51**, 2048 (1983)
1.91 H.E. Stanley, I. Majid, A. Margolina, A. Bunde: Phys. Rev. Lett. **53**, 1706 (1984)
1.92 S. Havlin, D. Ben-Avraham: Adv. Phys. **36**, 695 (1987);
A. Bunde: Adv. Solid State Phys. **26**, 113 (1986);
J.-P. Bouchaud, A. Georges: Phys. Rep. **195**, 127 (1990)
1.93 A.L. Ritzenberg, R.I. Cohen: Phys. Rev. B **30**, 4038 (1984)
1.94 K.M. Middlemiss, S.G. Whittington, D.C. Gaunt: J. Phys. A **13**, 1835 (1980)
1.95 Z. Alexandrowicz: Phys. Lett. A **80**, 284 (1980)
1.96 R. Pike, H.E. Stanley: J. Phys. A **14**, L169 (1981)
1.97 S. Havlin, R. Nossal: J. Phys. A **17**, L427 (1984);
S. Havlin, L. A. N. Amaral, S. V. Buldyrev, S. T. Harrington, H. E. Stanley: Phys. Rev. Lett. **74**, 4205 (1995)

1.98 H.J. Herrmann, H.E. Stanley: J. Phys. A **21**, L829 (1988);
      P. Grassberger: J. Phys. A **25**, 5475 (1992);
      P. Grassberger: J. Phys. A **25**, 5867 (1992)
1.99 S. Havlin, Z. Djordjevic, I. Majid, H.E. Stanley, G.H. Weiss: Phys. Rev. Lett. **53**, 178 (1984)
1.100 P. Meakin, I. Majid, S. Havlin, H.E. Stanley: J. Phys. A **17**, L975 (1984)
1.101 H.E. Stanley, D. Stauffer, J. Kertész, H.J. Herrmann: Phys. Rev. Lett. **59**, 2326 (1987);
      B. Derrida, G. Weisbuch: Europhys. Lett. **4**, 657 (1987);
      A. Coniglio, L. de Arcangelis, H.J. Herrmann, N. Jan: Europhys. Lett. **8**, 315 (1989);
      L. de Arcangelis, A. Coniglio, H.J. Herrmann: Europhys. Lett. **9**, 749 (1989)
1.102 P. Grassberger: J. Phys. A **18**, L215 (1985)
1.103 M. Barma: J. Phys. A **18**, L277 (1985)
1.104 H.E. Stanley: J. Phys. A **10**, L211 (1977);
      A. Coniglio: Phys. Rev. Lett. **46**, 250 (1981);
      A. Coniglio: J. Phys. A **15**, 3829 (1982)
1.105 S. Kirkpatrick: AIP Conf Proc **40**, 99 (1978);
      G. Shlifer, W. Klein, P.J. Reynolds, H.E. Stanley: J. Phys. A **12**, L169 (1979)
1.106 H.E. Stanley, R.J. Birgeneau, P.J. Reynolds, J.F. Nicoll: J. Phys. C **9**, L553 (1976)
1.107 H.J. Herrmann, H.E. Stanley: Phys. Rev. Lett. **53**, 1121 (1984)
1.108 G. Huber, H.E. Stanley: unpublished
1.109 H.J. Herrmann, D. Hong, H.E. Stanley: J. Phys. A **17**, L261 (1984)
1.110 F. Family, T. Vicsek: J. Phys. A **18**, L75 (1985);
      Y.C. Zhang: J. de Phys. **51**, 2113 (1990);
      J. Krug: J. de Phys. I **1**, 9 (1991);
      S.V. Buldyrev, S. Havlin, J. Kertész, H.E. Stanley, T. Vicsek: Phys. Rev. A **43**, 7113 (1991);
      S. Havlin, S.V. Buldyrev, H.E. Stanley, G. H. Weiss: J. Phys. A **24**, L925 (1991);
      S. Havlin, R. Selinger, M. Schwartz, H.E. Stanley, A. Bunde: Phys. Rev. Lett. **61**, 1438 (1988);
      S. Havlin, M. Schwartz, R. Blumberg Selinger, A. Bunde, H.E. Stanley: Phys. Rev. A **40**, 1717 (1989);
      R. Blumberg Selinger, S. Havlin, F. Leyvraz, M. Schwartz, H.E. Stanley: Phys. Rev. A **40**, 6755 (1989);
      C.K. Peng, S. Havlin, M. Schwartz, H.E. Stanley: Phys. Rev. A **44**, 2239 (1991);
      C.K. Peng, S. Havlin, M. Schwartz, H.E. Stanley, G.H. Weiss: Physica A **178**, 401 (1991);
      S. Prakash, S. Havlin, M. Schwartz, H. E. Stanley: Phys. Rev. A **46**, R1724 (1992)
1.111 H. Kunz, B. Souillard: J. Stat. Phys. **19**, 77 (1978)
1.112 B. Sapoval, M. Rosso, J.F. Gouyet: J. de Phys. Lett. **46**, L149 (1985);
      R.S. Voss: J. Phys. A **17**, L373 (1984)
1.113 H. Saleur, B. Duplantier: Phys. Rev. Lett. **58**, 2325 (1987)
1.114 A. Coniglio, H.E. Stanley: Phys. Rev. Lett. **52**, 1068 (1984)
1.115 A. Bunde, H.J. Herrmann, A. Margolina, H.E. Stanley: Phys. Rev. Lett. **55**, 653 (1985);
      A. Bunde, H.J. Herrmann, H.E. Stanley: J. Phys. A **18**, L523 (1985);
      H.J. Herrmann, H.E. Stanley: Z. Phys. **60**, 165 (1985)
1.116 T. Grossman, A. Aharony: J. Phys. A **19**, L745 (1986);
      T. Grossman, A. Aharony: J. Phys. A **20**, L1193 (1987);
      B. Duplantier, H. Saleur: Phys. Rev. Lett. **62**, 1368 (1989)
1.117 A. Coniglio, N. Jan, I. Majid, H.E. Stanley: Phys. Rev. B **35**, 3617 (1987);
      I. Majid, N. Jan, A. Coniglio, H.E. Stanley: Phys. Rev. Lett. **52**, 1257 (1984);
      B. Duplantier, H. Saleur: Phys. Rev. Lett. **59**, 539 (1987);
      P.H. Poole, A. Coniglio, N. Jan, H.E. Stanley: Phys. Rev. Lett. **60**, 1203 (1988);
      P.H. Poole, A. Coniglio, N. Jan, H.E. Stanley: Phys. Rev. B **39**, 495 (1989)
1.118 A.-L. Barabási, S.V. Buldyrev, S. Havlin, G. Huber, H.E. Stanley, T. Vicsek: in *Surface disordering: Growth, roughening and phase transitions*, edited by R. Jullien, J. Kertész, P. Meakin and D. E. Wolf (Nova Science, New York, 1992), pp. 193–204
1.119 L.-H. Tang, H. Leschhorn: Phys. Rev. A **45**, R8309 (1992)
1.120 S.V. Buldyrev, A.-L. Barabási, F. Caserta, S. Havlin, H.E. Stanley and T. Vicsek: Phys. Rev. A **45**, R8313 (1992)

1.121 S. V. Buldyrev, S. Havlin and H. E. Stanley, Physica A **200**, 200 (1993);

1.122 S. Havlin, A.-L. Barabási, S.V. Buldyrev, C.K. Peng, M. Scwartz, H.E. Stanley, T. Vicsek: in *Growth Patterns in Physical Sciences and Biology*, edited by E. Louis, L. Sander and P. Meakin, Proc. 1991 NATO Advanced Research Workshop, Granada, Spain (Plenum, New York, 1993), pp. 85–98

1.123 H. Leschhorn: Physica A **195**, 324 (1993)

1.124 H. Leschhorn, L.-H. Tang: Phys. Rev. E **49**, 1238 (1994)

1.125 Z. Olami, I. Procaccia, R. Zeitak: Phys. Rev. E **49**, 1232 (1994)

1.126 K. Sneppen: Phys. Rev. Lett. **69**, 3539 (1992)

1.127 L.-H. Tang, H. Leschhorn: Phys. Rev. Lett. **70**, 3832 (1993)

1.128 K. Sneppen, M.H. Jensen: Phys. Rev. Lett. **70**, 3833 (1993)

1.129 D. Ruelle: *Statistical Mechanics, Thermodynamic Formalism* (Addison-Wesley, Reading MA 1975);

M. J. Feigenbaum: J. Stat. Phys. **46**, 919 (1987)

1.130 H.E. Stanley, P. Meakin: Nature **335**, 405 (1988)

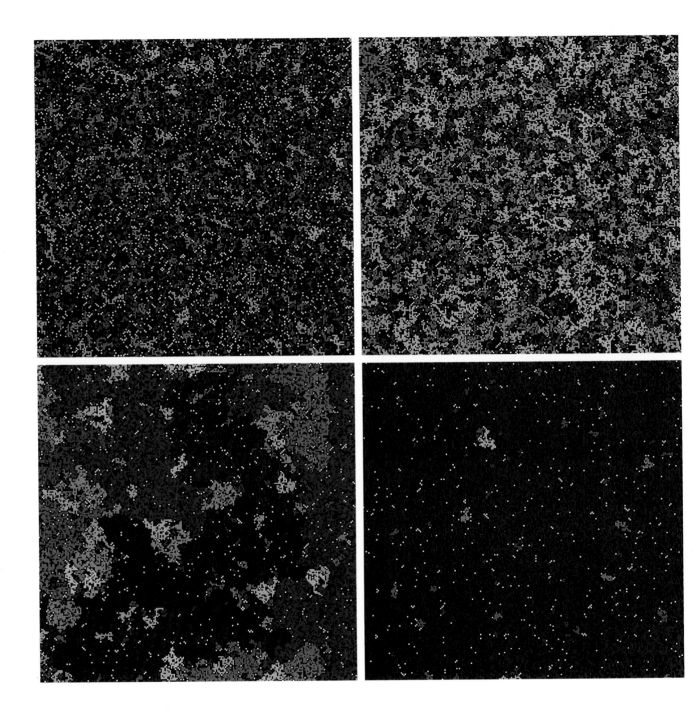

2 Percolation I

Armin Bunde and Shlomo Havlin

2.1 Introduction

Percolation represents the simplest model of a disordered system [2.1–10]. Consider a square lattice, where each site is occupied randomly with probability p or is empty with probability $1 - p$ (see Fig. 2.0). Occupied and empty sites may stand for very different physical properties. For simplicity, let us assume that the occupied sites are electrical conductors, the empty sites represent insulators, and that electrical current can flow only between nearest-neighbor conductor sites.

At low concentration p, the conductor sites are either isolated or form small clusters of nearest-neighbor sites. Two conductor sites belong to the same cluster if they are connected by a path of nearest-neighbor conductor sites, and a current can flow between them. At low p values, the mixture is an insulator, since no conducting path connecting opposite edges of our lattice exists. At large p values, on the other hand, many conduction paths between opposite edges exist, where electrical current can flow, and the mixture is a conductor.

At some concentration in between, therefore, a threshold concentration p_c must exist where for the first time electrical current can *percolate* from one edge to the other. Below p_c we have an insulator, above p_c we have a conductor. The threshold concentration is called the *percolation threshold,* or, since it separates two different phases, the *critical concentration.*

If the occupied sites are superconductors and the empty sites are conductors, p_c separates a normal-conducting phase below p_c from a superconducting phase

---

◄ **Fig. 2.0.** Site percolation clusters on a 510 × 510 square lattice, for p=0.4, 0.5, 0.593, 0.65. Different cluster sizes have different colors: empty sites are blue, clusters of one site are white. The color of the larger clusters depends on their size and varies from green (*small clusters*) via yellow and red to deep blue (infinite cluster at $p = 0.65$). Courtesy of S. Schwarzer

above p_c. Another example is a mixture of magnets and paramagnets, where the system changes at p_c from a paramagnet to a magnet.

In contrast to the more common thermal phase transitions, where the transition between two phases occurs at a critical temperature, the *percolation transition* described here is a *geometrical phase transition,* which is characterized by the geometric features of large clusters in the neighborhood of p_c. At low values of p only small clusters of occupied sites exist. When the concentration p is increased, the average size of the clusters increases. At the critical concentration p_c a large cluster appears which connects opposite edges of the lattice. We call this cluster the *infinite* cluster, since its size diverges when the size of the lattice is increased to infinity. When p is increased further, the density of the infinite cluster increases, since more and more sites become part of the infinite cluster. Accordingly, the average size of the *finite* clusters, which do not belong to the infinite cluster, decreases. At $p = 1$, trivially, all sites belong to the infinite cluster.

So far we have considered *site percolation,* where the sites of a lattice have been occupied randomly. When the bonds between the sites are randomly occupied, we speak of *bond percolation.* Two occupied bonds belong to the same cluster if they are connected by a path of occupied bonds (see Fig. 2.1). The critical concentration of bonds separates a phase of finite clusters of bonds from a phase with an infinite cluster.

Perhaps the most common example of bond percolation in physics is a *random resistor network,* where the metallic wires in a regular network are cut randomly with probability $q \equiv 1 - p$. Here q_c separates a conductive phase at low q from an insulating phase at large q.

Fig. 2.1. Bond-percolation clusters on a 20×20 square lattice

A possible application of bond percolation in chemistry is the polymerization process [2.11–13], where small branching molecules can form large molecules by activating more and more bonds between them. If the activation probability p is above the critical concentration, a network of chemical bonds spanning the whole system can be formed, while below p_c only macromolecules of finite size can be generated. This process is called a *sole-gel* transition. An example of this *gelation* process is the boiling of an egg, which at room temperature is liquid and upon heating becomes a more solid-like *gel*.

An example from biology concerns the spreading of an epidemic [2.14]. In its simplest form, the epidemic starts with one sick individual, which can infect its nearest-neighbors with probability p in one time step. After one time step, it dies and the infected neighbors in turn can infect their (so far) uninfected neighbors, and the process is continued. Here the critical concentration separates a phase at low p where the epidemic always dies out after a finite number of time steps from a phase where the epidemic can continue forever. The same process can be used as a model for forest fires [2.1,15,83,87], with the infection probability replaced by the probability that a burning tree can ignite its nearest-neighbor trees in the next time step.

In addition to these simple examples, percolation aspects have been found useful to describe a large number of disordered systems in physics and chemistry [2.3,11,16–32], such as porous and amorphous materials [2.20] (including thin films [2.23]), disordered ionic conductors (including mixed-alkali glasses, two-phase mixtures [2.24], and dispersed ionic conductors [2.24–27]), branched polymers [2.28], fragmentation [2.29] (including nuclear fragmentation [2.30] and earthquakes [2.9,10]), galactic structures [2.31], and supercooled water [2.32].

The definitions of site and bond percolation on a square lattice can easily be generalized to any lattice in d dimensions. In general, in a given lattice, a bond has more nearest-neighbors than a site. For example, in the square lattice one bond is connected to six nearest-neighbor bonds, while a site has only four nearest-neighbor sites. Thus, large clusters of bonds can be formed more effectively than large clusters of sites, and a lower concentration of bonds is needed to form a spanning cluster; i.e., on a given lattice the percolation threshold for bonds is smaller than the percolation threshold for sites (see Table 2.1).

So far, we have focused on either site or bond percolation, where either sites *or* bonds of a given lattice have been chosen randomly. If sites are occupied with probability p *and* bonds are occupied with probability q, we speak of *site–bond percolation*. Two occupied sites belong to the same cluster if they are connected by a path of nearest-neighbor occupied sites with occupied bonds in between. For $q = 1$, site–bond percolation reduces to site percolation, for $p = 1$ it reduces to bond percolation. In general, both parameters characterize the state of the system. Accordingly, a *critical line* in *p-q* space separates both phases, which for $p = 1$ and $q = 1$ takes the values of the critical bond and site concentrations,

Table 2.1. Percolation thresholds for several two- and three-dimensional lattices and the Cayley tree (see Sect. 2.4.2). [a]Exact [2.2,33,35]; [b]numerical method [2.34]; [c]Monte-Carlo [2.1]; [d]series expansion [2.36]; [e]Monte-Carlo [2.37]; [f]series expansion [2.38]; [g]numerical method [2.39]; [h]numerical method [2.35]

Lattice	Percolation of	
	sites	bonds
Triangular	$1/2^a$	$2\sin(\pi/18)^a$
Square	$0.5927460^{b,h}$	$1/2^a$
Honeycomb	0.6962^c	$1\text{-}2\sin(\pi/18)^a$
Face Centered Cubic	0.198^c	0.119^c
Body Centered Cubic	0.245^c	0.1803^d
Simple Cubic ($1^{st}nn$)	$0.31161^{e,g}$	0.248814^g
Simple Cubic ($2^{nd}nn$)	0.137^f	$-$
Simple Cubic ($3^{rd}nn$)	0.097^f	$-$
Cayley Tree	$1/(z-1)$	$1/(z-1)$

respectively. Site–bond percolations can be relevant for gelation and epidemic processes in dilute media (see Sect. 2.6).

The most natural example of percolation, perhaps, is *continuum percolation,* where the positions of the two components of a random mixture are not restricted to the discrete sites of a regular lattice. As a simple example, consider a sheet of conductive material, with circular holes punched randomly in it (Fig. 2.2). The relevant quantity now is the fraction p of remaining conductive material. Compared with site and bond percolation, the critical concentration is further decreased: $p_c \cong 0.312 \pm 0.005$ [2.40] for $d = 2$, when all circles have the same radius. This picture can easily be generalized to three dimensions, where spherical voids are generated randomly in a cube, and $p_c \cong 0.034 \pm 0.007$ [2.40,41]. Due to its similarity to Swiss cheese, this model of continuous percolation is called the Swiss cheese model. Similar models, where also the size of the spheres can vary, are used to describe sandstone and other porous materials.

Let us end this introductory section with some historical remarks. The first work introducing the concept of percolation was performed by Flory and Stockmayer about 50 years ago, when studying the gelation process [2.42]. The name percolation was suggested by Broadbent and Hammersley in 1957 when studying the spreading of fluids in random media [2.43]. They also introduced the relevant geometrical and probabilistic concepts. The developments of phase transition theory in the following years, in particular the series expansion method by Domb [2.5] and the renormalization group theory by Wilson, Fisher, and Kadanoff [2.44], stimulated tremendously the research activities on the geometric percolation transition. The fractal concepts introduced by Mandelbrot [2.45], have made available powerful new tools, which, together with the development of large-scale computers, have contributed significantly to our present understanding of percolation.

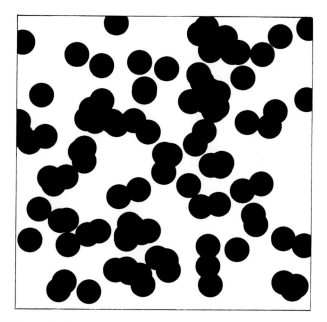

Fig. 2.2. Continuum percolation: Swiss cheese model

In this chapter we present mainly the static aspects of percolation, while dynamical properties associated with the percolation transition are discussed in Chap. 3.

2.2 Percolation as a Critical Phenomenon

The percolation transition is a simple example of a phase transition phenomenon. It is a geometrical phase transition where the critical concentration p_c separates a phase of finite clusters $(p < p_c)$ from a phase where an infinite cluster is present $(p > p_c)$. More common examples of phase transitions are thermal phase transitions such as the solid/liquid transition, where an ordered phase (the solid) changes into a disordered phase (the liquid) at some *critical* temperature T_c [2.46,47].

An illustrative example is the magnetic phase transition. At low temperatures some materials exhibit a spontaneous magnetization $m > 0$, without any external field (ferromagnetic phase). When the temperature increases the spontaneous magnetization decreases continuously and vanishes at the *critical* temperature T_c. Above T_c, in the paramagnetic phase, $m = 0$.

A magnetic material is composed of elementary magnetic moments (spins). The interactions between them favor an ordered state where all spins are parallel, while the thermal energy favors a disordered state where the spins have random orientation. At low temperatures, the interaction dominates and long-range order occurs, which is reflected by the nonzero spontaneous magnetization m. Since m describes the order in the system, it is called the *order parameter*. With increasing temperature, $m(T)$ decreases and close to T_c follows a power law, $m(T) \sim (T_c - T)^\beta$. Above T_c, the thermal energy dominates, only finite clusters of temporarily aligned spins can exist, and their random orientation leads to zero magnetization.

In percolation, the concentration p of occupied sites plays the same role as the temperature in thermal phase transitions. We will see later that as with thermal transitions, long-range correlations control the percolation transition and the relevant quantities near p_c are described by power laws and critical exponents.

The percolation transition is characterized by the geometrical properties of the clusters near p_c. An important quantity is the probability P_∞ that a site (or a bond) belongs to the infinite cluster. For $p < p_c$, only finite clusters exist, and $P_\infty = 0$. For $p > p_c$, P_∞ behaves similarly to the magnetization below T_c, and increases with p by a power law

$$P_\infty \sim (p - p_c)^\beta. \tag{2.1}$$

Similarly to the magnetization, P_∞ describes the order in the percolation system and can be identified as the *order parameter*.

The linear size of the *finite* clusters, below and above p_c, is characterized by the *correlation length* ξ. The correlation length is defined as the mean distance between two sites on the same finite cluster. When p approaches p_c, ξ increases as

$$\xi \sim |p - p_c|^{-\nu}, \tag{2.2}$$

with the same exponent ν below and above the threshold. The mean number of sites (mass) of a finite cluster also diverges,

$$S \sim |p - p_c|^{-\gamma}, \tag{2.3}$$

again with the same exponent γ above and below p_c. To obtain ξ and S, averages over all finite clusters in the lattice are required.

Analogs of the quantities P_∞ and S in magnetic systems are the magnetization m and the susceptibility χ (see Fig. 2.3 and Table 2.2).

The exponents β, ν, and γ describe the critical behavior of typical quantities associated with the percolation transition, and are called the *critical exponents*. The exponents are universal and depend neither on the structural details of the lattice (e.g., square or triangular) nor on the type of percolation (site, bond, or continuum), but only on the dimension d of the lattice.

This universality property is a general feature of phase transitions, where the order parameter vanishes continuously at the critical point (second-order phase transition). For example, the magnetization in all three-dimensional magnetic materials is described by the same exponent β, irrespective of the crystalline structure or the type of interactions between the spins, as long as they are of short range.

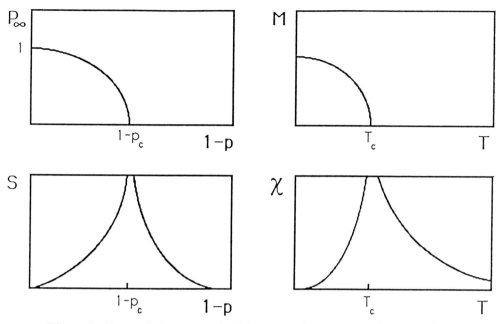

Fig. 2.3. P_∞ and S compared with magnetization m and susceptibility χ

The exponents β, ν, and γ are not the only critical exponents characterizing the percolation transition. The size distribution of percolation clusters, for example, is described by other exponents α, τ, and σ. However, as we show in Sect. 2.5 there exist relations between these exponents, and all of them can be obtained from the knowledge of just two of them.

In Table 2.2, the values of the critical exponents β, ν, and γ in percolation are listed for two, three, and six dimensions. They are compared with the analogous exponents of the magnetic phase transition.

The exponents considered here describe the geometrical properties of the percolation transition. The physical properties associated with the transition also show power-law behavior near p_c and are characterized by critical exponents.

Table 2.2. Exact values and best estimates for the critical exponents in percolation and magnetism. [a]Exact [2.48,49]; [b]numerical simulations [2.37]; [c]exact [2.4,6]

Percolation	d=2	d=3	d$\geq$ 6
Order parameter P_∞: β	$5/36$[a]	0.417 ± 0.003[b]	1[c]
Correlation length ξ: ν	$4/3$[a]	0.875 ± 0.008[b]	$1/2$[c]
Mean cluster size S: γ	$43/18$[a]	1.795 ± 0.005[b]	1[c]
Magnetism	d=2	d=3	d$\geq$ 6
Order parameter m: β	$1/8$	0.32	$1/2$
Correlation length ξ: ν	1	0.63	$1/2$
Susceptibility χ: γ	$7/4$	1.24	1

Examples are the conductivity in a random resistor or random superconducting network, or the spreading velocity of an epidemic disease near the critical infection probability. Presumably their "dynamical" exponents cannot be generally related to the geometric exponents discussed above. A review on the transport properties of percolation is given in Chap. 3.

At the end of this section we note that all quantities described above are defined in the thermodynamic limit of large systems. In a finite system, P_∞, for example, is not strictly zero below p_c. An approach to finite-size effects will be discussed in Sect. 2.5.

2.3 Structural Properties

a) **The fractal dimension d_f.** As was first noticed by Stanley [2.51], the structure of percolation clusters can be well described by the fractal concept [2.45]. We begin by considering the infinite cluster at the critical concentration p_c. A representative example of the infinite cluster is shown in Fig. 2.4.

As seen in the figure, the infinite cluster contains holes of all sizes, similarly to the Sierpinski gasket (see Chap. 1 and Sect. 2.8). The cluster is self-similar on all length scales (larger than the unit size and smaller than the lattice size), and can be regarded as a fractal. The fractal dimension d_f describes how, on average, the mass M of the cluster within a sphere of radius r scales with r,

$$M(r) \sim r^{d_f}. \tag{2.4}$$

As explained in Chap. 1, in random fractals $M(r)$ represents an average over many different cluster configurations or, equivalently, over many different centers of spheres on the same infinite cluster.

Below and above p_c, the mean size of the *finite* clusters in the system is described by the correlation length ξ. At p_c, ξ diverges and holes occur in the infinite cluster on all length scales. Above p_c, ξ also represents the linear size of the holes in the infinite cluster. Since ξ is finite above p_c, the infinite cluster can be self-similar only on length scales smaller than ξ. We can interpret $\xi(p)$ as a typical length up to which the cluster is self-similar and can be regarded as a fractal. For length scales larger than ξ, the structure is not self-similar and can be regarded as homogeneous. The crossover from the fractal behavior at small length scales to a homogeneous behavior at large length scales is best illustrated by a lattice composed of Sierpinski gasket unit cells of size ξ (see Fig. 2.5).

If our length scale is smaller than ξ, we see a fractal structure. On length scales larger than ξ, we see a homogeneous system which is composed of many unit cells of size ξ. Mathematically, this can be summarized as

$$M(r) \sim \begin{cases} r^{d_f}, & r \ll \xi, \\ r^d, & r \gg \xi. \end{cases} \tag{2.5}$$

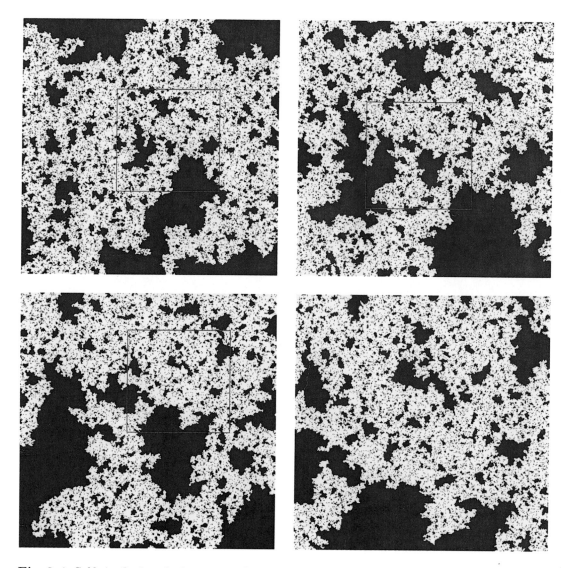

Fig. 2.4. Self-similarity of a large percolation cluster at the critical concentration. The cluster was grown by the Leath method (see Sect. 2.7.2) and consists of 20 000 shells. Courtesy of M. Meyer

Figure 2.6 shows this crossover in a $2d$ percolation system for p above p_c. One can relate the fractal dimension d_f of percolation clusters to the exponents β and ν. The probability that an arbitrary site within a circle of radius r smaller than ξ belongs to the infinite cluster, is the ratio between the number of sites on the infinite cluster and the total number of sites,

$$P_\infty \sim \frac{r^{d_f}}{r^d}, \qquad r < \xi. \tag{2.6}$$

This equation is certainly correct for $r = a\xi$, where a is an arbitrary constant smaller than 1. Substituting $r = a\xi$ in (2.6) yields

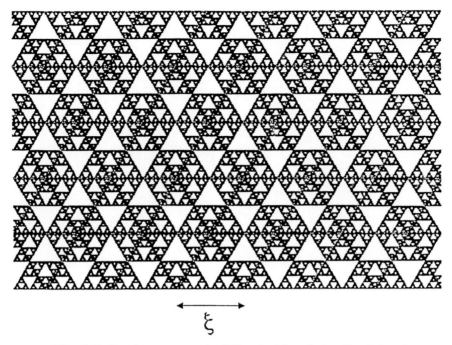

Fig. 2.5. Lattice composed of Sierpinski gasket cells of size ξ

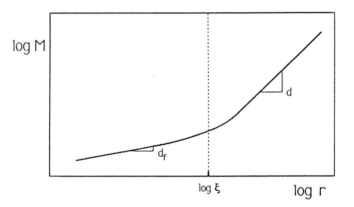

Fig. 2.6. Schematic plot of the crossover of $M(r)$ above the percolation threshold. For $r \ll \xi$ the slope is d_f while for $r \gg \xi$ the slope is d

$$P_\infty \sim \frac{\xi^{d_f}}{\xi^d}. \qquad (2.7)$$

Both sides are powers of $p - p_c$. Substituting (2.1) and (2.2) into (2.7) we obtain [2.4,40,41],

$$d_f = d - \frac{\beta}{\nu}. \qquad (2.8)$$

Thus the fractal dimension of the infinite cluster at p_c is not a new independent exponent but depends on β and ν. Since β and ν are universal exponents, d_f

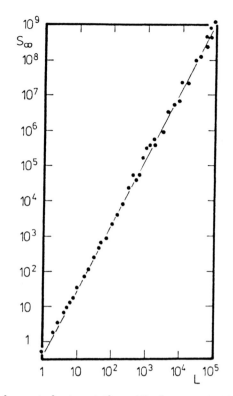

Fig. 2.7. The size of the largest cluster at the critical concentration $p_c = 1/2$ of the triangular lattice as a function of the linear size L of the lattice. For large L, the slope of this log–log plot is very close to the predicted value $d_f = 91/48$. (After [2.1])

is also universal. The actual dependence of $M(L)$ on the system linear size L for site percolation on a triangular lattice is shown in Fig. 2.7 [2.1]. At large values of L, the curve in the double logarithmic plot approaches a straight line with slope $d_f \cong 91/48$, in agreement with (2.8).

It can be shown that (2.8) also represents the fractal dimension of the finite clusters at p_c and below p_c, as long as their linear size is smaller than ξ. Below p_c there also exist (though very rarely) clusters with a linear size larger than ξ. These clusters are called *lattice animals* and their fractal dimension is smaller than d_f [2.53,54].

b) The graph dimensions d_{min} and d_ℓ. The fractal dimension, however, is not sufficient to fully characterize a percolation cluster. This becomes evident for example, on comparing pictures of diffusion limited aggregates (DLA) (see Chaps. 1, 4 and 10) and percolation clusters. The two clusters look very different. The percolation cluster has loops on all length scales, while the aggregate has practically no loops. In $d = 3$, both structures retain their characteristic differences, but their fractal dimensions are nearly the same, $d_f \cong 2.5$.

For a further intrinsic characterization of a fractal we consider the shortest path between two arbitrary sites A and B on the cluster (see Figs. 2.8–10) [2.14,55–59]. The structure formed by the sites of this path is also self-similar

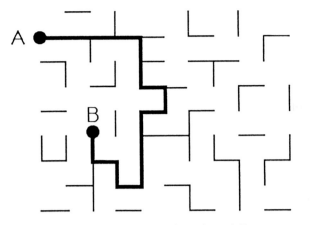

Fig. 2.8. Shortest path between two sites A and B on a percolation cluster

and is described by a fractal dimension $d_{\min}$ [2.60,61]. Accordingly, the length ℓ of the path, which is often called the "chemical distance", scales with the "Euclidean distance" r between A and B as

$$\ell \sim r^{d_{\min}}. \tag{2.9a}$$

The inverse relation

$$r \sim \ell^{1/d_{\min}} \equiv \ell^{\tilde{\nu}} \tag{2.9b}$$

tells how r scales with ℓ.

Closely related to $d_{\min}$ and d_f is the "chemical" dimension d_ℓ, which describes how the cluster mass M within the chemical distance ℓ from a given site scales with ℓ,

$$M(\ell) \sim \ell^{d_\ell}. \tag{2.10}$$

While the fractal dimension d_f characterizes how the mass of the cluster scales with the "Euclidean" distance r, the graph dimension d_ℓ characterizes how the mass scales with the chemical distance ℓ. The different arrangement of the cluster sites in ℓ and r space is demonstrated in Fig. 3.0. Combining (2.4), (2.9a), and (2.10) we obtain the relation between $d_{\min}$, d_ℓ, and d_f

$$d_\ell = \frac{d_f}{d_{\min}}. \tag{2.11}$$

To measure d_f, an arbitrary site is chosen on the cluster and one determines the number $M(r)$ of all sites within a distance r from this site. To measure d_ℓ, an arbitrary site is chosen on the cluster and one determines the number $M(\ell)$ of all sites which are connected to this site by a shortest path with length smaller or equal to ℓ. Finally, to measure $d_{\min}$, two arbitrary sites at Euclidean distance r are chosen on the cluster and one determines the length $\ell(r)$ of the shortest path connecting them. As for $M(r)$, averages have to be performed for $M(\ell)$ and $\ell(r)$ over many realizations. In regular "Euclidean" lattices, both d_ℓ and d_f coincide with the Euclidean space dimension d and $d_{\min} = 1$.

The chemical dimension d_ℓ (or $d_{min} = 1/\tilde{\nu}$) is an important tool for distinguishing between different fractal structures which may have a similar fractal dimension. In $d = 3$ for example, DLA clusters and percolation clusters have approximately the same fractal dimension $d_f \cong 2.5$, but have different $\tilde{\nu}$: $\tilde{\nu} = 1$ for DLA [2.62] but $\tilde{\nu} \cong 0.73$ for percolation [2.39,72,75,77].

While d_f has been related to the (known) critical exponents, (2.8), no such relation has been found (yet) for d_{min} or d_ℓ. The values of d_ℓ or d_{min} are known only from approximate methods, mainly numerical simulations.

The concept of the chemical distance also plays an important role in the description of dynamic phenomena in disordered systems, such as the spreading of forest fires or epidemics, that propagate along the shortest path from the seed. In Sect. 2.6 it is shown that the velocity with which the fire front or the epidemic propagates is related to the exponent $\tilde{\nu}$. Other examples where the concept of the chemical distance is essential include the diffusion of particles and localized vibrational excitation (see Sects. 3.2 and 3.7).

c) Probability densities. The relations between M, r, and ℓ are fully characterized by the probability densities $\phi(M \mid r)$, $\phi(M \mid \ell)$, and $\phi(r \mid \ell)$ [2.63,65]. $\phi(M \mid r)$ is the probability that within a circle of radius r there exist M cluster sites. Accordingly, $\phi(M \mid \ell)$ is the probability of finding M cluster sites within ℓ chemical shells, and $\phi(r \mid \ell)$ is the probability that two cluster sites separated by chemical distance ℓ are at (Euclidean) distance r from each other. The mean quantities $M(r)$, $M(\ell)$, and $r(\ell)$ discussed in (2.4), (2.9), and (2.10) represent the *first moments* of these distributions, e.g., $M(r) = \int M' \phi(M' \mid r) dM'$.

It is expected that the probability densities scale as [2.63] (see also Sect. 2.5)

$$\phi(M \mid r) \sim \frac{1}{M} f\left(\frac{M}{r^{d_f}}\right). \tag{2.12}$$

Of particular importance for the dynamical properties of percolation clusters (see Sects. 3.2 and 3.7) are the probability density $\phi(\ell \mid r)$, which is the probability that two cluster sites at an Euclidean distance r are separated by the chemical distance ℓ, and the analogous quantity $\phi(r \mid \ell)$.

Numerically, $\phi(r \mid \ell)$ can be obtained as follows. First one chooses one cluster site as a center site and counts the number $N(\ell)$ of all sites that are at a given chemical distance ℓ from this site. Among these $N(\ell)$ sites, there are $N(r, \ell)$ sites at an Euclidean distance r from the center (see Fig. 2.9). The fraction $N(r, \ell)/N(\ell)$, averaged over many configurations, can be identified with $\phi(r \mid \ell)$. The related probability density $\phi(\ell \mid r)$ is obtained in a similar way by calculating the number of sites $N(r)$ that are at an Euclidean distance r from the center, and averaging over the fraction $N(r, \ell)/N(r)$. Since by definition $\sum_r N(r, \ell) = N(\ell)$ and $\sum_\ell N(r, \ell) = N(r)$, the probability densities satisfy, in the continuum limit, the normalization condition $\int \phi(r \mid \ell) dr = \int \phi(\ell \mid r) d\ell = 1$.

Fig. 2.9. A percolation cluster at the critical concentration in a square lattice. The sites located between $r = 9.5$ and 10.5 from the center site ($\bigcirc$) are in blue, the sites located between $\ell = 14.5$ and 15.5 from the center site are in red, and those sites that are between $r = 9.5$ and 10.5 *and* between $\ell = 14.5$ and 15.5 are in red. Courtesy of J. Dräger and M. Meyer

Since $N(\ell)$ and $N(r)$ scale as

$$N(\ell) \sim \ell^{d_\ell - 1}, \quad N(r) \sim r^{d_f - 1}, \tag{2.13}$$

both probability densities are related by

$$\phi(\ell \mid r) \propto \phi(r \mid \ell)\, \ell^{d_\ell - 1}/r^{d_f - 1}. \tag{2.14}$$

For a percolation cluster, $\phi(r \mid \ell)$ cannot be calculated analytically. To obtain an idea about the specific functional form of $\phi(r \mid \ell)$, let us first consider a much simpler random fractal structure: the trace of a random walker in a high-dimensional lattice $(d \geq 4)$ (see Chap. 1 and Sect. 3.2 with Fig. 3.3). For $(d \geq 4)$, intersections of the trace are very rare, and the structure can be considered as linear $(d_\ell = 1)$. By construction, the length ℓ of this random-walk chain is identical to the number of steps t performed by the random walker. Hence, $\phi(r \mid \ell)$ for the random-walk chain has the same form as the well-known probability $P(r, t)$ of finding the random walker at time step t a distance r from his starting point, i.e., (see Sect. 3.2)

$$\phi(r \mid \ell) \propto \ell^{-d/2} \exp(-dr^2/2\ell). \tag{2.15}$$

There is numerical evidence [2.63] that a similar functional form holds also for percolation clusters,

$$\phi(r \mid \ell) = (C_1/r)(r/\ell^{\tilde{\nu}})^{g+d_f} \exp[-C_2(r/\ell^{\tilde{\nu}})^{\tilde{\delta}}], \quad \tilde{\delta} = (1 - \tilde{\nu})^{-1}, \qquad (2.16)$$

where C_1 and C_2 are (nonuniversal) constants and g is a new exponent, $g \approx 1.32$ $(d = 2)$ and $g \approx 1.5$ $(d = 3)$[2.63, 64]. For $d = d_c = 6$ it is expected that $g = 2$ [2.65]. For very small values of $r/\ell^{\tilde{\nu}}$, the values of g tend to be smaller [2.65]. Similar scaling forms also hold for $\phi(M \mid \ell)$ and $\phi(M \mid r)$.

It is obvious, however, that for fixed r, (2.15,16) do not hold for arbitrary small ℓ-values, since there exists a minimum chemical distance $\ell_{\min}(r)$ that a cluster site at Euclidean distance r from the center site can have. This quantity plays an important role when transport and vibrational properties of percolation clusters are studied. In contrast to the other quantities discussed in this chapter, $\ell_{\min}(r)$ is not self-averaging, but depends (logarithmically) on the number N of configurations averaged [2.64],

$$\ell_{\min}(r) \equiv \ell_{\min}(r, N) = \begin{cases} r, & r \ll r_c(N), \\ r_c(N)^{1-d_{\min}} r^{d_{\min}}, & r \gg r_c(N), \end{cases} \qquad (2.17)$$

where the "crossover" length $r_c(N)$ is given by $r_c(N) = (\ln z + \ln N)/\ln(1/p_c)$, and z is the coordination number of the underlying lattice; $z = 4$ for the square and $z = 6$ for the simple cubic lattice.

d) Fractal substructures. Next we show that d_f and d_ℓ are not the only exponents characterizing a percolation cluster at p_c. As is illustrated in Fig. 2.10, a percolation cluster is composed of several fractal substructures, which are described by other exponents. Imagine applying a voltage difference between two sites at opposite edges of a metallic percolation cluster: The *backbone* of the cluster consists of those sites (or bonds) which carry the electric current. The *dangling ends* are those parts of the cluster which carry no current and are connected to the backbone by a single site only. The *red bonds* (or singly connected bonds) [2.51,66] are those bonds that carry the total current; when they are cut the current flow stops. In analogy to red bonds we can define anti-red bonds [2.67]. If an anti-red bond is added to a nonconducting percolation system below p_c, the current will be able to flow in the system. The *blobs*, finally, are those parts of the backbone that remain after the red bonds have been removed.

Further substructures of the cluster are the *external perimeter* (which is also called the *hull*), the *skeleton*, and the *elastic backbone*. The hull consists of those sites of the cluster which are adjacent to empty sites and are connected with infinity via empty sites. In contrast, the *total* perimeter also includes the holes in the cluster. The external perimeter is an important model for random fractal interfaces (see Chaps. 7 and 8). The skeleton is defined as the union of

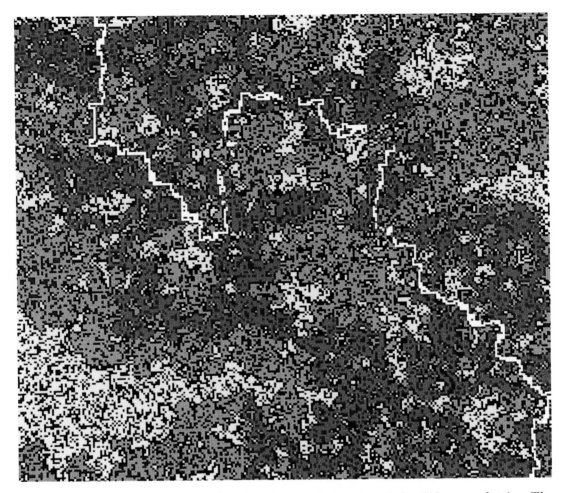

Fig. 2.10. Percolation system at the critical concentration in a 510×510 square lattice. The finite clusters are in yellow. The substructures of the infinite percolation cluster are shown in different colors: the shortest path between two points at opposite sites of the system is shown in white, the single connected sites (*"red" sites*) in red, the loops in blue, and the dangling ends in green. Courtesy of S. Schwarzer

all shortest paths from a given site to all sites at a chemical distance ℓ [2.68]. The elastic backbone is the union of all shortest paths between two sites [2.69].

The fractal dimension d_B of the backbone is smaller than the fractal dimension d_f of the cluster (see Table 2.3). This reflects the fact that most of the mass of the cluster is concentrated in the dangling ends, which is seen clearly in Fig. 2.11. The value of the fractal dimension of the backbone is known only from numerical simulations [2.70]. A recent conjecture yields $d_B = 13/8$ [2.75].

Fig. 2.11. Percolation cluster at the critical concentration in a simple cubic lattice. The ▶ backbone between two (*green*) cluster sites is shown in red, the gray sites represent the dangling ends. Courtesy of M. Porto

Table 2.3. Fractal dimensions of the substructures composing percolation clusters. Exact results and best estimates are presented: [a]exact [2.48,49]; [b]numerical simulations [2.37]; [c]numerical simulations [2.61,63]; [d]numerical simulations [2.70a]; [e]exact [2.66,71]; [f]exact [2.72,73]; [g]numerical simulations [2.37]; [h]numerical simulations [2.39]; [i]numerical simulations [2.84]; [k]numerical simulations [2.70b]. For other estimates see Sects. 2.7 and 2.8

Fractal dimensions	Space dimension		
	d=2	d=3	d≥ 6
d_f	$91/48^a$	2.524 ± 0.008^b	4
d_ℓ	1.678 ± 0.005^c	1.84 ± 0.02^i	2
$d_{\min}$	1.13 ± 0.004^c	1.374 ± 0.004^h	2
d_{red}	$3/4^e$	1.143 ± 0.01^b	2
d_h	$7/4^f$	2.548 ± 0.014^g	4
d_B	1.62 ± 0.02^d	1.855 ± 0.015^k	2
d_ℓ^B	1.45 ± 0.02^i	1.37 ± 0.03^i	1

Note that also the graph dimension d_ℓ^B of the backbone is smaller than that of percolation. In contrast, $\tilde{\nu}$ is the *same* for both the percolation cluster and its backbone, indicating the more universal nature of $\tilde{\nu}$. This can be understood since every two sites on a percolation cluster can be located on the corresponding backbone.

The fractal dimensions of the red bonds d_{red} and the hull d_h are known from exact analytical arguments. It has been proven by Coniglio [2.66,71] (see Sect. 2.8) that the mean number of red bonds varies with p as

$$n_{red} \sim (p - p_c)^{-1} \sim \xi^{1/\nu} \sim r^{1/\nu}, \tag{2.18}$$

and the fractal dimension of the red bonds is therefore $d_{red} = 1/\nu$. The fractal dimension of the skeleton is very close to $d_{min} = 1/\nu$, supporting the assumption that percolation clusters at criticality are finitely ramified [2.68] (see Sect. 2.8).

The hull of the cluster in $d = 2$ has the fractal dimension $d_h = 7/4$, which was first found numerically by Sapoval, Rosso, and Gouyet [2.72] and proven rigorously by Saleur and Duplantier [2.73]. If the hull is defined slightly differently and next-nearest neighbors of the perimeter are regarded as connected, many "fjords" are removed from the hull. According to Grossmann and Aharony [2.74], the fractal dimension of this modified hull is close to $4/3$, the fractal dimension of self-avoiding random walks. In three dimensions, in contrast, the mass of the hull seems to be proportional to the mass of the cluster, and both have the same fractal dimension (see also Chap. 7).

In Table 2.3 the values of the fractal dimension d_f and the graph dimension d_ℓ of the percolation cluster and its fractal substructures are summarized. The values for six dimensions are known rigorously from considerations given in the following sections.

2.4 Exact Results

Only very few results for the percolation problem can be obtained exactly. Some special cases can be treated exactly [2.1,2,6] and will be discussed here. From these examples, one easily becomes acquainted with the common definitions, techniques, and concepts used in percolation theory.

2.4.1 One-Dimensional Systems

Consider a one-dimensional chain, where each site is occupied randomly with probability p. Clusters are groups of neighboring sites. The sites neighboring the left and right end sites of a cluster must be empty. Since an infinite cluster can occur only if *all* sites in the chain are occupied, it follows that $p_c = 1$. Thus, only quantities below p_c, such as the correlation length or the mean number of sites (mass) of a cluster, can be calculated here.

We begin with the correlation length ξ and the correlation function $g(r)$, which is the mean number of sites on the same cluster at distance r from an arbitrary occupied site. In order for two sites separated a distance r to belong to the same cluster, all $r - 1$ sites in between must be occupied. Since the occupation is a random process with probability p,

$$g(r) = 2p^r. \tag{2.19}$$

The factor 2 comes from the fact that the site can be either to the left or to the right of the origin. The correlation length ξ is defined as the mean distance between two sites on the same cluster,

$$\xi^2 = \frac{\sum_{r=1}^{\infty} r^2 g(r)}{\sum_{r=1}^{\infty} g(r)} = \frac{\sum_{r=1}^{\infty} r^2 p^r}{\sum_{r=1}^{\infty} p^r}. \tag{2.20}$$

The sums can be performed easily and one obtains

$$\xi^2 = \frac{1+p}{(1-p)^2} \equiv \frac{1+p}{(p_c - p)^2}. \tag{2.21}$$

Thus we find $\nu = 1$ in one dimension. Using this result, one can express the correlation function $g(r)$ near p_c by an exponential:

$$\ln g(r) \sim -r/\xi, \tag{2.22}$$

where the correlation length ξ represents the decay radius of the correlation function.

Next we consider the mean mass S of the finite clusters. The mean mass S can be written as

$$S = 1 + \sum_{r=1}^{\infty} g(r), \tag{2.23}$$

where the 1 comes from the site at the origin, which was assumed to be occupied. From (2.23) it follows that

$$S = \frac{1+p}{1-p} \sim (p_c - p)^{-1}, \tag{2.24}$$

and hence $\gamma = 1$ in one dimension.

A quantity which plays a central role in the description of the cluster statistics is the probability that a chosen lattice site belongs to a cluster of s sites. This probability is given by $sp^s(1-p)^2$. The factor s is due to the fact that the chosen site can be any of the s sites in the cluster. The factor $(1-p)^2$ is due to the fact that every cluster must be surrounded by perimeter sites which are empty. In $d = 1$, every cluster has two perimeter sites. It is convenient to define the corresponding probability per cluster site, n_s, which in this case is

$$n_s = p^s(1-p)^2. \tag{2.25}$$

This quantity can also be interpreted as the number of clusters of size s divided by the total number of sites in the system. By definition, $\sum_{s=1}^{\infty} sn_s = p$, which can be easily checked using (2.25). The mean cluster mass S is related to n_s by

$$S = \sum_{s=1}^{\infty} s \left(\frac{sn_s}{\sum_{s=1}^{\infty} sn_s} \right). \tag{2.26}$$

Here the factor $(sn_s / \sum sn_s)$ is the probability that an occupied site belongs to a cluster of s sites. The sum in (2.26) can be performed easily, and of course gives the same result as (2.24).

The drawback of the one-dimensional chain is that there is no concentration range above p_c and thus the critical properties above p_c cannot be studied. Below p_c, however, the system shows critical (power-law) behavior, similar to the behavior in higher dimensions.

2.4.2 The Cayley Tree

A second system in which the percolation problem can be solved rigorously is the Cayley tree (also called the Bethe lattice) [2.1,6,62]. The Cayley tree has the advantage that the critical concentration is smaller than 1 and the regime above p_c can also be studied.

The Caylee tree is a structure without loops which is generated as follows. We start with a central site from which z branches (of unit length) emanate. The end of each branch is another site, and we obtain z sites which constitute the first shell of the Cayley tree. From each site $z - 1$ new branches grow out, generating $z(z - 1)$ sites in the second shell. This process is continued (see Fig. 2.12) and an infinite Cayley tree with z branches emanating from each site is generated. For $z = 2$, the tree reduces to the one-dimensional chain.

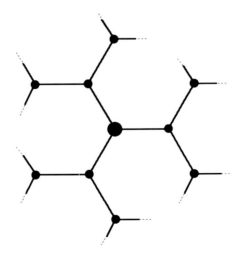

Fig. 2.12. Two shells of a Cayley tree, with $z = 3$

There are no loops in the system, since any two sites are connected by *only one* path. The Euclidean distance r has no meaning here, the lattice being described solely by the chemical distance ℓ between two sites. For example, the chemical distance between the central site and a site on the ℓth shell is exactly ℓ.

The ℓth shell of the tree consists of $z(z-1)^{\ell-1}$ sites, increasing exponentially with ℓ. In a d-dimensional Euclidean lattice, with d finite, the number of sites at distance ℓ increases as ℓ^{d-1}. Since the exponential dependence can be considered as a power-law behavior with an infinite exponent, the Cayley tree can be regarded as an infinite-dimensional lattice. From the universality property we can expect that the critical exponents derived for percolation on the Cayley tree will be the same as for percolation on *any* infinite-dimensional lattice. Moreover, as will be shown below, the upper critical dimension for percolation is $d_c = 6$ [2.77–79], i.e., the critical exponents are the same for all dimensions $d \geq 6$. Thus we expect that the exponents for percolation on the Cayley tree are the same as in $d \geq 6$ dimensions.

As for the one-dimensional case, we begin with the correlation function $g(\ell)$, which is defined as the mean number of sites on the same cluster at distance ℓ from an arbitrary occupied site. In order that two sites separated a distance ℓ belong to the same cluster, all $\ell - 1$ sites in between must be occupied. Since the occupation is a random process with probability p, and each shell contains $z(z-1)^{\ell-1}$ sites, it follows that

$$g(\ell) = z(z-1)^{\ell-1}p^\ell \equiv \frac{z}{z-1}[(z-1)p]^\ell. \tag{2.27}$$

For $z = 2$, the Cayley tree reduces to a linear chain and (2.27) reduces to (2.19). From (2.27) the critical concentration p_c can be easily derived. For ℓ approaching infinity, the correlation function tends to zero exponentially for $p(z-1) < 1$, and diverges for $p(z-1) > 1$. Accordingly, an infinite cluster can be generated only if $p \geq 1/(z-1)$. Hence

$$p_c = \frac{1}{z-1} \tag{2.28}$$

for the Cayley tree. From (2.27) we can easily calculate the correlation length ξ_ℓ in ℓ space,

$$\xi_\ell^2 = \frac{\sum_{\ell=1}^{\infty} \ell^2 g(\ell)}{\sum_{\ell=1}^{\infty} g(\ell)}. \tag{2.29}$$

As for the linear chain, the sums can be performed easily to give

$$\xi_\ell^2 = p_c \frac{p_c + p}{(p_c - p)^2}, \quad p < p_c. \tag{2.30}$$

Thus the correlation exponent in ℓ space is 1, as in one dimension.

As in $d = 1$, the mean mass S of the finite clusters can be written as

$$S = 1 + \sum_{\ell=1}^{\infty} g(\ell). \qquad (2.31)$$

Substituting (2.27) into (2.31) we find

$$S = p_c \frac{1+p}{p_c - p}, \quad p < p_c, \qquad (2.32)$$

which is a simple generalization of the one-dimensional result (2.24). Accordingly, we find $\gamma = 1$ for the Cayley tree.

Next we consider $n_s(p)$, the probability that a chosen site on the Cayley tree belongs to a cluster of s sites, divided by s. In the one-dimensional chain, n_s was simply the product of the probability p^s that s sites are occupied and the probability $(1-p)^2$ that the (two) perimeter sites are empty. For one s cluster there exists only one realization.

In general, the probability of finding a cluster with s sites having t perimeter sites is $p^s(1-p)^t$, and there exist more than one realization for a cluster of s sites. Clusters with the same s may have different t. On a square lattice, for example, a cluster of three sites can have either eight or seven perimeter sites, depending on whether the three sites are in a line or form a right angle, respectively. Thus one has to define a quantity $g_{s,t}$, which gives the number of configurations for a cluster of s sites and t perimeter sites. For a cluster of three sites on the square lattice, $g_{3,7} = 4$ and $g_{3,8} = 2$. The general expression for n_s is therefore

$$n_s = \sum_t g_{s,t} p^s (1-p)^t. \qquad (2.33)$$

On a Cayley tree, in contrast to the square lattice, there exists a unique relation between s and t. A cluster of one site is surrounded by z perimeter sites and a cluster of two sites is surrounded by $z + (z-2)$ perimeter sites. In general, a cluster of s sites has $z - 2$ more perimeter sites than a cluster of $s - 1$ sites. Denoting the number of perimeter sites of an s cluster by $t(s)$, we obtain

$$t(s) = z + (s-1)(z-2). \qquad (2.34)$$

Thus for the Cayley tree, (2.33) reduces to

$$n_s(p) = g_s p^s (1-p)^{2+(z-2)s} \equiv g_s(1-p)^2 [p(1-p)^{z-2}]^s, \qquad (2.35)$$

where g_s is simply the number of configurations for an s-site cluster. Note that (2.35) is general and holds for all values of p.

We are interested in the behavior of n_s near the critical concentration. To this end, we expand $p(1-p)^{z-2}$ around $p_c = 1/(z-1)$. To lowest order in $(p - p_c)$, we obtain

$$n_s(p) \sim n_s(p_c) f_s(p), \qquad (2.36)$$

where $f_s(p) = \left(1 - [(p - p_c)^2/2p_c^2(1 - p_c)]\right)^s$ decays exponentially for large s, $f_s(p) = \exp(-cs)$ and $c \sim (p - p_c)^2$. Accordingly, close to p_c, $f_s(p)$ is a function of *only* the *combined variable* $(p - p_c)s^\sigma$, with $\sigma = 1/2$. From the derivation it is clear that this result is valid for p approaching p_c from below or above. The exponent σ describes how fast the number of clusters of size s decreases as a function of s if we are above or below the threshold. At the threshold, this decay is described by the prefactor of $f_s(p)$ in (2.36). In principle, for the Cayley tree, it is possible to exactly evaluate g_s and hence $n_s(p_c)$ by combinatorial approaches. Instead of doing this, let us *assume* that $n_s(p_c)$ follows a power law [2.80]

$$n_s(p_c) \sim s^{-\tau}. \tag{2.37}$$

The exponent τ then follows from (2.26) and (2.36):

$$S = \frac{1}{p} \sum_s s^2 n_s(p) \sim \sum_s s^{2-\tau} e^{-cs} \sim \int_1^\infty s^{2-\tau} e^{-cs} ds. \tag{2.38a}$$

Substituting $z = cs$ we obtain

$$S \sim c^{\tau-3} \int_c^\infty z^{2-\tau} e^{-z} dz. \tag{2.38b}$$

For $\tau < 3$, the integral is nonsingular for c approaching zero and we obtain $S \sim |p - p_c|^{(\tau-3)/\sigma}$, or

$$\gamma = (3 - \tau)/\sigma. \tag{2.39}$$

Since $\gamma = 1$ and $\sigma = 1/2$, we obtain $\tau = 5/2$, consistent with the assumption $\tau < 3$.

Next we consider P_∞. There exists a simple relation between n_s and P_∞. Every site in the lattice is either (a) empty with probability $1 - p$, or (b) occupied *and* on the infinite cluster with probability pP_∞, or (c) occupied but not on the infinite cluster with probability $p(1 - P_\infty) \equiv \sum_s sn_s$. This leads to the *exact* relation

$$P_\infty = 1 - \frac{1}{p} \sum_s sn_s, \tag{2.40}$$

which we will use to determine the exponent β. Below p_c, $\sum_s sn_s = p$ and $P_\infty = 0$.

If, as we did in (2.38), we express the sum in (2.40) by an integral, the integral is singular for c approaching zero and cannot be treated as a constant. In order to get rid of this artificial singularity, we write (2.40) in the form

$$P_\infty = \frac{1}{p} \sum_s s(n_s(p_c) - n_s(p)) + (p - p_c)/p. \tag{2.41}$$

The sum is proportional to $c^{\tau-2} \int_c^\infty z^{1-\tau}[1 - \exp(-z)]dz$. Since for small z

$[1 - \exp(-z)] \sim z$, the integral is no longer singular, and

$$P_\infty \sim c^{\tau - 2} + \text{const} \times (p - p_c). \qquad (2.42)$$

With c from (2.36) we obtain

$$\beta = (\tau - 2)/\sigma, \qquad (2.43)$$

yielding $\beta = 1$ for the Cayley tree. Note that, for the Cayley tree, the two terms in (2.42) are of the same order in $(p - p_c)$. In regular lattices, the same treatment can be used to relate the exponents. Since there $(\tau - 2)/\sigma$ is less than 1, the first term is dominant.

As indicated by (2.26) and (2.40), the mean cluster mass, S, and P_∞ can be regarded as the second and first moments of the cluster distribution function $n_s(p)$. The zeroth moment $M_0 \equiv \sum_s n_s$ represents the mean number of clusters per lattice site. To leading order in $|p - p_c|$, M_0 scales as

$$M_0 \equiv \sum_s n_s \sim |p - p_c|^{2 - \alpha}, \qquad (2.44)$$

which defines the exponent α. The sum here can be calculated in the same way as above, and one easily obtains

$$2 - \alpha = (\tau - 1)/\sigma, \qquad (2.45)$$

yielding $\alpha = -1$ for the Cayley tree.

It should be noticed that, in contrast to the critical concentration p_c, the exponents α, β, γ, σ, and τ are *independent* of z, a fact which reflects the universality property of the critical exponents.

As mentioned above, we expect that these exponents will be the same in all lattices with dimension $d \geq d_c = 6$. A special role is played by the exponent ν, which here characterizes the correlation length ξ_ℓ in chemical space. Since the Euclidean distance r has no meaning on the Cayley tree, we cannot determine the correlation length in r space. But the behavior of ξ in r space for $d \geq d_c$ can be obtained from the following arguments. Above the critical dimension, correlations between any two sites in the lattice are no longer relevant, and *any* path on a cluster will behave like a random walk (see,i.e., Sect. 3.2).

$$r^2 \sim \ell. \qquad (2.46)$$

From (2.46) it follows immediately that $\tilde{\nu} = 1/d_{\min} = d_\ell/d_f = 1/2$ and hence $\xi \sim \xi_\ell^{1/2}$. Since $\xi_\ell \sim |p - p_c|^{-1}$, we have $\nu = 1/2$ in r space, for all lattices with $d \geq 6$. From the relation $d_f = d - \beta/\nu$ (2.8) and $\beta = 1$ we obtain $d_f = 4$ in $d = 6$, which then yields $d_\ell = 2$. Since $d_c = 6$, we expect the same fractal dimension for *all* $d \geq 6$. Thus the infinite percolation cluster at p_c, even in $d = \infty$, can be embedded in a four-dimensional space.

For a different approach for studying percolation on a Cayley tree, using the generating function method, see Appendix 2.A.

2.5 Scaling Theory

In the previous section, we introduced the quantity $n_s(p)$, which describes the distribution of s clusters per lattice site. It turned out that $n_s(p)$ plays a central role in the description of the geometric percolation transition. From $n_s(p)$ we obtained the mean cluster mass S, the correlation length ξ as well as the probability P_∞ that an occupied site belongs to the infinite cluster.

In general, in contrast to percolation on the Cayley tree and in $d = 1$, the cluster size distribution $n_s(p)$ is not known exactly. However, as we will show below (see also [2.1,6,67]), not all the information on $n_s(p)$ is needed in order to calculate the relevant quantities. What is important is the *scaling* form of $n_s(p)$, from which a scaling theory can be applied to determine the relations between the different critical exponents. First we discuss, as in the sections before, percolation on an infinite lattice, and later we consider effects of finite lattice size.

2.5.1 Scaling in the Infinite Lattice

We are interested in the region around the critical concentration (*critical region*) where the quantities related to the percolation transition show power-law behavior. When considering the Cayley tree we saw that, to lowest order in $(p - p_c)$, $n_s(p)$ was also described by power laws with the exponents σ and τ,

$$n_s(p) \sim s^{-\tau} f_\pm \left(|p - p_c|^{1/\sigma} s \right), \tag{2.47}$$

where $\tau = 5/2$, $\sigma = 1/2$, and $f_\pm(z) = \exp(-z)$. The subscripts $+$ and $-$ refer to $p > p_c$ and $p < p_c$, respectively. In the Cayley tree, f was the same above and below p_c, but this is generally not the case. We will *assume* now that $n_s(p)$ retains the form of (2.47) on regular lattices, with the same exponents τ and σ below and above p_c. Like the other critical exponents, σ and τ are universal and depend only on the lattice dimension, not on the lattice structure. This is a consequence of the self-similarity in the neighborhood of p_c. In contrast to the exponents, the form of $f(z)$ need not be universal.

If we accept the *scaling ansatz* (2.47), we can calculate the mean cluster mass S and the probability P_∞ in the same way as in Sect. 2.4. In deriving the relations between σ, τ, β, and γ for the Cayley tree we did not use the analytic form of $n_s(p)$ explicitly, but rather the general scaling form (2.47). Consequently, the relations (2.39), (2.43), and (2.45) are not restricted to the Cayley tree, but hold for all cases where the scaling ansatz (2.47) is valid, i.e., for general d-dimensional systems.

To relate the correlation length exponent ν to σ and τ note that ξ is defined as the root mean square (rms) distance between occupied sites on the same finite cluster, averaged over all finite clusters, where each distance is given the same weight. For clusters with s sites, the rms distance between all pairs of sites on each cluster, averaged over all clusters of size s, is simply

$$R_s^2 = \frac{2}{s(s-1)} \sum_{i=1}^{s} \sum_{j=1}^{i} \overline{(\mathbf{r_i} - \mathbf{r_j})^2}, \tag{2.48}$$

and to find ξ we have to average over all cluster sizes.

Since every site on the lattice has probability sn_s of belonging to a cluster of s sites, and every site is connected with s sites (including the connection with itself), we have

$$\xi^2 = \sum_{s=1}^{\infty} R_s^2 s^2 n_s / \sum_{s=1}^{\infty} s^2 n_s. \tag{2.49a}$$

The factor s^2 here is due to the fact that the same weight is given to each pair of sites. Close to p_c, the large clusters dominate the sum in (2.49a). Since they are fractals, their mass s is related to R_s by [see (2.4)] $R_s \sim s^{1/d_f}$, and we obtain from (2.49a)

$$\xi^2 \sim \sum_{s=1}^{\infty} s^{2/d_f+2-\tau} f_{\pm}\left(|p-p_c|^{1/\sigma} s\right) / \sum_{s=1}^{\infty} s^{2-\tau} f_{\pm}\left(|p-p_c|^{1/\sigma} s\right). \tag{2.49b}$$

To calculate the sums we transform them into integrals, as in Sect. 2.4. Since $2-\tau$ is greater than -1, the integrations are over nonsingular integrands, and to lowest order in $(p-p_c)$ the lower integration limit does not contribute to the integral. Following the same procedure as in (2.38) we obtain

$$\xi^2 \sim |p-p_c|^{-2/(d_f\sigma)}, \tag{2.50}$$

which yields the desired relation between ν, σ, and τ,

$$\nu = \frac{1}{d_f\sigma} = \frac{\tau-1}{d\sigma}. \tag{2.51}$$

Note that (2.51), together with (2.39), (2.43), and (2.45) (which hold also for general lattices, as will be shown below), represent *four* relations between the *six* exponents (α, β, γ, σ, τ, and ν), thus we have only two independent exponents.

A second quantity which characterizes the size of a finite cluster is the mean square cluster radius R^2, defined as

$$R^2 = \sum_{s=1}^{\infty} R_s^2 s n_s / \sum_{s=1}^{\infty} s n_s. \tag{2.52a}$$

Here the same weight is given to each site of the cluster, and not to each pair of sites as in (2.49a). Following the treatment of (2.49) we obtain

$$R^2 \sim |p-p_c|^{-2\nu+\beta}. \tag{2.52b}$$

We will show in Chap. 3 that R^2 is an important quantity for describing diffusion in percolation systems. We show now that the correlation length defined in (2.49a) is the only characteristic length scale of percolation.

Using (2.51), the argument $z = |p - p_c|^{1/\sigma} s$ of the scaling function $f_\pm(z)$ can be written as $z = s/\xi^{d_f}$, and (2.47) becomes

$$n_s(p) \sim s^{-\tau} f_\pm(s/\xi^{d_f}), \tag{2.53a}$$

or equivalently, using (2.51), in terms of ξ,

$$n_s(p) \sim \xi^{-\tau d_f}(s/\xi^{d_f})^{-\tau} f_\pm(s/\xi^{d_f}) = \xi^{-d-d_f} F_\pm(s/\xi^{d_f}), \tag{2.53b}$$

where $F_\pm(z) = z^{-\tau} f_\pm(z)$. Equations (2.53a,b) show that the correlation length ξ represents the only characteristic length scale near the percolation threshold: the cluster distribution function $n_s(p)$ depends on s via only the ratio s/ξ^{d_f} or, on replacing s by $R_s^{d_f}$, on only the ratio R_s/ξ.

It can easily be checked that the sums calculated so far are special cases of the more general expression

$$M_k = \sum_{s=1}^{\infty} s^k n_s(p) \sim \sum_{s=1}^{\infty} s^{k-\tau} f_\pm(s/\xi^{d_f}), \tag{2.54a}$$

from which all relations between the exponents can be obtained. The sum is transformed into the integral

$$M_k \sim \int_1^\infty s^{k-\tau} f_\pm(s/\xi^{d_f}) ds \sim \xi^{d_f(k-\tau+1)} \int_{\xi^{-d_f}}^\infty z^{k-\tau} f_\pm(z) dz. \tag{2.54b}$$

As long as the integrand is nonsingular, $k - \tau > -1$, the lower integration limit can be extended to zero, yielding

$$M_k \sim \xi^{d_f(k-\tau+1)} \sim |p - p_c|^{(\tau-1-k)/\sigma}. \tag{2.55}$$

For $k < \tau - 1$ this procedure does not work, since the lower limit dominates the integral. In this case, one can consider derivatives of M_k with respect to ξ^{-d_f}:

$$\frac{d^n M_k}{(d\xi^{-d_f})^n} \sim \xi^{d_f(k-\tau+n+1)} \int_{\xi^{-d_f}}^\infty z^{k-\tau+n} \frac{d^n f(z)}{dz^n} dz, \tag{2.56a}$$

where n is the smallest integer greater than $\tau - k - 1$. The integrand in (2.56a) is nonsingular and hence

$$\frac{d^n M_k}{(d\xi^{-d_f})^n} \sim \xi^{d_f(k-\tau+n+1)}. \tag{2.56b}$$

From (2.56b), we can obtain by simple integration, up to lowest-order terms,

$$M_k \sim \xi^{d_f(k-\tau+1)} \sim |p - p_c|^{(\tau-1-k)/\sigma}, \tag{2.57}$$

which shows that (2.55) is also valid for $k > \tau - 1$. From (2.57) we obtain $M_0 \sim |p - p_c|^{(\tau-1)/\sigma} \sim |p - p_c|^{2-\alpha}$ and $M_1 \sim |p - p_c|^{(\tau-2)/\sigma} \sim |p - p_c|^\beta$, which yields relations (2.45) and (2.43) between the exponents. Similarly, we obtain for $k = 2$ relation (2.39), $\gamma = (3-\tau)/\sigma$. These three relations constitute, together with (2.51), $d\nu = (\tau-1)/\sigma$, four relations between the six exponents. From (2.51), (2.39), and (2.43) one obtains

$$d\nu = 2\beta + \gamma, \qquad (2.58)$$

a relation which has been found useful by Toulouse [2.77] to obtain the upper critical dimension d_c for percolation. At d_c, $\nu = 1/2$, $\beta = 1$, and $\gamma = 1$ (see Sect. 2.2), and hence $d_c = 6$. The same argument leads to $d_c = 4$ in Ising systems, where $\beta = 1/2$ at the critical dimension.

2.5.2 Crossover Phenomena

From the scaling form (2.53a) we learned that the correlation length ξ is the only characteristic length scale in percolation. Above p_c, ξ is finite, and we expect different behavior on length scales $r < \xi$ and $r > \xi$. In the following we present a scaling theory that describes the crossover behavior in several quantities such as P_∞, $M(r)$, $M(\ell)$, and $R(\ell)$. First we consider the probability P_∞ that an occupied site belongs to the infinite cluster. To determine P_∞ we choose subsystems of linear size r and determine the fraction of sites in this system that belong to the infinite cluster. If $r \gg \xi$, the system behaves as if $r = \infty$, and P_∞ is independent of r and described by $P_\infty \sim (p - p_c)^\beta$. If $r \ll \xi$, on the other hand, the number of sites of the infinite cluster within a circle of radius r is proportional to r^{d_f}. The total number of occupied sites within r is proportional to pr^d, and hence $P_\infty \sim r^{d_f - d}$. Since ξ is the only characteristic length, we can assume that P_∞ depends on r and ξ via only the ratio r/ξ. This leads, in close analogy to (2.53), to the scaling ansatz

$$P_\infty \sim (p - p_c)^\beta G(r/\xi) \sim \xi^{-\beta/\nu} G(r/\xi). \qquad (2.59)$$

The scaling function G describes the crossover from $r/\xi \ll 1$ to $r/\xi \gg 1$. In order to obtain the expected results in the two limits, we must require that

$$G(x) \sim \begin{cases} x^{d_f - d}, & x \ll 1, \\ \text{const}, & x \gg 1. \end{cases} \qquad (2.60)$$

From (2.60) we can determine how the mean mass M of the infinite cluster scales with r and ξ above p_c. Since $M \sim r^d P_\infty(r, \xi)$, the mean mass of the infinite cluster scales as

$$M \sim r^{d - \beta/\nu} H(r/\xi), \qquad H(x) = x^{\beta/\nu} G(x), \qquad (2.61)$$

and we recover (2.5) and (2.8). Equations (2.60) and (2.61) generalize (2.4) for $p \geq p_c$.

Next we consider how M scales with the chemical length above p_c. Substituting (2.10), $r \sim \ell^{\tilde{\nu}}$, into (2.61) gives [2.50]

$$M \sim \ell^{\tilde{\nu}(d-\beta/\nu)} g(\ell/\xi_\ell), \tag{2.62}$$

where ξ_ℓ is the correlation length in ℓ space. From (2.10), (2.61), and (2.62) we obtain the scaling function $g(x)$,

$$g(x) \sim \begin{cases} \text{const,} & x \ll 1, \\ x^{d-d_f \tilde{\nu}}, & x \gg 1. \end{cases} \tag{2.63}$$

Equations (2.62) and (2.63) generalize (2.9).

Finally we study how the average Euclidean distance r between sites separated by a fixed chemical distance ℓ scales above p_c. By scaling arguments similar to those given above we expect that r scales as

$$r \sim \ell^{\tilde{\nu}} f(\ell/\xi_\ell), \tag{2.64}$$

where

$$f(x) \sim \begin{cases} \text{const,} & x \ll 1, \\ x^{1-\tilde{\nu}}, & x \gg 1. \end{cases} \tag{2.65}$$

Equations (2.64) and (2.65) generalize (2.10) for $p > p_c$. Accordingly, at large Euclidean distances $(r > \xi)$ the chemical distance and the Euclidean distance have the same metric, and [2.14,83]

$$r \sim \xi_\ell^{\tilde{\nu}-1} \ell \sim \xi^{(\tilde{\nu}-1)/\tilde{\nu}} \ell \sim (p-p_c)^{\nu(1/\tilde{\nu}-1)} \ell. \tag{2.66}$$

Equations (2.64–66) will be shown to be useful for describing spreading phenomena in Sect. 2.6.

Scaling assumptions can also be used to treat various quantities at criticality. Examples are the probability densities $\phi(M \mid r)$, $\phi(M \mid \ell)$, and $\phi(r \mid \ell)$, which at p_c scale as $\phi(M \mid r) \sim (1/M) f(M/r^{d_f})$, etc., see (2.12). Similar quantities have also been studied for the percolation backbone [2.84] and scaling theory has been successfully applied to study the blob-size distribution [2.85]. More examples (related to transport properties) are given in Chap. 3.

So far, we have considered percolation on infinitely extended lattices. In the next subsection we will use the above results to study the effect of *finite* lattices.

2.5.3 Finite-Size Effects

Consider a finite system, e.g., a triangular lattice of $L \times L$ sites, where each bond has unit length $a = 1$. We expect that the relevant quantities depend on the magnitude of L. Since the characteristic length of the system is the correlation length ξ (which is defined for the infinite system), we expect different behavior for $L/\xi \gg 1$ and for $L/\xi \ll 1$.

First consider again P_∞. For $L \gg \xi$, the system behaves as if $L = \infty$, i.e., P_∞ is independent of L and described by $P_\infty \sim (p - p_c)^\beta$. For $\xi \gg L \gg 1$, the number of sites of the infinite cluster in the $L \times L$ "window" is proportional to L^{d_f}. We obtain P_∞ by dividing this result by the total number of occupied sites in the window, which is pL^d, hence $P_\infty \sim L^{d_f - d}$. Since ξ is the only characteristic length, P_∞ depends on L via only the ratio L/ξ, which in analogy to (2.59) leads to the scaling ansatz

$$P_\infty \sim (p - p_c)^A G(L/\xi) \sim \xi^{-A/\nu} G(L/\xi). \qquad (2.67)$$

The scaling function $G(x)$ describes the crossover from $L/\xi \ll 1$ to $L/\xi \gg 1$. In order to obtain the expected results in both regimes, we must have $A = \beta$ and

$$G(x) \sim \begin{cases} x^{d_f - d}, & x \ll 1, \\ \text{const}, & x \gg 1. \end{cases} \qquad (2.68)$$

To see the consequences of these relations, assume that we are going to perform computer simulations of P_∞ on the triangular lattice, where the critical concentration is exactly $1/2$ (for site percolation on the infinite lattice). We choose a large lattice, say $L = 1000$, and occupy randomly all sites with probability p. Next we analyze the clusters and determine the size of the infinite cluster (if it exists). We repeat the calculations say for 5000 lattice configurations, and average P_∞ over all configurations. For p well above the percolation threshold, where ξ is considerably smaller than $L = 1000$, we do not notice the finite size of the lattice. Accordingly, if we approach p_c more closer, P_∞ will decrease as $(p - p_c)^\beta$, as long as $L \gg \xi$. If we approach p_c more and more closely we will reach a crossover concentration p^*, where $L \sim \xi$. Between p^* and p_c, $L < \xi$ and P_∞ remains roughly constant at a finite value.

This behavior can be understood as follows. In the *finite* system of 10^6 sites a small change in concentration is equivalent to adding or substracting, on average, only a few occupied sites, which leaves the system practically unchanged. In our lattice we cannot see a difference in P_∞ for $p = p_c + 10^{-6}$ and $p = p_c + 10^{-12}$. The small change in concentration becomes more effective if we take larger systems, and for the infinite system drastic changes also occur very close to p_c. In the above example, P_∞ decreases by a factor $10^{6\beta}$ which is of the order of 10. It is obvious that the finite-size effects described here for P_∞ occur also in the cluster site distribution $n_s(p)$ and the other related quantities.

It is convenient to write the scaling relation (2.67) also in a slightly different way. Multiplying and dividing (2.67) by $L^{-\beta/\nu}$ we obtain

$$P_\infty \sim L^{-\beta/\nu} H(L/\xi), \qquad (2.69)$$

where $H(x) = G(x) x^{\beta/\nu}$, as in (2.61).

From the finite-size scaling we can learn that in principle there are two ways to determine a critical exponent x of some quantity X, $X \sim (p - p_c)^{-x}$, by computer simulations. Instead of determining X, e.g., P_∞ or S, directly as a function of $(p - p_c)$, one can calculate X right at p_c for various system sizes. This way one obtains $X \sim L^{x/\nu}$, and if ν is known, one finds x.

Using extensive computer simulations, Ziff [2.35] studied how the critical concentration depends on the linear size L of the system. He finds for large L in $d = 2$ that

$$p_c(L) = p_c(\infty) + \text{const} \times L^{-1-1/\nu}. \qquad (2.70)$$

Hence, for obtaining accurate estimates for $p_c \equiv p_c(\infty)$ one has to plot $p_c(L)$ versus $x = L^{-1-1/\nu}$ and to extrapolate to $x = 0$.

A further interesting result is that the probability $R_L(p_c)$ for finding a spanning cluster at p_c (in the site percolation) approaches $1/2$ for L approaching infinity. Since this result is the same as for bond percolation, it suggests that $R_L(p_c)$ is a universal quantity that depends on only the dimension d of the system, not on the type of the lattice considered. Further numerical support for this hypothesis was obtained recently by Stauffer et al. [2.86] for 3d percolation, where $R_L(p_c) \cong 0.42$.

2.6 Related Percolation Problems

In this section we consider modifications of the percolation problem which are useful for describing several physical, chemical, and biological processes, such as the spreading of epidemics or forest fires, gelation processes, and the invasion of water into oil in porous media, which is relevant for the process of recovering oil from porous rocks. In some cases the modification changes the universality class of percolation. We begin with an example in which the universality class does not change.

2.6.1 Epidemics and Forest Fires

The simplest model for an epidemic disease or a forest fire is the following [2.12,15] (see also Chap. 4). On a large square lattice, each site represents an individual which can be infected with probability p and which is immune with probability $(1 - p)$. At an initial time $t = 0$ the individual at the center of the lattice (seed) is infected. We assume now that in one unit of time this infected site infects all nonimmune nearest-neighbor sites. In the second unit of time, these infected sites will infect all their nonimmune nearest-neighbor sites, and so on. In this way, after t time steps all nonimmune sites in the ℓth chemical shell around the seed are infected, i.e., the maximum length ℓ of the shortest path between the infected sites and the seed is $\ell \equiv t$. The process of randomly infecting individuals is exactly the same as that of randomly occupying sites in the site percolation discussed in Sect. 2.1, but instead of many clusters beeing generated on a lattice only single clusters are formed from each seed (see also the Leath method, Sect. 2.7). Below p_c, only finite clusters can be generated and the disease stops spreading after a finite number of time steps. Above p_c, there exists a finite probability that the disease continues for ever.

Accordingly, at p_c the mean number of infected individuals at time t is determined by the graph dimension d_ℓ (see Sect. 2.3),

$$M(t) \sim t^{d_\ell}, \qquad t \gg 1. \tag{2.71}$$

The spreading radius of the disease is related to $M(t)$ by d_f, $R(t) \sim M(t)^{1/d_f}$.

Above p_c, for $R \gg \xi$, $d_\ell = d_f = 2$ and the spreading of the epidemic is characterized by the velocity v of the epidemic front; by definition $v \equiv dR/dt \sim t^{d_\ell/d_f - 1}$. It is clear from (2.71) and $d_\ell < d_f$ that $v \to 0$ at p_c and $v > 0$ above p_c. To derive the dependence of v on $p - p_c$ note that $R(t)$ can be written as [see (2.64)]

$$R(t) \sim t^{d_\ell/d_f} f(t/t_\xi), \tag{2.72}$$

where t_ξ is the characteristic time scale in the process. Since the correlation length ξ is the only characteristic length scale here, we can associate t_ξ with the time needed to generate a cluster of size ξ. For $r < \xi$ the cluster is self-similar, and we obtain from (2.71) $t_\xi \sim \xi^{d_f/d_\ell}$. For $t \ll t_\xi$, $R(t)$ is proportional to t^{d_ℓ/d_f} and f is constant. For $t \gg t_\xi$, the system is homogeneous and $R(t)$ is linear in t. In order to satisfy this relation, we require that the right-hand side of (2.72) is also linear in t, which yields $f(t/t_\xi) \sim (t/t_\xi)^{1-d_\ell/d_f}$ and $R(t) \sim t_\xi^{d_\ell/d_f - 1} t \sim \xi^{1-d_f/d_\ell} t$. Hence the spreading velocity of the disease is also described by a critical exponent [2.83],

$$v \sim \xi^{1-d_f/d_\ell} \sim (p - p_c)^{(1/\tilde\nu - 1)\nu}. \tag{2.73}$$

Since in $d = 2$ the exponent $(1/\tilde\nu - 1)\nu$ is much smaller than 1, a relatively large velocity occurs just above p_c, and the spreading transition at p_c is abrupt rather than smooth.

In the above model, an infected site infects *all* its nonimmune nearest-neighbor sites. In an improved version of the model [2.14,87] (see also [2.15]), an infected site can infect its nearest-neighbors only with probability q and within a certain time interval τ which we take as unit time (see Fig. 2.13). It is easy to see that in the absence of immune sites, i.e., $p = 1$, this model corresponds to bond percolation, since the process of randomly infecting individuals is equivalent to the process of randomly occupying bonds (and the sites connected by them). For $p < 1$, the model is equivalent to site–bond percolation (see Sect. 2.1) where only those bonds connecting occupied sites can be occupied.

We would like to note that this simple model already combines essential features of real epidemics and emphasizes the importance of the protection, here characterized by the parameter q. For small q below the percolation threshold, the epidemic cannot spread and dies out. In AIDS, for example, the virus is

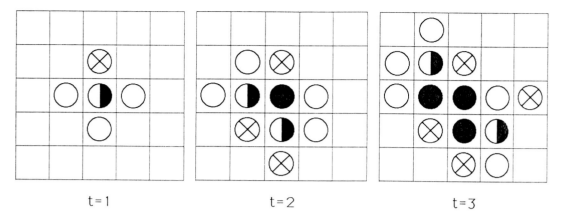

Fig. 2.13. The first three steps of the spreading of an "epidemic". The open circles represent "nonimmune" individuals, the crosses represent the "immune" individuals, the half-full circles are the "infected" sites which are still "contagious", and the full circles denote those infected individuals which are no longer contagious

mainly transfered by blood contacts, and q can be decreased drastically by choosing the proper type of protection.

There are several other models for spreading phenomena in dilute media. In a variant of the Eden model [2.88], one starts with a lattice consisting of immune and infectable individuals as before, with a seed in the center of the lattice. The infection spreads from the seed by randomly choosing one of the nonimmune nearest-neighbor sites of the spreading cluster ("growth sites") and converting it into a cluster site. This process is repeated up to the desired number of time steps. Here again, the structure of the spreading clusters is described by percolation, but the dynamical features depend on how time is associated with this process. There are two choices: (a) if a new site is added to the cluster, the time is increased by one unit; (b) if a new site is added to the cluster, the time is increased by $\Delta t = 1/P$, where P is the number of growth sites in the previous step, before the new site was chosen. In the second way, in one unit of time all available growth sites can, on average, be occupied, a process that is similar to the epidemic model above. Note, however, that here an infected site is contagious for ever, i.e., it can still infect an infectable nearest-neighbor individual even after many time steps. If this condition is relaxed, and an infected site is contagious only for a finite time range τ, the dynamical process is changed completely. There exists a *critical* lifetime $\tau_c(p)$ which depends explicitly on the concentration p of the infectable species. For $p \geq p_c$ and $\tau > \tau_c$, percolation clusters are generated, while for $p \geq p_c$ and $\tau < \tau_c$ the generated clusters belong to the universality class of self-avoiding walks, with $d_f = 4/3$ in $d = 2$ [2.89].

For related models for forest fires, which exhibit features of self-organized criticality (Chap. 2 in [2.22]) we refer the reader to [2.90,91].

Next we discuss further models for spreading phenomena in dilute media, which to some extent can be also related to percolation theory.

2.6.2 Kinetic Gelation

The gelation process describes the transition from a solution containing only small molecules to a gel state where a large molecule of the size of the system is formed [2.12,13,17]. This transition is called a sol-gel transition and separates two phases with very different physical properties. The shear viscosity diverges when the transition is approached from the sol phase. Gels occur in biological systems (e.g., eye humor), and play an important role in chromatography, in the fabrication of glues and cosmetics, and in food technology (yoghurts, gelatine, etc.).

During the gelation process, chemical bonds between neighboring molecules become activated and connect the molecules to form larger and larger clusters. There exist different mechanisms for activating the bonds. Here we concentrate on the so-called addition polymerization, where unsaturated electrons hop between the molecules and create bonds.

For this process, the following kinetic gelation model was suggested by Manneville and de Seze [2.92], which is related to the percolation problem. In this model, one considers, for example, a square lattice where sites are occupied randomly by monomers with probability p and by solvent molecules with probability $(1-p)$. Each monomer has a certain "functionality" f which states how many of its bonds can be activated (f=1,2,3,4 on the square lattice). Next a small concentration of initiators, modeling unsaturated electrons, are placed randomly on the monomers. The role of these initiators is to activate the available bonds between monomers by a random walk process (see Sect. 3.2). A randomly chosen initiator attempts to jump randomly to a nearest-neighbor site. If this site is a monomer with functionality $f \geq 1$, the jump is successful, the functionality is reduced by 1, and the bond between both monomers is activated. If the attempted site is a monomer with $f = 0$ or a solvent molecule, the jump is rejected. In both cases the time is increased by $1/N_i(t)$, where $N_i(t)$ is the number of initiators present at this time t. If two initiators meet at the same monomer, they annihilate. Accordingly, the concentration of initiators decreases monotonically with time t, but in general it will not reach zero since initiators surrounded by monomers with functionality zero are trapped.

The concentration $p(t)$ of occupied bonds increases with t. There exists a critical concentration p_c at which a spanning cluster exists for the first time and the gel is formed. As in normal percolation, the critical behavior of this geometric transition is described by the exponents ν and β.

Using Monte Carlo simulations it has been shown [2.93] that ν and β are the same as for normal percolation. This implies that d_f is also unchanged. In contrast, the fractal dimension of the backbone of the gel at p_c is $d_B \cong 2.22$ in $d = 3$ and differs considerably from the value 1.74 found in ordinary percolation [2.94,95].

2.6.3 Branched Polymers

The polymerization process, i.e., the generation of long polymers from mono-mers, has become an important topic in both applied and basic research [2.11,12,42,96–99]. It is commonly accepted that long polymer chains can be modelled by kinetic growth walks (KGW), where at each step a random walker moves only to those neighboring sites that have not been visited before [2.12]. Very recently, in order to generate branched polymer structures, Lucena *et al.* [2.102] generalized this KGW model to include branching (branched polymer growth model – (BPGM)).

The BPGM generates a branched polymer from a seed in a self-avoiding manner similar to the KGW, but allows for the possibility of branching with bifurcation probability b. To be specific, consider a square lattice where at $t = 0$ the center of the lattice is occupied by a polymer "seed". There are four empty nearest-neighbor sites of the seed, where the polymer is allowed to grow. At step $t = 1$, two of these four growth sites are chosen randomly: one of them is occupied by the polymer with probability 1, the other one is occupied with probability b. This process is continued. At step $t+1$, the polymer can grow from each of the sites added at the step t to empty nearest-neighbor sites (growth sites) either in a linear fashion or by bifurcation with probability b, provided there are enough growth sites left; otherwise the polymer stops growing. It is clear that large polymers can be generated only for q below the percolation threshold q_c of the considered lattice ($q_c \approx 0.40723$ on the square lattice).

The structure generated by this procedure depends crucially on the branching parameter b and the impurity concentration q. For small b and large q, the structures are finite and belong to the universality class of KGW, while for large b and small q, compact Eden clusters (see Sect. 2.6.1) are generated. There exists a critical line $b_c(q)$ [2.38,102] that separates both phases (see Fig. 2.14). Numerical results suggest [2.28] that at the critical line the structures belong to the universality class of percolation, which is controversial with the earlier common belief that branched polymers are in the universality class of lattice animals [2.101].

Three interesting points should be mentioned. (i) In the considered square lattice, in the absence of impurities, the phase transition between both struc-tures occurs at a *finite* value of the branching probability ($b > b_c(0) \cong 0.055$) [2.102]. For higher dimensions, numerical results suggest that $b_c(0) = 0$ [2.103]. (ii) For $b = 1$, the phase transition occurs at $q'_c < q_c$ (see Fig. 2.14), since an additional finite concentration of sites ($q_c - q'_c$) are blocked when the branch-ing (here two branches at each step) is below the maximum possible branching (here three branches) on the considered lattice. (iii) Since the generated clus-ters are loopless, the dynamical exponents are not the same as those of lattice percolation (see Sect. 3.6).

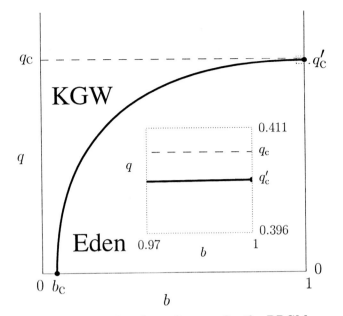

Fig. 2.14. The phase diagram for the BPGM

2.6.4 Invasion Percolation

Consider the flow of water into a porous rock filled with oil. Water and oil are two incompressible immiscible fluids and hence the oil is displaced by the water when the water invades the rock. The porous medium can be considered as a network of pores which are connected by narrow throats. The flow rate is kept constant and very low so that viscous forces can be neglected compared to capillary forces. The invasion percolation was introduced as a model to describe this dynamical process [2.104–106] (see also Sects. 4.3.3 and 7.4.2).

In the model, we again consider a regular lattice of size $L \times L$, which here represents the oil (the "defender"). Water (the "invader") is initially placed along one edge of the lattice. To describe the different resistances of the throats to the invasion of the water we assign to each site of the lattice a random number between 0 and 1. The invading water follows the path of lowest resistance. At each time step that perimeter site of the invader which has the lowest random number is occupied by the water and the oil is displaced (see Fig. 2.15).

Since oil and water are incompressible and immiscible fluids, oil that is surrounded completely by water cannot be replaced and oil can be trapped in some regions of the porous medium. Thus, if by the Monte Carlo process a closed loop is generated the invader can no longer enter this region.

The fractal dimension of the clusters of water was found numerically to be $d_f \cong 1.82$ [2.106,108]. Experiments on air slowly invading a $d = 2$ network of ducts with glycerol performed by Lenormand and Zarcone [2.109] yielded $d_f \cong 1.8$, in good agreement with the numerical results. This value of d_f suggests that the invasion percolation model is in a univerality class different from regular

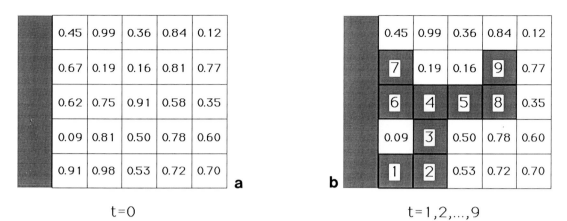

Fig. 2.15. Invasion percolation on a 5 × 5 square lattice (a) at the initial time and (b) after 15 time steps

percolation. Support for this conclusion comes from recent numerical results for the chemical dimension yielding $d_\ell = 1.40 \pm 0.07$ (or $\tilde{\nu} = 1/d_{\min} \cong 0.77$) [2.110]. This value for d_ℓ is significantly smaller than that of percolation (see Table 2.3).

On the other hand, if the defender is compressible, closed regimes can continue to grow and the fractal dimension of the invading fluid is larger [2.106], $d_f \cong 1.89$ $(d = 2)$ and $d_f \cong 2.5$ $(d = 3)$, in close agreement with percolation. For a further discussion and the relation to aggregation and viscous fingering, we refer the reader to Chaps. 4 and 7. For a recent application of invasion percolation to the invasion of nonwetting fluids through porous media see [2.107].

2.6.5 Directed Percolation

Directed percolation can be viewed as a model for a forest fire under the influence of a strong wind in one direction. Consider bond percolation on a square lattice at concentration p, and assign an arrow to each bond such that vertical bonds point in the positive x direction and horizontal bonds point in the positive y direction (see Fig. 2.16). To illustrate directed percolation let us now assume that each bond is a conductor where electric current is transmitted along *only* the directions of the arrows. There exists a critical concentration p_c that separates an insulating phase from a conducting phase. The critical concentration is larger than for ordinary percolation (see Fig. 2.16), $p_c \cong 0.479$ for the triangular lattice [2.111] and $p_c \cong 0.644701$ for the square lattice [2.112].

The structure of the clusters is now strongly anisotropic and two characteristic correlation lengths $\xi_\perp$ and $\xi_\parallel$ exist, perpendicular and parallel to the main direction, here the x–y direction. This anisotropy is due to the fact that the clusters at p_c are self-affine (see Sect. 7.2.1) rather than self-similar objects. The critical behavior of both correlation lengths is described by

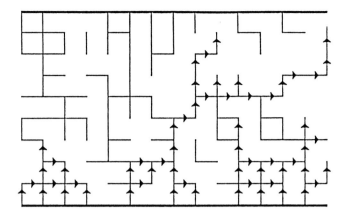

Fig. 2.16. Directed percolation on a square lattice below the percolation threshold. Although a path exists between the top and bottom lines, a current *cannot* go through. Hence the percolation threshold in directed percolation is larger than p_c in ordinary percolation

$$\xi_\perp \sim |p_c - p|^{-\nu_\perp}$$

and

$$\xi_\parallel \sim |p_c - p|^{-\nu_\parallel}, \tag{2.74}$$

with $\nu_\perp \neq \nu_\parallel$, and both are different from the exponent ν. Using scaling theory the following relation can be derived:

$$\gamma + 2\beta = (d-1)\nu_\perp + \nu_\parallel, \tag{2.75}$$

which is a generalization of (2.58). Substituting the mean field exponents $\gamma = \beta = \nu_\parallel = 2\nu_\perp = 1$ in (2.75) yields the upper critical dimension $d_c = 5$ for directed percolation [2.113]. The best estimates for the critical exponents have been achieved using 35 terms in the series expansion method [2.112] (see Sect. 2.8.2), yielding for $d = 2$, $\nu_\parallel = 1.7334 \pm 0.001$, $\nu_\perp = 1.0972 \pm 0.0006$, and $\gamma = 2.2772 \pm 0.0003$, suggesting the rational values $\nu_\parallel = 26/15$, $\nu_\perp = 79/72$, and $\gamma = 41/18$. Also β and the other static exponents differ from ordinary percolation.

Recently, progress has been made in relating directed percolation to self-organized criticality [2.114], and in calculating the spectrum of its transfer matrix [2.115] and its fractal dimension [2.116]. It has also been shown that in directed percolation hyperscaling is violated [2.118]. A detailed review on directed percolation in $d = 2$ has been given in [2.111]. An application of directed percolation to explain interface roughening in porous media [2.117] is presented in detail in Sect. 1.15. For directed percolation in $d = 3$ see [2.119].

2.7 Numerical Approaches

In this section we present several numerical approaches to generating percolation clusters and their substructures such as the backbone and the external
perimeter. Percolation clusters can be generated either by the Leath method
[2.120], where the percolation cluster grows in the way a simple epidemic
spreads (see Sect. 2.6), or by the Hoshen-Kopelman algorithm [2.121], where
all sites in the percolation system belonging to the same cluster are identified.
Substructures such as the backbone and the skeleton are obtained from the
percolation cluster, while the external perimeter of the cluster (in $d = 2$) is
generated directly using specific self-avoiding-walk rules.

2.7.1 Hoshen-Kopelman Method

In the Hoshen-Kopelman algorithm [2.121], all sites in the percolation system
are labeled in such a way that sites with the same label belong to the same
cluster and different labels are assigned to different clusters. If the same label
occurs at opposite sides of the system, an infinite cluster exists. In this way the
critical concentration can be determined. By counting the number of clusters
with s sites, we obtain the cluster distribution function.

The algorithm is quite tricky and we use a simple example to demonstrate
it. Consider the 5×5 percolation system in Fig. 2.17a which we want to analyse.
Beginning at the upper left corner and ending at the lower right corner, we
assign cluster labels to the occupied sites. The first occupied site gets the label
1, the neighboring site gets the same label because it belongs to the same
cluster. The third site is empty and the fourth one is labeled 2. The fifth site
is empty.

In the second line the first site is connected to its neighbor at the top and
is therefore labeled 1. The second site is empty and the third one is labeled 3.

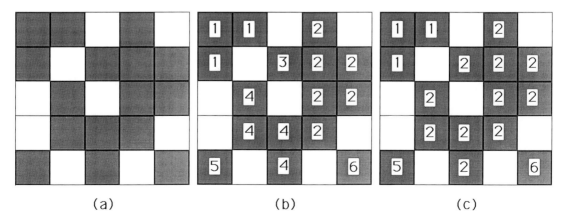

(a)   (b)   (c)

Fig. 2.17a-c. Demonstration of the Hoshen-Kopelman algorithm

The fourth site is now the neighbor of two sites, one labeled 2 and the other labeled 3. All three sites belong to the same cluster, which was first labeled 2. Accordingly, we assign the label 2 also to the new site, but we have to keep track that clusters 2 and 3 are connected. This is achieved by defining a new array $N_L(k)$: $N_L(3) = 2$ tells us that the correct label of cluster 3 is 2. If we continue the labeling we end up with Fig. 2.17b, with $N_L(3) = 2$ and $N_L(4) = 2$. Sites labeled 1, 2, 5, and 6 are not connected with sites with lower labels, and we define $N_L(1) = 1$, $N_L(2) = 2$, $N_L(5) = 5$ and, $N_L(6) = 6$.

In the second step (Fig. 2.17c) we change the improper labels [where $N_L(k) < k$] into the proper ones, beginning with the lowest improper label (here $k = 3$) and ending with the largest improper label (here $k = 4$).

The Hoshen-Kopelman algorithm is useful when investigating the distribution of cluster sizes as well as the largest cluster in any disordered system, not necessarily a percolation system. The method can also be used to determine p_c and to generate the infinite percolation cluster in percolation, but for these purposes nowadays more efficient algorithms are available, which will be described next.

2.7.2 Leath Method

In the Leath method [2.120] (see also [2.56,122]), single percolation clusters are generated in the same way as with the epidemic spreading (see Sect. 2.6). In the first step the origin of an empty lattice is occupied, and its nearest-neighbor sites are either occupied with probability p or blocked with probability $1 - p$. In the second step the empty nearest-neighbors of those sites occupied in the step before are occupied with probability p and blocked with probability $1 - p$. In each step, a new chemical shell is added to the cluster.
The process continues until no sites are available for growth or the desired number of shells has been generated. Thus, percolation clusters are generated with a distribution of cluster sizes, $s \cdot sn_s(p)$. The factor s comes from the fact that each site of the cluster has the same chance of being the origin of the cluster, and thus exactly the same cluster can be generated in s ways, enlarging the distribution $sn_s(p)$ by a factor of s. Figure 2.18 shows a cluster grown up to four shells.

The Leath method is particularly useful for studying structural and physical properties of single percolation clusters. If a large cluster is generated, the substructures like the backbone [2.70] or the skeleton [2.68] can be obtained by efficient algorithms.

2.7.3 Ziff Method

A direct method for generating the external perimeter of a percolation cluster in $d = 2$ (the hull) was suggested by Ziff et al. [2.123]. The method is particularly important because it also represents the most efficient algorithm

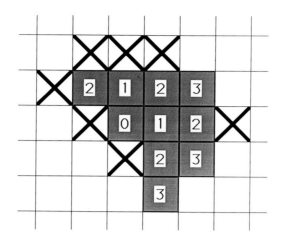

Fig. 2.18. The first four steps of the Leath cluster growth method

for determining p_c. One starts, for example, with a square lattice, where each square represents a site. First, two horizontal neighboring sites are chosen. The right one is occupied and the left is blocked (see Fig. 2.19a).

The bond between the two sites is directed along the positive y direction. This arrow represents the first step of the self-avoiding walk (SAW) by which the hull is generated. In the second step the two neighboring upper sites are considered. We begin with the right one, which is occupied with probability p and blocked with probability $1 - p$. If this site is blocked, the walk steps to the right. If the site is occupied, the left site is tested. If this is blocked the walker steps upwards; if it is occupied the walker steps to the left (see Fig. 2.19b). A new arrow is drawn accordingly and the process continues in the same way.

The method is particularly useful in $d = 2$ where only a relatively small part of the cluster is generated. It is used for determining p_c. Below p_c, finite

(a)

(b)

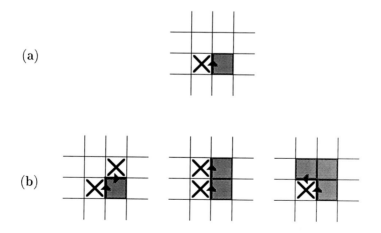

Fig. 2.19. (a) The initial step for generating the perimeter of a percolation cluster on a square lattice. (b) The three generators describing the next step. The full squares represent occupied sites and $\times$ represents blocked sites

clusters of occupied sites dominate and closed perimeter loops in the clockwise direction will mostly be generated. Above p_c, finite clusters of empty sites dominate and closed loops of counterclockwise direction are mostly generated. At $p = p_c$ the numbers of clockwise and counterclockwise loops are equal. In this way, very accurate values of p_c can be obtained, as well as several exponents characterizing the cluster perimeter at criticality. It was shown [2.124] that these exponents are related to the correlation length exponent ν. The fractal dimension of the external perimeter is related to ν by $d_h = 1 + 1/\nu$ [2.72,73], and the distribution of clusters with perimeter length u scales as $n(u) \sim u^{-\tau}$ with $\tau = 1 + 2\nu/(1 + \nu)$. The mean perimeter length of a finite cluster diverges on approaching p_c according to $\langle u \rangle \sim (p - p_c)^{-\gamma}$ with $\gamma = 2$ [2.125].

2.8 Theoretical Approaches

Here we discuss some deterministic fractal structures which are useful for *qualitative* understanding of the fractal nature of percolation clusters, and describe the main theoretical approaches (series expansions and renormalization techniques), by which a *quantitative* description of the percolation transition can be achieved.

2.8.1 Deterministic Fractal Models

a) Sierpinski gasket. The Sierpinski gasket (see Figs. 1.1,2 in Chap. 1) was suggested [2.126] as a model for the backbone of the infinite percolation cluster at p_c. Both structures contain loops of all length scales and are *finitely ramified* [2.45], i.e., parts of them of *any* size can be isolated from the rest by cutting only a *finite* number of bonds.

The fractal dimension of the gasket is $d_f = \ln 3/\ln 2 = 1.585$ for $d = 2$ and $d_f = \ln(d+2)/\ln 2$ for general d. In $d = 2$ and $d = 3$, d_f is thus very close to the fractal dimension of the percolation backbone, see Table 2.3. However, there the similarity ends. Other exponents, such as the graph dimension $d_\ell \, (= d_f)$ and the fractal dimension of the red bonds $d_{\text{red}} (= 0)$, show that the structure of the Sierpinski gasket is very different from the structure of the percolation backbone. Moreover, while in percolation the effect of loops decreases with increasing dimension, the effect of loops increases in the gasket.

The main advantage of the Sierpinski gasket is that many physical properties can be calculated exactly, since the structure is finitely ramified. These exact solutions help to provide insight onto the anomalous behavior of physical properties on fractals, as will be shown in Chap. 3.

b) Mandelbrot-Given fractal. A more refined model for percolation clusters and its substructures has been introduced by Mandelbrot and Given [2.127]. The first four generations of the Mandelbrot-Given fractal are shown in Fig. 2.20.

Fig. 2.20. Three generations of the Mandelbrot-Given fractal

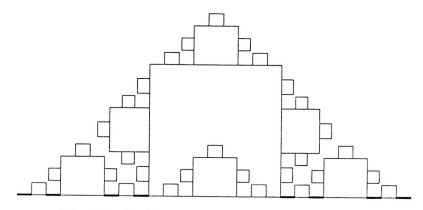

Fig. 2.21. Backbone of the Mandelbrot-Given fractal

The structure contains loops, branches, and dangling ends of all length scales. At each generation, each segment of length a is replaced by eight segments of length $a/3$. Accordingly, the fractal dimension is $d_f = \ln 8/\ln 3 \cong$ 1.893, which is very close to $d_f = 91/46 \cong 1.896$ for percolation. Note, however, that in contrast to percolation $d_{\min} = 1$ or $d_\ell = d_f$.

The backbone of this fractal can be obtained easily by eliminating the dangling ends when generating the fractal (see Fig. 2.21). It is easy to see that the fractal dimension of the backbone is $d_B = \ln 6/\ln 3 \cong 1.63$. The red bonds are all located along the x axis of the figure and they form a Cantor set with the fractal dimension $d_{\mathrm{red}} = \ln 2/\ln 3 \cong 0.63$.

c) Modified Koch curve. The modified Koch curve was suggested in Chap. 1 of [2.22] as a deterministic fractal structure with a nontrivial chemical distance metric, where d_ℓ is neither 1 (as in linear fractals) nor d_f (as in the Sierpinski

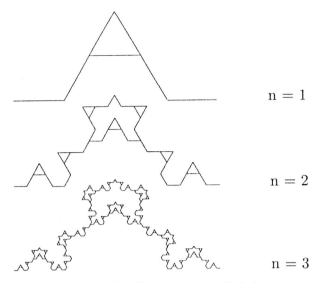

$n = 1$

$n = 2$

$n = 3$

Fig. 2.22. The first iterations of a modified Koch curve, which has a nontrivial chemical distance metric

gasket or the Mandelbrot-Given fractal). The chemical distance ℓ is defined as the shortest path on the fractal between two sites of the fractal. From Fig. 2.22 we see that if we reduce ℓ by a factor of 5, the mass of the fractal within the reduced chemical distance is reduced by a factor of 7, i.e., $M(\frac{1}{5}\ell) = \frac{1}{7}M(\ell)$, yielding $d_\ell = \ln 7/\ln 5 \cong 1.209$. It is easy to see that the fractal dimension of the structure is $d_f = \ln 7/\ln 4 \cong 1.404$. From d_ℓ and d_f one obtains $d_{\min} = \ln 5/\ln 4 \cong 1.161$.

d) Hierarchical model. The *hierarchical structure* (Fig. 2.23) was suggested [2.128] as a model for the percolation backbone at criticality. One starts at step $n = 0$ with a single bond of unit length. At $n = 1$ this bond is replaced

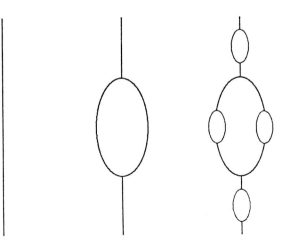

Fig. 2.23. Three generations of the hierarchical model

by four unit bonds, two in parallel and two in series. This structure, shown in Fig. 2.23, is the *unit cell*. At $n = 2$ each bond is replaced by the unit cell and this process continues indefinitely. This self-similar structure consists of blobs and singly connected bonds. The total mass of the structure is $M = 4^n$ and the number of red bonds is $n_{red} = 2^n$. In percolation $n_{red} \sim L^{1/\nu}$, see (2.18), and thus $M \sim L^{2/\nu}$ with $d_B = 1.5$ for $d = 2$ compared with $d_B = 1.62$ (Table 2.3). This model, however, gives better estimates of the dynamical exponents and, in particular, may explain the seemingly log–normal distribution in the voltage drops found in $d = 2$ percolation [2.128] (see Sect. 3.7).

Finally, we mention the *node-link-blob model* [2.51,71] for the backbone, which is essentially based on the ideas of Skal and Shklowskii [2.129] and de Gennes [2.130]. In this model, the backbone represents a superlattice network of nodes. Neighboring nodes are separated by a distance ξ, and are connected by links which include both singly connected bonds and blobs.

2.8.2 Series Expansion

The series expansion method [2.5,131] is one of the earliest techniques to determine the critical exponents for percolation. In this method, the moments M_k of the cluster distribution function [see (2.54)] are approximated by a finite power series in p (or $1 - p$),

$$M_k = \sum_{\ell=1}^{\ell_{max}} a_\ell(k) p^\ell. \tag{2.76}$$

If we combine (2.33) and (2.54), M_k can be written as $M_k = \sum_{s,t} g_{s,t} s^k p^s (1 - p)^k$, and the coefficients $a_\ell(k)$ can be expressed in terms of $g_{s,t}$ which can be calculated up to certain cluster sizes s by exactly enumerating all possible configurations. This way, for example, the first 20 terms in the expansion of the mean cluster size $S = \sum_\ell a_\ell p^\ell$ can be determined. When $p \to p_c$, S diverges and one can determine p_c and the critical exponent γ, by best fitting the series to the form $S \sim |p - p_c|^{-\gamma}$. The simplest (but not most efficient) way to do this is by expanding $(p - p_c)^{-\gamma}$ in powers of p, $(p - p_c)^{-\gamma} = \sum_{\ell=1}^{\infty} b_\ell p^\ell$. The ratio of two consecutive terms in the expansion is $b_{\ell+1}/b_\ell = (1/p_c) + [(\gamma - 1)/p_c](1/\ell)$. Thus by plotting $a_{\ell+1}/a_\ell$ as as a function of $1/\ell$ one expects a straight line, the slope yields $(\gamma - 1)/p_c$ and the intercept is $1/p_c$.

This method, however, is only useful if a large number of coefficients a_ℓ are known. The reason for this is that $S(p)$ is not simply proportional to $(p - p_c)^{-\gamma}$ but also includes corrections with less divergent terms. Only for large ℓ_{max} does the dominant contribution come from the most singular term and the approach become accurate. To improve the convergency for the (realistic) case that only few terms in the expansion are known, several quite sophisticated extrapolation techniques have been developed.

In the Padè approximation, $d\log S/dp \sim [-\gamma/(p_c - p)]$ is calculated from (2.76) by expanding $\log S(p)$, $d\log S/dp = \sum_{\ell=0}^{N} c_\ell p^\ell$. The polynomial is then approximated by a ratio of two polynomials, $\sum_{\ell=0}^{m} a_\ell p^\ell / \sum_{\ell=0}^{n} b_\ell p^\ell$, where $m + n = N$ is the number of the known terms in (2.76) and $b_0 = 1$. The remaining $N + 1$ coefficients a_ℓ and b_ℓ are determined from the $N + 1$ linear equations, $\sum_{\ell=0}^{n} b_\ell c_{j-\ell} = a_j$, where $j = 0, 1, 2, ..., N$.

For different sets of n and m one obtains different Padè approximants for $d\log S/dp$. Since they are expected to behave as $-\gamma/(p_c - p)$, the polynomials in the denominator, $\sum_{\ell=1}^{n} b_\ell p^\ell$, must vanish at p_c. In general, the polynomial will have several zeros from which the most likely one is chosen as p_c ($0 < p_c < 1$). Following standard algebra, the polynomial then can be replaced by $(p_c - p)\sum_{\ell=0}^{n-1} d_\ell p^\ell$ and γ is identified as $\gamma = -\sum_{\ell=1}^{m} a_\ell p_c^\ell / \sum_{\ell=0}^{n-1} d_\ell p_c^\ell$. By plotting γ versus p_c for all pairs (m, n) it appears that many pairs concentrate near one point, usually around $n \cong m$, which then gives the best estimate of γ and p_c. This way, quite accurate results can be obtained. For example, one of the earliest estimations of the exponent α by Domb and Pearce [2.132] yielded $\alpha = -0.668 \pm 0.004$ for $d = 2$, which is to be compared with the exact value $\alpha = -2/3$ found several years later [2.131] (see Sect. 2.8.4). For bond percolation on the cubic lattice, the method yields $p_c = 0.2488 \pm 0.0002$. Quite accurate estimates based on 15 terms in the series expansion have been obtained also for the critical exponents in $d = 3$, $\gamma = 1.805 \pm 0.02$, $\nu = 0.872 \pm 0.023$, and $\beta = 0.405 \pm 0.025$ (see Table 2.2). For a recent pedagogical review on the method see [2.133].

2.8.3 Small-Cell Renormalization

The scaling theory discussed in Sect. 2.5 just provides relations between the critical exponents. Wilson's renormalization technique offers a method where the values of the exponents can also be studied systematically. There exist basically two variants of the method, the ϵ expansion, which is performed in reciprocal q space, and the small-cell (real-space) renormalization, which is performed in real space. In this section we will concentrate on the small cell renormalization and briefly review the ϵ expansion. The ϵ expansion is also useful for studying dynamical quantities near the percolation threshold, which will be discussed in Chap. 3.

On small scales $r < \xi$ the clusters are self-similar. Thus, if one cuts a small piece out of a cluster, and magnifies it, and compares it with the original, then one cannot decide which is the original and which is the magnification, as long as the size of the original is below ξ. At p_c, ξ diverges and we can "renormalize" the system by not looking at each site separately but by averaging over regimes of size $b \ll \xi$. If we then change the length scale accordingly, we cannot distinguish between the original and the renormalized system. This averaging procedure can be repeated again and again since $\xi = \infty$ at p_c. In this respect

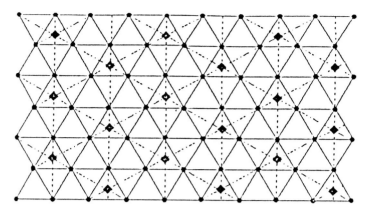

Fig. 2.24. Renormalization of a triangular lattice. The full lines represent the original lattice and the dashed lines are the renormalized one

the critical point p_c is called the mathematical "fixed point" of the renormalization transformation. If the transformation is done in the real lattice space as described here, the method is called the real-space renormalization group transformation [2.134].

Figure 2.24 shows the renormalization group transformation on a triangular lattice. Three sites form a "supersite". The supersite is regarded as empty if the majority of the three sites (i.e. two or three sites) are empty, otherwise it is regarded as occupied. The supersites also form a triangular lattice, the lattice constant of the new lattice being $b = 3^{1/2}$. Here we assume that the lattice constant of the original lattice is 1.

In the original lattice, the correlation length behaves as

$$\xi \sim \xi_0 |p - p_c|^{-\nu}, \tag{2.77a}$$

where ξ_0 is of the order of the lattice constant. In the new superlattice, the correlation length can be written as

$$\xi' \sim \xi_0' |p' - p_c|^{-\nu}, \tag{2.77b}$$

where $\xi_0' = b\xi_0$ and p' is the probability that a site in the superlattice is occupied. Since the actual value of the correlation length remains unchanged by the transformation, we have $\xi' = \xi$ and hence $b|p' - p_c|^{-\nu} = |p - p_c|^{-\nu}$ or

$$\nu = \frac{\ln b}{\ln(\frac{p' - p_c}{p - p_c})}. \tag{2.78}$$

Since a site in the superlattice is occupied if either all three sites in the normal lattice are occupied (with probability p^3) or if two of them are occupied and one is empty (with probability $3p^2(1-p)$), the probability of having a supersite occupied is simply

$$p' = p^3 + 3p^2(1 - p). \tag{2.79}$$

At the percolation threshold we expect $p_c = p'_c (= 1/2)$, which is satisfied by (2.79). Inserting (2.79) into (2.78) we obtain

$$\nu = \frac{\ln 3^{1/2}}{\ln 3/2} \cong 1.355,$$

in close agreement with the exact value $4/3$.

At first glance it seems surprising that the result for ν is not exact. The reason lies in the fact that the renormalization transformation as described does not preserve the connectivity between the sites. Two sites which belong to the same cluster before the transformation may belong to different clusters after the transformation. Hence different transformation schemes may lead to different values of the exponents, which is a serious drawback of the renormalization group method in percolation.

Despite this, renormalization arguments have been used successfully by Coniglio [2.66,71] to determine the fractal dimension of the "red bonds". To calculate d_{red}, consider again the original lattice with lattice constant 1 and a superlattice with lattice constant $b \gg 1$. In the superlattice, bonds between two sites are occupied if the sites are connected in the original lattice. This rule is better suited to treating percolation problems than the majority rule discussed above. At the percolation threshold, the bond concentrations in the two lattices are the same.

If we cut a small fraction q of bonds in the original lattice, a certain fraction q' of renormalized bonds becomes disconnected. The concentrations of bonds in both lattices are $p = p_c(1-q)$ and $p' = p_c(1-q')$, respectively, hence $q = (p_c - p)/p_c$ and $q' = (p_c - p')/p_c$.

The relation between q and q' can be obtained as follows: consider a "renormalized" bond connecting two sites on the superlattice, which by definition belong to the same cluster in the original lattice. If we cut a small fraction q of the bonds, the probability that both sites become disconnected is

$$q' = n_{\mathrm{red}} q, \quad (q \to 0), \tag{2.80}$$

where $n_{\mathrm{red}} \sim b^{d_{\mathrm{red}}}$ is the number of red bonds on the backbone connecting the two sites. Equation (2.80) reflects the fact that cutting only one red bond between the two sites is sufficient to disconnect them. The probability of cutting blobs is at least of order q^2, which can be neglected for $q \to 0$.

From (2.80) we obtain immediately $(p'_c - p) \sim b^{d_{\mathrm{red}}} (p_c - p)$. Comparing this result with (2.78) we obtain the exact relation

$$d_{\mathrm{red}} = 1/\nu. \tag{2.81}$$

2.8.4 Potts Model, Field Theory, and ϵ Expansion

There exist several ways to map the percolation model onto the q-state Potts model in the limit $q \to 0$ [2.135]. A common way [2.7] is to start with the dilute Ising Hamiltonian,

$$H = - \sum_{<i,j>} J_{ij} S_i S_j, \qquad (2.82)$$

where $S_i = \pm 1$ and J_{ij} are nearest-neighbor interactions which can take the value J with probability p and 0 with probability $1 - p$. The partition function is

$$Z\{J_{ij}\} = \text{Tr} \left\{ \exp \left(\beta \sum_{<i,j>} J_{ij} S_i S_j \right) \right\}, \qquad (2.83)$$

and the average free energy is

$$F = -\frac{kT}{N} \langle \ln Z\{J_{ij}\} \rangle. \qquad (2.84)$$

The average $\langle \cdots \rangle$ is over the quenched distribution of J_{ij}. Using the replica assumption

$$\ln Z = \lim_{n \to 0} \frac{Z^n - 1}{n}, \qquad (2.85)$$

one obtains

$$\langle Z^n \rangle = \prod_{<ij>} \exp(h_{ij}), \qquad (2.86)$$

where

$$\exp(h_{ij}) = p \exp \left(\beta J \sum_{\alpha=1}^{n} \left(S_i^\alpha S_j^\alpha - 1 \right) \right) + 1 - p. \qquad (2.87)$$

The trace here is over the n replicated spins S_i^α, $\alpha = 1, \ldots, n$. For $\beta J \to \infty$ (corresponding to $T \to 0$), the rhs of (2.86) takes only two values, $\exp(h_{ij}) = 1$ if $S_i^\alpha = S_j^\alpha$ for all α, and $\exp(h_{ij}) = 1 - p$ otherwise. We can therefore express h_{ij}, for $\beta J \to \infty$, by

$$h_{ij} = - \ln(1 - p) \left(\prod_{\alpha=1}^{n} \delta_{S_i^\alpha, S_j^\alpha} - 1 \right). \qquad (2.88)$$

Equation (2.88) represents the pair interaction between sites i and j in the Hamiltonian of the q-state Potts model with $q = 2^n$ states and a coupling constant $- \ln(1 - p)$. The essential point is that by taking the limit $n \to 0$ one eliminates the property of spins, and only the geometrical connectivity is left. Thus the exponents derived for the Potts model in the limit of $n \to$

0 also represent the exponents for the pure percolation problem. From the renormalization treatment of the Potts model (which gives exact values only in $d = 2$) one finds for percolation in $d = 2$, $\nu = 4/3$, $\beta = 5/36$, $\gamma = 43/18$ and $d_f = 91/48$ [2.48,49].

The fractal dimensions of the percolation substructures like d_B and the graph dimension d_ℓ could not be derived by similar analogies.

For dimensions $d \geq 3$, the critical exponents can be estimated from the ϵ expansion. In this method, a Potts model field theory is derived using the momentum space representation of the Potts Hamiltonian, and standard momentum space renormalization methods are applied where large momentum values (corresponding to small distances) are traced out. The critical exponents are expanded in $\epsilon = 6 - d$ around the critical dimension $d = 6$. The method is tedious and only expansions up to second order in ϵ have been performed. The results are

$$\nu = \frac{1}{2} + \frac{5}{84}\epsilon + \frac{589}{37044}\epsilon^2 + O(\epsilon^3), \qquad (2.89a)$$

$$\beta = 1 - \frac{1}{7}\epsilon - \frac{61}{12348}\epsilon^2 + O(\epsilon^3), \qquad (2.89b)$$

and

$$d_f = d - \frac{\beta}{\nu} = 4 - \frac{10}{21}\epsilon + \frac{103}{9261}\epsilon^2 + O(\epsilon^3), \qquad (2.89c)$$

which gives reasonable results for high dimensions. For example, in $d = 3$ (2.89c) yields $d_f \cong 2.67$, which is to be compared with the best numerical value 2.52.

The exponent $d_{\min} = 1/\tilde{\nu}$ was calculated by Cardy and Grassberger [2.59] up to the order of ϵ. Janssen [2.136] calculated $d_{\min} = 1/\tilde{\nu}$ up to the order of ϵ^2, $d_{\min} = 2 - \frac{1}{6}\epsilon - [\frac{937}{588} + \frac{45}{49}(\ln 2 - \frac{9}{10}\ln 3)](\frac{\epsilon}{6})^2 + O(\epsilon^3)$, yielding $d_{\min} \cong 1.1695$, which is to be compared with the best numerical value 1.374 (see Table 2.3).

2.A Appendix: The Generating Function Method

In this appendix we present an alternative method to derive the percolation exponents on a Cayley tree. We show how the generating-function approach works in the case of classical percolation on the Cayley tree with coordination number $z = 3$, where the bonds are randomly disconnected with probability p. For each site we introduce a probability $F(s)$ that this site is connected to the branch of s bonds via a given bond. Actually $F(s)$ can be identified with sn_s defined in Sect. 2.5. Since all branches and sites on the Cayley tree are equivalent, we can write

$$F(s) = (1 - p) \sum_{s_1 + s_2 = s - 1} F(s_1)F(s_2), \qquad F(0) = p. \qquad (2.90)$$

To simplify this convolution sum one can use the generating function

$$f(x) = \sum_{s=0}^{\infty} F(s) x^s. \tag{2.91}$$

As shown below, simple relations exist between this generating function and the quantities of interest such as P_{∞} and $\langle s \rangle$. Multiplying both sides of (2.90) by x^s and summing over s yields the quadratic equation,

$$f(x) = p + (1 - p) x f^2(x), \tag{2.92}$$

with two real solutions. Taking into account that if $x \leq 1$ then $f(x) \leq 1$, we select the smaller of the two solutions,

$$f(x) = \frac{1 - \sqrt{1 - 4p(1 - p)x}}{2(1 - p)x}. \tag{2.93}$$

The order parameter is $P_{\infty} = 1 - f(1)$. At $x = 1$, the positive value of the square root is equal to $|2p - 1|$. If $p > 1/2$ then $|2p - 1| = 2p - 1$ which gives $f(1) = 1$. For p below $1/2$ we have $f(1) = p/(1 - p) < 1$ and the order parameter is different from zero. Hence $p_c = 1/2$ and

$$P_{\infty} = 1 - f(1) = \frac{1 - 2p}{1 - p} \sim (p_c - p)^{\beta}, \tag{2.94}$$

with $\beta = 1$. The function $f(x)$ has a singularity at point $x_0 = 1/4p(1 - p)$ of the type

$$f(x) = A + B|x - x_0|^a, \tag{2.95}$$

with $a = 1/2$. According to the Tauberian theorems [2.81], which relate singularities in $f(x)$ to singularities in $F(s)$, (2.95) suggests that the Taylor expansion of the function $f(x)$ near $x = 0$ behaves as

$$F(s) = s^{-\tau+1} \phi(s^{\sigma} |p - p_c|) \sim s^{-\tau+1} \exp(-s/s_c) \tag{2.96}$$

where

$$\tau = 2 + a, \qquad s_c = 1/\ln(x_0). \tag{2.97}$$

Thus, in our case,

$$\tau = 5/2 \tag{2.98}$$

and

$$s_c \approx \frac{1}{x_0 - 1} = \frac{p(1 - p)}{(p - p_c)^2}, \tag{2.99}$$

which gives

$$\sigma = 1/2. \tag{2.100}$$

In this simple case the function $F(s)$ can be computed according to the binomial expansion of $\sqrt{1 - y}$

$$F(s) = 2p\frac{(2s-1)!!}{(2s+2)!!}(4p(1-p))^s, \tag{2.101}$$

where the ratio of factorials has the power-law asymptotic behavior $s^{-3/2}$, and the term $(4p(1-p))^s$ gives an exponential cutoff.

We can also compute the average cluster size

$$\langle s \rangle = \sum_{s=0}^{\infty} sF(s) = df(x)/d\ln x|_{x=1}$$

and get

$$\langle s \rangle = \frac{(1-p)}{2|p-p_c|} \qquad p > p_c, \tag{2.102a}$$

$$\langle s \rangle = \frac{p}{2|p-p_c|} \qquad p < p_c, \tag{2.102b}$$

which means that $\gamma = 1$ on both sides.

Acknowledgements. We like to thank D. Ben-Avraham, I. Dayan, J. Dräger, H. Bolterauer, P. Maass, M. Meyer, U. A. Neumann, R. Nossal, M. Porto, S. Rabinovich, H. E. Roman, S. Schwarzer, H. E. Stanley, D. Stauffer, H. Taitelbaum, and G. H. Weiss for valuable discussions and helpful remarks.

References

2.1   D. Stauffer: *Introduction to Percolation Theory* (Taylor and Francis, London 1985)
2.2   A mathematical approach can be found in H. Kesten: *Percolation Theory for Mathematicians* (Birkhauser, Boston 1982);
        G.R. Grimmet: *Percolation* (Springer, New York 1989)
2.3   A collection of review articles: *Percolation Structures and Processes*, ed. by G. Deutscher, R. Zallen, J. Adler (Adam Hilger, Bristol 1983)
2.4   S. Kirkpatrick: in *Le Houches Summer School on Ill Condensed Matter*, ed. by R. Maynard, G. Toulouse (North Holland, Amsterdam 1979)
2.5   C. Domb: in Ref. [2.3];
        C. Domb, E. Stoll, T. Schneider: Contemp. Phys. **21**, 577 (1980)
2.6   J.W. Essam: Rep. Prog. Phys. **43**, 843 (1980)
2.7   A. Aharony: in *Directions in Condensed Matter Physics*, ed. by G. Grinstein, G. Mazenko (World Scientific, Singapore 1986)
2.8   A. Bunde: Adv. Solid State Phys. **26**, 113 (1986)
2.9   M. Sahimi: *Application of Percolation Theory* (Taylor and Francis, London 1993)
2.10   D.L. Turcotte: *Fractals and Chaos in Geology and Geophysics* (Cambridge University Press, Cambridge 1992)
2.11   P.G. de Gennes: *Scaling Concepts in Polymer Physics* (Cornell University Press, Ithaca 1979)
2.12   H.J. Herrmann: Phys. Rep. **136**, 153 (1986)
2.13   F. Family, D. Landau, eds.: *Kinetics of Aggregation and Gelation* (North Holland, Amsterdam 1984);

For a recent review on gelation see M. Kolb, M.A.V. Axelos: in *Correlatios and Connectivity: Geometric Aspects of Physics, Chemistry and Biology*, ed. by H.E. Stanley, N. Ostrowsky, (Kluwer, Dordrecht 1990) p. 225

2.14   P. Grassberger: Math. Biosci., **62**, 157 (1986); J. Phys. A **18**, L215 (1985); J. Phys. A **19**, 1681 (1986)

2.15   G. Mackay, N. Jan: J. Phys. A **17**, L757 (1984)

2.16   J. Feder: *Fractals* (Plenum, New York 1988)

2.17   T. Viscek: *Fractal Growth Phenomena* (World Scientific, Singapore 1989)

2.18   H.E. Stanley, N. Ostrowsky, eds.: *On Growth and Form: Fractal and Non-Fractal Patterns in Physics* (Nijhoff, Dordrecht 1985);
       H.E. Stanley, N. Ostrowsky, eds.: *Random Fluctuations and Pattern Growth; Experiments and Models* (Kluwer, Dordrecht 1988);
       H.E. Stanley, N. Ostrowsky, eds.: *Correlatios and Connectivity: Geometric Aspects of Physics, Chemistry and Biology* (Kluwer, Dordrecht 1990)

2.19   D. Avnir, ed.: *The Fractal Approach to Heterogeneous Chemistry* (John Wiley, New York 1989)

2.20   R. Zallen: *The Physics of Amorphous Solids* (John Wiley, New York 1983)

2.21   J.F. Gouyet: *Physics and Fractal Structures* (Springer, New York 1995)

2.22   A. Bunde, S. Havlin, eds.: *Fractals in Science* (Springer, Heidelberg 1994)

2.23   G. Dumpich, St. Friedrichowski, A. Plewnia, P. Ziemann: Phys. Rev. B **48**, 15332 (1993)

2.24   U. Lauer, J. Maier: Ber. Bunsenges. Phys. Chem. **96**, 111 (1992);
       J. Maier: Prog. Solid St. Chem. **23**, 171 (1995)

2.25   M.D. Ingram: Phys. Chem. Glasses **28**, 215 (1987)

2.26   H. Harder, A. Bunde, W. Dieterich: J. Chem. Phys. **85**, 4123 (1986);
       M.D. Ingram, M.A. Mackenzie, W. Müller, M. Torge: Solid State Ionics **40&41**, 671 (1990);
       A. Bunde, M.D. Ingram, P. Maass: J. Non-Cryst. Solids **173-174**, 1222 (1994)

2.27   A. Bunde, W. Dieterich, H.E. Roman: Phys. Rev. Lett. **55**, 5 (1985)

2.28   A. Bunde, S. Havlin, M. Porto: Phys. Rev. Lett. **74** 2714 (1995)

2.29   R. Engelman, Z. Jaeger, eds.: *Fragmentation, Form and Flow in Fractured Media*, (Adam Hilger, Bristol 1986)

2.30   M. Plogzajczak, A. Tucholski: Phys. Rev. Lett. **65**, 1539 (1990)

2.31   P.E. Seiden, L.S. Schulman: Adv. Phys. **39**, 1 (1990)

2.32   H.E. Stanley, R.L. Blumberg, A. Geiger: Phys. Rev. B **28**, 1626 (1983);
       R.L. Blumberg, H.E. Stanley, A. Geiger, P. Mausbach: J. Chem. Phys. **80**, 5230 (1984);
       F. Sciortino, P.H. Pool, H.E. Stanley, S. Havlin: Phys. Rev. Lett. **64**, 1686 (1990)

2.33   J.W. Essam, D.S. Gaunt, A.J. Guttmann: J. Phys. A **11**, 1983 (1978)

2.34   R.M. Ziff, B. Sapoval: J. Phys. A **19**, L1169 (1987)

2.35   R.M. Ziff: Phys. Rev. Lett. **69**, 2670 (1992)

2.36   M.F. Sykes, M.K. Wilkinson: J. Phys. A **19**, 3415 (1986);
       J. Adler, Y. Meir, A.B. Harris, A. Aharony: Bull. Isr. Phys. Soc. **35**, 102 (1989)

2.37   R.M. Ziff, G. Stell: unpublished
       P.N. Strenski, R.M. Bradley, J.M. Debierre: Phys. Rev. Lett. **66**, 133 (1991)

2.38   C. Domb: Proc. Phys. Soc. **89**, 859 (1966)

2.39   P. Grassberger: J. Phys. A **25**, 5867 (1992)

2.40   T. Viscek, J. Kertesz: J. Phys. A **14**, L31 (1981);
       J. Kertesz: J. Physique **42** L393 (1981)

2.41   W.T. Elam, A.R. Kerstein, J.J. Rehr: Phys. Rev. Lett. **52**, 1515 (1984)

2.42   P.J. Flory: *Principles of Polymer Chemistry* (Cornell University, New York 1971);
       P.J. Flory: J. Am. Chem. Soc. **63**, 3083, 3091, 3096 (1941);
       W.H. Stockmayer: J. Chem. Phys. **11**, 45 (1943)

2.43   S.R. Broadbent, J.M. Hammersley: Proc. Camb. Phil. Soc., **53**, 629 (1957)

2.44   See e.g., S.K. Ma: *Modern Theory of Critical Phenomena* (Benjamin, Reading 1976)

2.45   B.B. Mandelbrot: *Fractals: Form, Chance and Dimension* (Freeman, San Francisco 1977);
       B.B. Mandelbrot: *The Fractal Geometry of Nature* (Freeman, San Francisco 1982)

2.46   H.E. Stanley: *Introduction to Phase Transition and Critical Phrenomena* (Oxford University, Oxford 1971)

2.47 C. Domb, J.L. Lebowitz, eds.: *Phase Transitions and Critical Phenomena* (Academic Press, New York 1972)

2.48 M.P.M. den Nijs: J. Phys. **12**, 1857 (1979)

2.49 B. Nienhuis: J. Phys. A **15**, 199 (1982)

2.50 S. Havlin, D. Ben-Avraham: Adv. in Phys. **36**, 693 (1987)

2.51 H.E. Stanley: J. Phys. A **10**, L211 (1977)

2.52 R.J. Harrison, G.H. Bishop, G.D. Quinn: J. Stat. Phys. **19**, 53 (1978)

2.53 B.H. Zimm, W.H. Stockmayer: J. Chem. Phys. **17**, 1301 (1949)

2.54 T.C. Lubensky, J. Isaacson: Phys. Rev. A **20**, 2130 (1979)

2.55 K.M. Middlemiss, S.G. Whittington, D.C. Gaunt: J. Phys. A **13**, 1835 (1980)

2.56 Z. Alexandrowicz: Phys. Lett. A **80**, 284 (1980)

2.57 R. Pike, H.E. Stanley: J. Phys. A **14**, L169 (1981)

2.58 S. Havlin, R. Nossal: J. Phys. A **17**, L427 (1984)

2.59 J.L. Cardey, P. Grassberger: J. Phys A **18**, L267 (1985)

2.60 H.E. Stanley: J. Stat. Phys. **36**, 843 (1984)

2.61 H.J. Herrmann, H.E. Stanley: J. Phys. A **21**, L829 (1988)

2.62 P. Meakin, I. Majid, S. Havlin, H.E. Stanley: J. Phys A **17**, L975 (1984)

2.63 S. Havlin, B. Trus, G.H. Weiss, D. Ben-Avraham: J. Phys. A **18**, L247 (1985); U.A. Neumann, S. Havlin: J. Stat. Phys. **52**, 203 (1988)

2.64 A. Bunde, J. Dräger: Phil. Mag. B **71**, 721 (1995); Phys. Rev. E **52**, 53 (1995)

2.65 H.E. Roman: Phys. Rev. E **51**, 5422 (1995)

2.66 A. Coniglio: J. Phys. A **15**, 3829 (1982)

2.67 J.F. Gouyet: Physica A **191**, 301 (1992)

2.68 S. Havlin, R. Nossal, B. Trus, G.H. Weiss: J. Stat. Phys. A **17**, L957 (1984)

2.69 H.J. Herrmann, D.C. Hong, H.E. Stanley: J. Phys. A **17**, L261 (1984)

2.70 (a) H.J. Herrmann, H.E. Stanley: Phys. Rev. Lett. **53**, 1121 (1984); (b) M.D. Rintoul, H. Nakanishi: J. Phys. A **27**, 5445 (1994)

2.71 A. Coniglio: Phys. Rev. Lett. **46**, 250 (1982)

2.72 B. Sapoval, M. Rosso, J.F. Gouyet: J. Physique Lett. **46** L149 (1985)

2.73 H. Saleur, B. Duplantier: Phys. Rev. Lett. **58**, 2325 (1987)

2.74 T. Grossman, A. Aharony: J. Phys. A **20**, L1193 (1987)

2.75 G. Huber, H.E. Stanley: unpublished

2.76 S. Havlin, J.E. Kiefer, F. Leyvraz, G.H. Weiss: J. Stat. Phys. **47**, 173 (1987)

2.77 G. Toulouse: Nuovo Cimento B **23**, 234 (1974)

2.78 S. Kirkpatrick: Phys. Rev. Lett. **36**, 69 (1976)

2.79 M.E. Sykes, D.S. Gaunt, H. Ruskin: J. Phys. A **9**, 1899 (1976)

2.80 M.E. Fisher: Physics **3**, 255 (1967)

2.81 G.H. Weiss: *Aspects and Applications of the Random Walk* (North Holland, Amsterdam 1994)

2.82 D. Stauffer: Phys. Rep. **54**, 1 (1979)

2.83 A.L. Ritzenberg, R.I. Cohen: Phys. Rev. B **30**, 4036 (1984); M. Barma: J. Phys. A **18**, L277 (1985)

2.84 M. Porto, A. Bunde, S. Havlin, H.E. Roman: Phys. Rev. E (1996)

2.85 M.F. Guyer, M.V. Ferer, B.F. Edwards, G. Huber: Phys. Rev. E **51**, 2632 (1995)

2.86 D. Stauffer, J. Adler, A. Aharony: J. Phys. A **27**, L475 (1994)

2.87 W. von Niessen, A. Blumen: Canadian Journal of Forest Research **18**, 805 (1988)

2.88 M. Eden: Proc. 4th Berkeley Symp. on Math. Stat. and Prob., **4**, 223 (1961); A. Bunde, H.J. Herrmann, A. Margolina, H.E. Stanley: Phys. Rev. Lett. **55**, 653 (1985)

2.89 A. Bunde, S. Miyazima: J. Phys. A **21**, L345 (1988)

2.90 B. Drossel, F. Schwabl: Phys. Rev. Lett. **69**, 1628 (1992); B. Drossel, S. Clar, F. Schwabl: Phys. Rev. Lett. **71** 3739 (1993)

2.91 E.V. Albano: J. Phys. A **27**, L881 (1994)

2.92 P. Manneville, L. de Seze: in *Numerical Methods in the Study of Critical Phenomena*, ed. by I. Della Dora, J. Demongeot, B. Lacolle (Springer, Heidelberg 1981)

2.93 H.J. Herrmann, D.P. Landau, D. Stauffer: Phys. Rev. Lett. **49**, 412 (1982); N. Jan, A. Coniglio, H.J. Herrmann, D.P. Landau, F. Leyvraz, H.E. Stanley: J. Phys. A **19**, L399 (1986)

2.94 A. Chhabra, D.P. Landau, H.J. Herrmann: in *Fractals in Physics*, ed. by L. Pietronero, E. Tossati (North-Holland, Amsterdam 1986)

2.95 D. Stauffer, A. Coniglio, A. Adam: Adv. Polymer Science **44**, 103 (1982)

2.96 P. Munk: *Introduction to Macromolecular Science* (Wiley, New York 1989)

2.97 M. Doi, S.F. Edwards: *The Theory of Polymer Dynamics* (Oxford University Press, Oxford 1986)

2.98 A.Yu. Grossberg, A. Khokhlov: *Statistical Physics of Macromolecules* (AIP Press, New York 1994)

2.99 M. Daoud, in [2.22]

2.100 I. Majid, N. Jan, A. Coniglio, H.E. Stanley: Phys. Rev. Lett. **52**, 1257 (1984);
K. Kremer, J.W. Lyklema: Phys. Rev. Lett. **55**, 2091 (1985);
S. Havlin, B. Trus, H.E. Stanley: Phys. Rev. Lett. **53**, 1288 (1984)

2.101 T.C. Lubensky, J. Isaacson: Phys. Rev. Lett. **41**, 829 (1978);
S. Havlin, Z.V. Djordjevic, I. Majid, H.E. Stanley: Phys. Rev. Lett. **53**, 178 (1984)

2.102 L.A. Lucena, J.M. Araujo, D.M. Tavares, L.R. da Silva, C. Tsallis: Phys. Rev. Lett. **72**, 230 (1994)

2.103 M. Porto, A. Shehter, A. Bunde, S. Havlin: preprint

2.104 R. Lenormand, S. Boris: C. R. Acad. Sci. (Paris) **291**, 279 (1980)

2.105 R. Chandler, J. Koplik, K. Lerman, J. Willemsen: J. Fluid Mech. 119, 249 (1982)

2.106 D. Wilkinson, J. Willemsen: J. Phys. A **16**, 3365 (1983)

2.107 P. Meakin, G. Wagner, J. Feder, T. Jøssang: Physica A **200**, 241 (1993)

2.108 L. Furuberg, J. Feder, A. Aharony, I. Jøssang: Phys. Rev. Lett. **61**, 2117 (1988)

2.109 R. Lenormand, C. Zarcone: Phys. Rev. Lett. **54**, 2226 (1985)

2.110 S. Havlin, A. Bunde, J.E. Kiefer, unpublished (1991)

2.111 W. Kinzel, in: [2.3]

2.112 S. Redner: Phys, Rev. B **25**, 3242 (1982)

2.113 J.W. Essam, K. De'Bell, J. Adler: Phys. Rev. B **33**, 1982 (1986);
J.W. Essam, A.J. Guttmann, K. De'Bell: J. Phys. A **21**, 3815 (1988)

2.114 S.P. Obukov: Phys, Rev. Lett. **65**, 1395 (1990);
D.E. Wolf, J. Kertesz, S.S. Manna, S.P. Obukhov: Phys. Rev. Lett. **68**, 546 (1992)

2.115 M. Henkel, H.J. Herrmann: J. Phys. A **23**, 3719 (1990)

2.116 B. Hede, J. Kertesz, T. Viscek: J. Stat. Phys. **64**, 829 (1991)

2.117 L.A.N. Amaral, A.-L. Barabási, S.V. Buldyrev, S.T. Harrington, S.Havlin, R. Sadr-Lahijany, and H.E. Stanley: Phys. Rev. E **51**, 4655 (1995)

2.118 M. Henkel, V. Privman: Phys. Rev. Lett. **65**, 1777 (1990)

2.119 P. Grassberger: J. Phys. A **22**, 3673 (1989);
J. Krug, J. Kertesz, D.E. Wolf: Europhys. Lett. **12**, 113 (1990)

2.120 P.L. Leath: Phys. Rev. B **14**, 5046 (1976)

2.121 J. Hoshen, R. Kopelman: Phys. Rev. B **14**, 3428 (1976)

2.122 J.M. Hammersley: Meth. in Comp. Phys. **1**, 281 (1963)

2.123 R.M. Ziff, P.T. Cummings, G. Stell: J. Phys. A **17**, 3009 (1984);
R.M. Ziff: J. Stat. Phys. **28**, 838 (1982)

2.124 R.M. Ziff: Phys. Rev. Lett. **56**, 545 (1986)

2.125 A. Weinreb, S. Trugman: Phys. Rev. B **31**, 2993 (1985)

2.126 Y. Gefen, A. Aharony, B.B. Mandelbrot, S. Kirkpatrick: Phys. Rev. Lett. **47**, 1771 (1981)

2.127 B.B. Mandelbrot, J. Given: Phys. Rev. Lett. **52**, 1853 (1984)

2.128 L. de Arcangelis, S. Redner, A. Coniglio: Phys. Rev. B **31**, 4725 (1985); Phys. Rev. A **34**, 4656 (1986)

2.129 A.S. Skal, B.I. Shklowskii: Sov. Phys. Semicond., **8**, 1029 (1975)

2.130 P.G. de Gennes: La Recherche **7**, 919 (1976)

2.131 D.S. Gaunt, A.J. Guttmann, in: *Phase Transitions and Critical Phenomena*, ed. by C. Domb, M.S. Green (Academic Press, New York 1974) p. 181

2.132 C. Domb, C.J. Pearce: J. Phys. A **9**, L137 (1976)

2.133 J. Adler: Computers in Physics **8**, 287 (1994)

2.134 P.J. Reynolds, H.E. Stanley, W. Klein: Phys. Rev. B **21**, 1223 (1980)

2.135 P.W. Kasteleyn, C.M. Fortain: J. Phys. Soc. Jap. Suppl. **26**, 11 (1969)

2.136 H.K. Janssen: Z. Physik B **58**, 311 (1985)

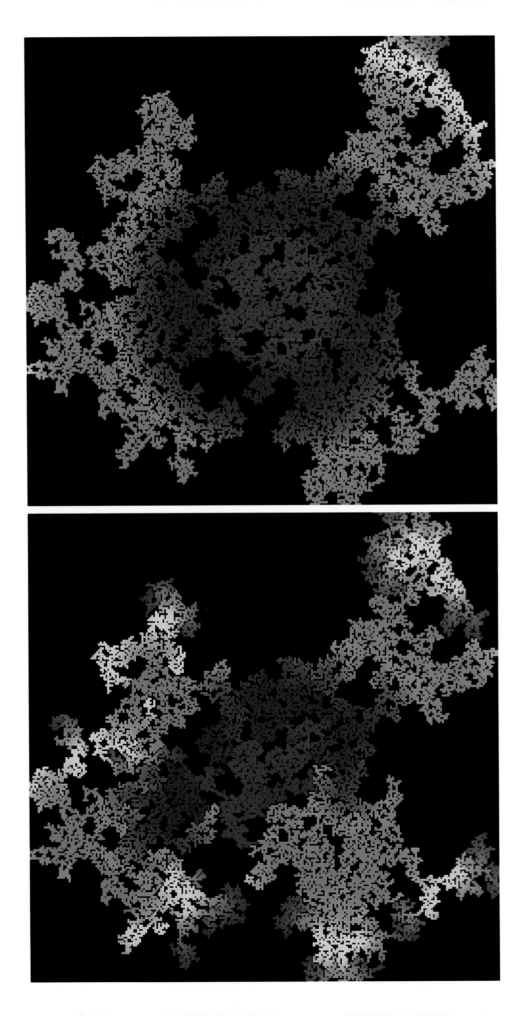

3 Percolation II

Shlomo Havlin and Armin Bunde

3.1 Introduction

In the previous chapter we discussed the structural properties of percolation systems, in particular close to the percolation threshold. To describe the percolation transition we studied geometrical quantities such as the cluster size distribution n_s, the mean size S of a finite cluster, the correlation length ξ, and the probability P_∞ that an occupied site belongs to the infinite cluster. These quantities are characterized by the critical exponents σ and τ, γ, ν, and β, respectively, which are not independent of each other. If one knows two exponents, the others follow.

The correlation length ξ describes the linear extension of the finite clusters below and above p_c and represents the only characteristic length in the system. On length scales smaller than ξ both the finite clusters and the infinite cluster (if it exists) are self-similar and are characterized by the fractal dimension d_f and the graph dimension d_ℓ. While d_f is related to the static exponents ($d_f = d - \beta/\nu$), a similar relation for d_ℓ is not known. Above p_c on length scales larger than ξ, the infinite cluster is homogeneous and has the dimension d of the lattice (see Sect. 2.3).

In this chapter we discuss *dynamical* properties of percolation systems, where to each site or bond a physical property such as conductivity or elasticity is assigned. We show that due to the fractal nature of percolation near p_c, the physical laws of dynamics are changed essentially and become *anomalous*.

For example, we discuss transport properties of percolation systems by assuming that occupied sites are conductors, empty sites are insulators, and electric current can flow only between nearest-neighbor conductor sites. The spe-

◄ Fig. 3.0. Large percolation cluster at the critical concentration. The colors of the sites characterize (a) their "Euclidean" distance and (b) their chemical distance from a site chosen as the origin of the cluster. Courtesy of B.L. Trus

cific conductivity is no longer constant but depends anomalously on the size of the system. Vibrational properties will be discussed in the context of bond percolation by assuming that occupied bonds can be represented by springs with a finite spring constant, while empty bonds are disconnected. If all bonds in the lattice are occupied, one has normal phonons. When the system is diluted randomly, localized modes occur for large frequencies, which have been introduced and called *fractons* by Alexander and Orbach [3.1], and their density of states shows an anomalous frequency behavior.

As was emphasized in the previous chapter, the fact that the percolation transition is a critical phenomenon implies that physical quantities related to the transition can be characterized by power laws of $|p - p_c|$. The same applies to the dynamical properties. The first empirical evidence for power-law behavior was given by Last and Thouless [3.2] in 1971 when studying a $2d$ diluted conducting material. Their data suggest that above $p_c \cong 0.6$, in the critical regime, the conductivity σ behaves as

$$\sigma \sim (p - p_c)^\mu \qquad (3.1)$$

with μ greater than 1. Figure 3.1 shows the results of a more recent conductivity measurement obtained by Lauer and Maier [3.3] for a $3d$ mixture of AgCl (moderate Ag$^+$ conductivity) at concentration $p = \phi$ and AgI (high Ag$^+$ conductivity) at concentration $1 - \phi$. The curve shows a nice power-law with an exponent of about 2.3 (see Sect. 3.6.4).

The idea that transport properties of percolation systems can be efficiently studied by means of diffusion was suggested by de Gennes [3.4] (see also Kopel-

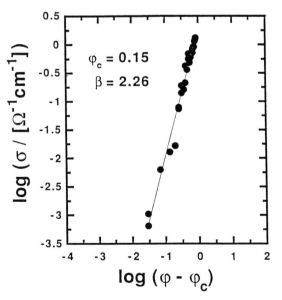

Fig. 3.1. Double logarithmic plot of the ionic conductivity for two phase mixures AgCl/α-AgI at $T = 182^o$ C. After [3.3]

man [3.5]). The diffusion process can be modeled by random walkers, which can jump between nearest-neighbor occupied sites in the lattice. For such a random walker moving in a disordered environment including bottlenecks, loops, and dead ends de Gennes coined the term *ant in the labyrinth*. By calculating the mean square displacement of the walker one obtains the diffusion constant, which according to Einstein is proportional to the dc conductivity.

As we will show, not only the conductivity and diffusion exponents above p_c, but also the exponents characterizing the size dependence of the dc conductivity and the time dependence of the mean square displacement of the random walker, are related. Since it is numerically more efficient to calculate the relevant transport quantities by simulating random walks than to determine the conductivity directly from Kirchhoff's equations, the study of random walks has improved our knowledge not only of diffusion but also of transport process in percolation in general. For reviews on transport in disordered systems and in particular on percolation see [3.6–15].

To introduce the various definitions, concepts, and methods related to transport on percolation clusters we begin in Sect. 3.2 with transport on fractals, including the Sierpinski gasket. In Sects. 3.3–5 we discuss the dynamical properties of percolation systems for $p \geq p_c$: transport in Sect. 3.3, fractons in Sect. 3.4, and ac conductivity in Sect. 3.5. In Sect. 3.6 we present theoretical and numerical methods by which the transport exponents can be obtained. In Sect. 3.7 we show that at p_c several intrinsic dynamical distributions, such as the distribution of the voltage drops along the bonds or the probability densities of random walks, have multifractal features. Section 3.8 is devoted to several other dynamical aspects of percolation such as biased diffusion, diffusion-controlled reactions, and dynamical percolation.

3.2 Anomalous Transport in Fractals

Due to self-similarity, the transport quantities are significantly modified for fractal substrates. We will consider here three representative examples: (1) the total resistance and the conductivity, (2) the mean square displacement and the probability density of random walks, (3) the density of states and the mean amplitudes of vibrational excitations. To see the anomalous behavior for fractals, we first discuss transport in regular lattices.

3.2.1 Normal Transport in Ordinary Lattices

a) **Total resistance.** Consider a metallic network of size L^d. At opposite faces of the network there are metallic bars with a voltage difference between them (see Fig. 3.2).

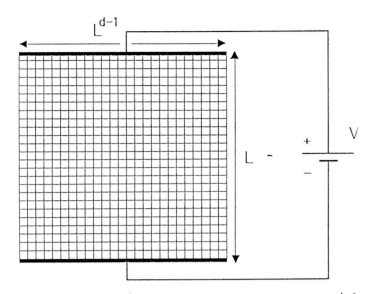

Fig. 3.2. A metallic network of size L^d between two metallic bars of size L^{d-1} with a voltage difference V

If we vary the linear size L of the system, the total resistance ρ varies as

$$\rho \sim \sigma^{-1} \frac{L}{L^{d-1}}, \tag{3.2}$$

where $\sigma \sim L^0 =$ const is the conductivity of the metal. Since σ does not depend on L, (3.2) states that the total resistance of the network depends on its linear size L via the power-law $\rho \sim L^{2-d} \equiv L^{\tilde{\zeta}}$, which defines the resistance exponent $\tilde{\zeta}$; here $\tilde{\zeta} = 2 - d$.

b) Diffusion. The diffusion process is commonly modeled by a simple random walk (see, e.g., [3.15,17]), which in one time unit advances one step of length a to a randomly chosen nearest-neighbor site on a given d-dimensional lattice. Assume that a random walker starts at time $t = 0$ at the origin of the lattice. After t time steps, the actual position of the walker is described by the vector (see Fig. 3.3)

$$\mathbf{r}(t) = a \sum_{\tau=1}^{t} \mathbf{e}_\tau, \tag{3.3}$$

where $\mathbf{e}_\tau$ denotes the unit vector pointing in the direction of the jump at the τth time step.

The mean distance the random walker has traveled after t time steps is described by the root mean square displacement $\langle r^2(t) \rangle^{1/2}$, where the average $\langle \cdots \rangle$ is over all random-walk configurations on the lattice. From (3.3) we obtain

$$\langle r^2(t) \rangle = a^2 \sum_{\tau,\tau'=1}^{t} \langle \mathbf{e}_\tau \cdot \mathbf{e}_{\tau'} \rangle = a^2 t + \sum_{\tau \neq \tau'} \langle \mathbf{e}_\tau \cdot \mathbf{e}_{\tau'} \rangle. \tag{3.4a}$$

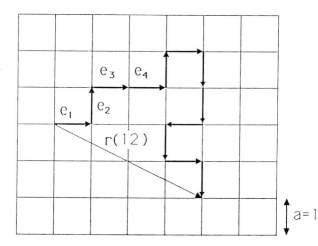

Fig. 3.3. A random-walk in a square lattice. The lattice constant $a = 1$ is equal to the jump length of the random walker

Since jumps at different steps τ and τ' are uncorrelated, we have $\langle \mathbf{e}_\tau \cdot \mathbf{e}_{\tau'} \rangle = \delta_{\tau\tau'}$, and we obtain Fick's law of diffusion

$$\langle r^2(t) \rangle = a^2 t. \tag{3.4b}$$

Note that (3.4) is independent of the dimension d of the lattice. In the general case when the random walker has a probability to stay in place, $\langle r^2(t) \rangle = 2dDt$, where D is the *diffusion constant*.

The mean square displacement can be obtained from the probability density $P(r, t)$, which is the probability of finding the walker after t time steps at a site within distance r from its starting point, via $\langle r^2(t) \rangle = \int d\mathbf{r}\, r^2 P(r, t)$.

The probability density can most easily be determined for a linear chain where a walker can only jump in two directions. Jumps to the right or to the left occur with probability $p = 1/2$. For simplicity we assume $a = 1$.

After t time steps, the random walker has jumped m times to the right and $t - m$ times to the left, so its actual position is at $x = m - (t - m) = 2m - t$. The probability that this happens is described by the binomial distribution,

$$P(m, t) = \binom{t}{m} p^m (1-p)^{t-m} = \binom{t}{m} \left(\frac{1}{2} \right)^t. \tag{3.5}$$

For large t, m, and $(t-m)$, the binomial coefficient can be expressed by the Stirling formula, $t! \cong (2\pi t)^{1/2} (t/e)^t$, etc. Since the actual distance x the walker has moved from its starting point is much smaller than the number of times it stepped to the right or to the left, the resulting expressions can be expanded in terms of $x/t \ll 1$, and we obtain finally

$$P(x, t) \cong \frac{1}{(2\pi t)^{1/2}} e^{-x^2/2t}. \tag{3.6}$$

Equation (3.6) represents a Gaussian with the width t, which is identical to $\langle x^2(t) \rangle$. Since $P(x, t)$ represents a probability, $\int_{-\infty}^{\infty} dx P(x, t) = 1$, which is satisfied by (3.6). In the more general case of a d-dimensional hypercubic lattice, (3.6) becomes simply

$$P(r, t) \propto \frac{1}{\langle r^2(t) \rangle^{d/2}} e^{-(d/2)r^2/\langle r^2(t) \rangle}, \tag{3.7}$$

where $\langle r^2(t) \rangle = 2dDt$, and the probability $P(0, t)$ that the random walker is at the origin after t time steps is proportional to $\langle r^2(t) \rangle^{-d/2}$.

c) **Lattice vibrations.** Consider L^d particles located at the sites of a regular d-dimensional lattice, where neighbor particles are coupled by springs. The particles can perform local vibrations around their equilibrium positions, which are the sites of the lattice. According to the translational invariance of the lattice, the vibrational excitations are waves characterized by the (discrete) reciprocal lattice vectors $\mathbf{q}$ and frequencies $\omega(\mathbf{q})$. For small wave vectors q, corresponding to large wavelengths λ, these excitations (phonons) are the ordinary sound waves with

$$\omega(q) = cq, \tag{3.8}$$

where c is the sound velocity of the lattice. From (3.8) one easily obtains the phonon density of states (see, e.g., [3.18]),

$$z(\omega) \sim \omega^{d-1}. \tag{3.9}$$

As we shall see later, $z(\omega)$ can be related to the Fourier transform of the probabiliy of being at the origin $P(0, t)$. We will use this analogy when considering $z(\omega)$ for fractals.

3.2.2 Transport in Fractal Substrates

a) **Total resistance.** Equation (3.2) is valid for homogeneous conductors where the density of the conducting material is constant. For fractal conductors, the density is proportional to $L^{d_f - d}$ and approaches zero for $L \to \infty$. This is a consequence of the fact that in fractal structures holes of all sizes up to the size of the system exist. If we increase L, we increase the size of the (nonconducting) holes as well, and by this we decrease the conductivity σ. Due to self-similarity, σ is decreased on all length scales, leading to the power-law dependence

$$\sigma \sim L^{-\tilde{\mu}}, \tag{3.10}$$

which defines the exponent $\tilde{\mu}$. In the following section we shall see that for percolation clusters, $\tilde{\mu}$ is related to the exponent μ defined in (3.1) and the correlation exponent ν by $\tilde{\mu} = \mu/\nu$.

As a consequence of (3.10) and (3.2), the total resistance behaves as

$$\rho \sim L^{\tilde{\zeta}}, \tag{3.11}$$

where now $\tilde{\zeta} = 2 - d + \tilde{\mu}$ is greater than the value of $2 - d$ for homogeneous conductors.

It is instructive to calculate $\tilde{\zeta}$ for the Sierpinski gasket [3.19]. The method is very simple and can be used in a straightforward way to obtain the resistance exponents of other finitely ramified deterministic fractals, such as the Mandelbrot-Given fractal [3.20] (see Sect. 2.8).

Consider a voltage difference between the top of the Sierpinski gasket and the two edge points in the bottom line (see Fig. 3.4) and compare the end-to-end resistances of a gasket of length L and a smaller gasket of length $L/2$. According to Kirchhoff's law, the end-to-end resistance (top to bottom) of the large system, $\rho(L)$, is the resistance of one small resistor, $\rho(L/2)$, plus the resistance of one small resistor in parallel with two small resistors, i.e.,

$$\rho(L) = \rho(L/2) + \left(\frac{1}{2\rho(L/2)} + \frac{1}{\rho(L/2)} \right)^{-1} = \frac{5}{3}\rho\left(\frac{L}{2} \right). \tag{3.12}$$

Using (3.11) we obtain [3.19]

$$\tilde{\zeta} = \ln(5/3)/\ln 2 \qquad [\text{Sierpinski gasket}]. \tag{3.13a}$$

For the Mandelbrot-Given fractal (see Sect. 2.8.1) one finds in an analogous way [3.20]

$$\tilde{\zeta} = \ln(11/4)/\ln 3 \qquad [\text{Mandelbrot} - \text{Given fractal}]. \tag{3.13b}$$

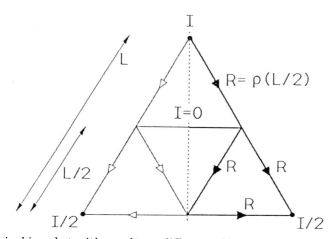

Fig. 3.4. Sierpinski gasket with a voltage difference V between the top site and the two sites at the bottom corners. The arrows describe the direction of the current. By symmetry, the total current along the horizontal direction is zero

b) Diffusion. Due to the presence of large holes, bottlenecks, and dangling ends in the fractal, the motion of a random walker is slowed down. Since due to self-similarity, holes, bottlenecks, and dangling ends occur on all length scales, the motion of the walker is slowed down on all length scales. Fick's law (3.4b) is no longer valid. Instead, the mean square displacement is described by a more general power-law,

$$\langle r^2(t)\rangle \sim t^{2/d_w}, \tag{3.14}$$

where the new exponent d_w ("diffusion exponent" or "fractal dimension of the random walk") is always greater than 2.

Both the resistance exponent $\tilde{\zeta}$ and the exponent d_w can be related by the Einstein equation (see also Sect. 3.5)

$$\sigma = \frac{e^2 n}{k_B T} D, \tag{3.15}$$

which relates the dc conductivity σ of the system to the diffusion constant $D = \lim_{t\to\infty}\langle r^2(t)\rangle/2dt$ of the random walk. In (3.15), e and n denote the charge and density of the mobile particles, respectively.

Simple scaling arguments can now be used to relate d_w to $\tilde{\zeta}$ and $\tilde{\mu}$. Since n is proportional to the density of the substrate, $n \sim L^{d_f-d}$, the right-hand side of (3.15) is proportional to $L^{d_f-d}t^{2/d_w-1}$. The left-hand side of (3.15) is proportional to $L^{-\tilde{\mu}}$. Since the time a random walker takes to travel a distance L scales as L^{d_w}, we find $L^{-\tilde{\mu}} \sim L^{d_f-d+2-d_w}$, from which the "Einstein relation" [3.1]

$$d_w = d_f - d + 2 + \tilde{\mu} = d_f + \tilde{\zeta} \tag{3.16}$$

follows. Since $d_f = \ln 3/\ln 2$ for the Sierpinski gasket and $d_f = \ln 8/\ln 3$ for the Mandelbrot-Given fractal, we obtain

$$d_w = \begin{cases} \ln 5/\ln 2 & \text{[Sierpinski gasket]}, \\ \ln 22/\ln 3 & \text{[Mandelbrot-Given fractal]}. \end{cases}$$

For random fractals, the determination of d_w is not that easy in general. The exceptions include topologically linear fractal structures ($d_\ell = 1$, see Chap. 1), which can be considered as nonintersecting paths. Along the path (in ℓ space), diffusion is normal and $\langle \ell^2(t)\rangle = t$. Since $\ell \sim r^{d_f}$, the mean square displacement in r space scales as $\langle r^2 \rangle \sim t^{1/d_f}$, leading to $d_w = 2d_f$ in this case. In percolation, d_w cannot be calculated exactly but upper and lower bounds can be derived, which are very close in $d \geq 3$ dimensions (see Sect. 3.6). A good estimate is $d_w \cong 3d_f/2$ (Alexander-Orbach conjecture [3.1]).

For fractals, the distribution function $\langle P(r,t)\rangle$, averaged over all starting points on the fractal is no longer Gaussian [3.6,21]. Following arguments similar to those used by Fisher [3.22] and Domb [3.23] (see also de Gennes [3.24]) to describe the distribution of self-avoiding random walks, one obtains that $\langle P(r,t)\rangle$ for $r/\langle r^2\rangle^{1/2} \gg 1$ is described by a *stretched Gaussian* [3.6,25–32] (for a sketch of the derivation see Sect. 3.7),

$$\langle P(r,t)\rangle / \langle P(0,t)\rangle \sim \exp\left[-\text{const} \times \left(r/\langle r^2(t)\rangle^{1/2}\right)^u\right], \qquad (3.17a)$$

where the exponent u is related to d_w by

$$u = \frac{d_w}{d_w - 1}. \qquad (3.17b)$$

There is strong numerical evidence that this equation is valid for a large class of random fractals, including percolation clusters [3.26]. For topologically linear fractal structures, (3.17) is exact [3.26]. Since for large times each site has the same probability of being visited, the probability of being at the origin $\langle P(0,t)\rangle$ appearing in (3.17a) is proportional to the inverse of the number of distinct visited sites $S(t)$. For fractals, $S(t)$ scales as [3.1] $\langle r^2(t)\rangle^{d_f/2}$, and thus

$$\langle P(0,t)\rangle \sim (1/\langle r^2(t)\rangle)^{d_f/2} \sim t^{-d_f/d_w}. \qquad (3.18)$$

While (3.17a,b) seem to describe correctly the asymptotic behavior of both regular and random fractals, the behavior in the short-distance regime $r < \langle r^2(t)\rangle^{1/2}$ may be different for regular and random fractals. For random walks on the Sierpinski gasket numerical results suggest [3.27]

$$\langle P(r,t)\rangle / \langle P(0,t)\rangle \sim \exp\left[-\text{const} \times \left(r/\langle r^2(t)\rangle^{1/2}\right)^{d_w}\right]. \qquad (3.19)$$

For random walks on nonintersecting random-walk structures (see Sects. 1.16 and 3.7) or walks on percolation, one can show analytically [3.28–30] that

$$\langle P(r,t)\rangle / \langle P(0,t)\rangle \cong 1 - \text{const} \times \left(r/\langle r^2(t)\rangle^{1/2}\right)^g. \qquad (3.20)$$

For random-walk structures $g = d - 2$ for $d \leq 6$ and $g = 4$ for $d \geq 6$, while for percolation structures g is the exponent appearing in the structural function $\phi(r \mid \ell)$ from (2.16).

For random-walk structures and percolation, the behavior of $\langle P(r,t)\rangle$ becomes more complicated if averages are performed over a finite (but large) number of configurations, N, which is always the case, e.g., in computer simulations. Equations (3.17a,b) and (3.20) are then valid only below a second crossover distance $r_2(N) \propto \langle r^2(t)\rangle^{1/2} r_c(N)^{1/u}$, with $r_c(N)$ from (2.17). Above $r_2(N)$, $\langle P(r,t)\rangle_N$ is described by [3.29]

$$\langle P(r,t)\rangle_N / \langle P(0,t)\rangle \sim \exp\left[-\text{const} \times \left(\frac{\ell_{\min}(r,N)}{\langle r^2(t)\rangle^{d_{\min}/2}}\right)^v\right] \qquad (3.21)$$

with $\ell_{\min}(r,N) \propto r_c(N)^{1-d_{\min}} r^{d_{\min}}$ from (2.17) and

$$v = d_w/(d_w - d_{\min}). \qquad (3.22)$$

For the meaning of the exponent v, we refer the reader to Sect. 3.7.

c) **Vibrational excitations.** Consider $N = L^{d_f}$ particles located at the sites of a fractal embedded in a d-dimensional hypercubic lattice, where neighbor particles are coupled by strings. If we denote the matrix of spring constants between nearest-neighbor particles i and j by $k_{ij}^{\alpha\beta}$, the equation of motion reads

$$\frac{d^2 u_i^\alpha(t)}{dt^2} = \sum_j \sum_\beta k_{ij}^{\alpha\beta} \left(u_j^\beta(t) - u_i^\beta(t) \right), \qquad (3.23)$$

where u_i^α is the displacement of the ith atom along the α coordinate.

For simplicity we assume that the coupling matrix $k_{ij}^{\alpha\beta}$ can be considered as a scalar quantity, $k_{ij}^{\alpha\beta} = k_{ij}\delta_{\alpha\beta}$. Then different components of the displacements decouple, and we obtain the *same* equation,

$$\frac{d^2 u_i(t)}{dt^2} = \sum_j k_{ij} \left(u_j(t) - u_i(t) \right), \qquad (3.24)$$

for *all* components $u_i^\alpha \equiv u_i$. This equation is identical to the diffusion equation when $u_i(t)$ is replaced by $P(\mathbf{i}, t)$ and the second time derivative is replaced by a first time derivative. In the diffusion equation, the matrix $k_{ij} = k_{ji}$ describes the jump probabilities between sites i and j and obeys the same symmetries as the matrix of the spring constants. The solution of (3.24) can be obtained following standard classical mechanics. The ansatz $u_i(t) = A_i \exp(-i\omega t)$ leads to a homogeneous system of equations for the N unknowns A_i, from which the N real eigenvalues $\omega_\alpha^2 \geq 0$, $\alpha = 1, 2, \ldots, N$, and the corresponding eigenvectors $(A_1^\alpha, \ldots, A_N^\alpha)$ can be determined. It is convenient to choose an orthonormal set of eigenvectors $(\psi_1^\alpha, \ldots, \psi_N^\alpha)$. Then the general solution of (3.24) becomes $u_i(t) = \mathrm{Re}\{\sum_{\alpha=1}^N c_\alpha \psi_i^\alpha \exp(-i\omega_\alpha t)\}$ where the complex constants c_α have to be determined from the initial conditions. If at $t = 0$ only the k_0th atom is displaced, i.e., $u_i(0) = u_{k_0}(0)\delta_{ik_0}$, we have

$$u_{i+k_0}(t) = u_{k_0}(0)\mathrm{Re}\left\{ \sum_{\alpha=1}^N (\psi_{k_0}^\alpha)^* \psi_{i+k_0}^\alpha \exp(-i\omega_\alpha t) \right\}. \qquad (3.25)$$

The solution of the corresponding diffusion equation can be found accordingly,

$$P(i + k_0, t) = \mathrm{Re}\left\{ \sum_{\alpha=1}^N (\psi_{k_0}^\alpha)^* \psi_{i+k_0}^\alpha \exp(-\epsilon_\alpha t) \right\}, \qquad (3.26)$$

where $\epsilon_\alpha = \omega_\alpha^2$. According to the initial condition $P(i, 0) = \delta_{ik_0}$, $P(k_0, t)$ denotes the probability of being at the origin of the walk. We obtain the average probability $\langle P(r, t) \rangle$ that the walker is at time t at a site separated by a distance r from the starting point by (a) averaging over all sites $i + k_0$, which are at distance r from k_0 and (b) choosing all sites of the fractal as starting points k_0, and averaging over all of them,

$$\langle P(r,t) \rangle = \mathrm{Re} \left\{ \sum_{\alpha=1}^{N} \psi(r,\alpha) \exp(-\epsilon_\alpha t) \right\}, \tag{3.27}$$

where

$$\psi(r,\alpha) = \frac{1}{N} \sum_{k_0=1}^{N} \frac{1}{N(r)} \sum_{i=1}^{N(r)} (\psi_{k_0}^\alpha)^* \psi_{i+k_0}^\alpha \tag{3.28}$$

and the inner sum here is over all $N(r)$ sites i, which are at distance r from k_0. For $r = 0$ we have simply $\psi(0,\alpha) = 1/N$ (since the eigenvectors are normalized) and thus

$$\langle P(0,t) \rangle = \frac{1}{N} \sum_{\alpha=1}^{N} \exp(-\epsilon_\alpha t). \tag{3.29a}$$

In the limit of $N \to \infty$ the sum over α can be transformed into an integral over ϵ by introducing the energy density of states $n(\epsilon)$,

$$\langle P(0,t) \rangle = \frac{1}{N} \int_0^\infty d\epsilon \, n(\epsilon) \exp(-\epsilon t). \tag{3.29b}$$

Accordingly, we can find $n(\epsilon)$ by the inverse Laplace transform of $\langle P(0,t) \rangle$. It is easy to verify that $\langle P(0,t) \rangle \sim t^{-d_f/d_w}$, (3.18), implies $n(\epsilon) \sim \epsilon^{d_f/d_w - 1}$. From $n(\epsilon)$ we obtain the vibrational density of states $z(\omega)$. Since $\epsilon_\alpha = \omega_\alpha^2$ and, by definition, $n(\epsilon)d\epsilon = z(\omega)d\omega$, we have [3.1]

$$z(\omega) \sim \omega^{2d_f/d_w - 1}. \tag{3.30}$$

The exponent $d_s \equiv 2d_f/d_w$ has been termed the fracton dimension [3.1] or spectral dimension [3.33], and replaces the Euclidean dimension d in the expression (3.9) for the phonon density of states. For percolation clusters, d_s is close to $4/3$ for all dimensions [3.1] (see Table 3.1 in Sect. 3.6.5).

The vibrational excitations in fractals are called *fractons* [3.1]. In contrast to regular phonons, fractons are strongly localized in space. From the above treatment it is easy to verify that

$$\langle P(r,t) \rangle = \int_0^\infty d\omega \, z(\omega) \langle \psi(r,\omega) \rangle \exp(-\omega^2 t), \tag{3.31}$$

where here $z(\omega)$ is normalized to unity. As in the foregoing subsection, the brackets $\langle \cdots \rangle$ denote an average over all configurations. For large distances r, the inverse Laplace transform of $\langle P(r,t) \rangle$, (3.17b), can be performed by the method of steepest descent, yielding [3.34] the leading exponential term

$$\langle \psi(r,\omega) \rangle / \langle \psi(0,\omega) \rangle \sim \mathrm{Im} \exp \left[-\mathrm{const} \times c(d_w) \, (r/\lambda(\omega))^{d_\phi} \right], \tag{3.32}$$

with

$$d_\phi = \frac{u d_w}{u + d_w}, \tag{3.33}$$

$$c(d_w) = \cos(\pi/d_w) + i\sin(\pi/d_w), \tag{3.34}$$

and

$$\lambda(\omega)^{-1} \sim \omega^{2/d_w}. \tag{3.35}$$

Using (3.17b) one obtains $d_\phi = 1$, i.e., the fractons are localized vibrations and decay by a simple exponential, with a localization length $\lambda(\omega)$. Equations (3.32–35) are general and also include the case of regular lattices. In this case, $d_w = 2$ and $c(d_w) = i$, and we recover the well-known result that harmonic vibrations in regular lattices are infinitely extended. For a different derivation of (3.33) see [3.6] and [3.35].

Equations (3.32–37) were derived from (3.17a), and are valid only in the asymptotic regime $r \gg \lambda(\omega)$ when the average is over all configurations. By using the approach sketched in Sect. 3.7 it is also possible to treat the decay of $\langle\psi(r,\omega)\rangle$ for percolation clusters in the short-distance regime below the localization length. One obtains [3.28], in close analogy to (3.20),

$$\langle\psi(r,\omega)\rangle/\langle\psi(0,\omega)\rangle \cong 1 - \text{const} \times (r/\lambda(\omega))^g, \tag{3.36}$$

where g again is the exponent occuring in the structural function $\phi(r \mid \ell)$ from Sect. 2.3. Hence, in the short-distance regime, the spatial decays of $\langle P(r,t)\rangle$ and $\langle\psi(r,\omega)\rangle$ are described by the same exponent g. Numerical simulations [3.28] show that most of the decay of $\langle\psi(r,\omega)\rangle$ occurs in the short-distance regime, and therefore one expects that this behavior can be detected by inelastic scattering experiments. In $d = 3$, one has $g = 1.5\pm0.1$, which is consistent with conclusions by Tsujimi *et al.* [3.62] who extracted a localization exponent of about 1.6 from their Raman scattering data in aerogels.

If one averages over only a finite number N of configurations, a more complicated behavior for $\langle\psi(r,\omega)\rangle_N$ is predicted. Similar to the case of random walks, (3.32–36) are then only valid below a second crossover distance $\tilde{r}_2(N) \propto \lambda(\omega) r_c(N)$. Above, $\tilde{r}_2(N)$, $\langle\psi(r,\omega)\rangle_N$ is described by [3.28–30]

$$\langle\psi(r,\omega)\rangle_N/\psi(0,\omega)\rangle \sim \exp\left[-\text{const} \times \left(\frac{\ell_{\min}(r,N)}{\lambda(\omega)^{d_{\min}}}\right)\right], \tag{3.37}$$

with $\ell_{\min}(r,N) \propto r_c(N)^{1-d_{\min}} r^{d_{\min}}$ from (2.17).

3.3 Transport in Percolation Clusters

In the previous section we discussed anomalous transport in fractal structures, which are self-similar on *all* length scales. Next we discuss transport in percolation systems above p_c where fractal structures appear only at length scales below the correlation length ξ. Above ξ, the infinite cluster can be regarded as homogeneous (see (2.5) in Chap. 2). Since the correlation length is the only relevant length scale we expect that, similar to the static properties, the transport properties can also be described by simple scaling laws.

We discuss separately diffusion (a) only on the infinite cluster and (b) in the whole percolation system, where finite clusters are also present, and relate the diffusion exponents to the conductivity exponents. We also show how transport in superconductor/conductor mixtures below p_c can be modeled by a simple random walk.

3.3.1 Diffusion in the Infinite Cluster

The long-time behavior of the mean square displacement of a random walker in the infinite percolation cluster is characterized by the diffusion constant D. It is easy to see that D is related to the diffusion constant D' of the whole percolation system: above p_c, the dc conductivity of the percolation system increases as $\sigma \sim (p-p_c)^\mu$, (3.1), so due to the Einstein equation (3.15) also the diffusion constant D' must increase this way. The mean square displacement (and hence D') is obtained by averaging over all possible starting points of a particle in the percolation system. It is clear that only those particles which start on the infinite cluster can travel from one side of the system to the other and thus contribute to D'. Particles that start on a finite cluster cannot leave the cluster, and thus do not contribute to D'. Hence D' is related to D by $D' = DP_\infty$, implying

$$D \sim (p - p_c)^{\mu-\beta} \sim \xi^{-(\mu-\beta)/\nu}. \tag{3.38}$$

If we combine (3.14) and (3.38), the mean square displacement on the infinite cluster can be written as [3.36–38]

$$\langle r^2(t) \rangle \sim \begin{cases} t^{2/d_w} & \text{if } t \ll t_\xi, \\ (p-p_c)^{\mu-\beta} t & \text{if } t \gg t_\xi, \end{cases} \tag{3.39a}$$

where

$$t_\xi \sim \xi^{d_w} \tag{3.39b}$$

describes the time scale the random walker needs, on average, to explore the fractal regime in the cluster. As $\xi \sim (p - p_c)^{-\nu}$ is the only length scale here, t_ξ is the only relevant time scale, and we can bridge the short-time regime and the long-time regime by a scaling function $f(t/t_\xi)$ (see also Sect. 2.5),

$$\langle r^2(t) \rangle = t^{2/d_w} f(t/t_\xi). \tag{3.40}$$

To satisfy (3.39), we require $f(x) \sim x^0$ for $x \ll 1$ and $f(x) \sim x^{1-2/d_w}$ for $x \gg 1$. The first relation trivially satisfies (3.39). The second relation gives $D = \lim_{t\to\infty} \langle r^2(t) \rangle / 2dt \sim t_\xi^{2/d_w-1}$, which in connection with (3.38) and (3.39b) yields a relation between d_w and μ [3.36–38]:

$$d_w = 2 + \frac{\mu - \beta}{\nu}. \tag{3.41}$$

Comparing (3.16) and (3.41), we can express the exponent $\tilde{\mu}$ by μ,

$$\tilde{\mu} = \mu/\nu. \tag{3.42}$$

3.3.2 Diffusion in the Percolation System

To calculate $\langle r^2(t) \rangle$ in the percolation system [3.36] (which consists of all clusters), we have to average over all starting points of the walkers that are uniformly distributed over all conductor sites. To do this, we average first over all random walks that start on clusters of *fixed size* s, and thus obtain the mean square displacement $\langle r_s^2(t) \rangle$ of a random walker on an s-site cluster. Then we average $\langle r_s^2(t) \rangle$ over all cluster sites. This is easy to do at p_c where the cluster-size distribution $n_s(p)$ is described by the power-law $n_s(p) \sim s^{-\tau}$ (see (2.47) in Chap. 2).

The mean radius R_s of all clusters of s sites is related to s by $s \sim R_s^{d_f}$. As long as the distance traveled by the random walkers is smaller than R_s, diffusion is anomalous and $\langle r_s^2(t) \rangle \sim t^{2/d_w}$. For very large times, however, since the random walker cannot escape the s cluster, $\langle r_s^2(t) \rangle$ is bounded by R_s^2. Hence we can write

$$\langle r_s^2(t) \rangle \sim \begin{cases} t^{2/d_w} & \text{if } t^{2/d_w} < R_s^2, \\ R_s^2 & \text{if } t^{2/d_w} > R_s^2. \end{cases} \tag{3.43}$$

From $\langle r_s^2(t) \rangle$ we obtain the total mean square displacement $\langle r^2(t) \rangle$ by averaging over all clusters,

$$\langle r^2(t) \rangle \sim \sum_{s=1}^{\infty} s n_s(p_c) \langle r_s^2(t) \rangle \sim \sum_{s=1}^{\infty} s^{1-\tau} \langle r_s^2(t) \rangle. \tag{3.44}$$

According to (3.43), there exists, for every *fixed* time t, a crossover size $S_\times(t) \sim R_s^{d_f} \sim t^{d_f/d_w}$: for $s < S_\times(t)$, $\langle r_s^2(t) \rangle \sim R_s^2$, while for $s > S_\times(t)$, $\langle r_s^2(t) \rangle \sim t^{2/d_w}$. Accordingly, (3.44) can be written as

$$\langle r^2(t) \rangle \sim \sum_{s=1}^{S_\times(t)} s^{1-\tau} s^{2/d_f} + \sum_{s=S_\times(t)}^{\infty} s^{1-\tau} t^{2/d_w}. \tag{3.45}$$

The first term in (3.45) is proportional to $[S_\times(t)]^{2-\tau+2/d_f}$ and the second term is proportional to $[S_\times(t)]^{2-\tau}t^{2/d_w}$. Since $S_\times(t) \sim t^{d_f/d_w}$, both terms scale the same and we obtain

$$\langle r^2(t)\rangle \sim t^{2/d'_w}, \tag{3.46}$$

with the effective exponent $d'_w = 2/[(d_f/d_w)(2-\tau+2/d_f)]$. Using the scaling relations $\tau = 1 + d/d_f$ and $d_f = d - \beta/\nu$ [(2.51) and (2.8)] we find [3.36,37]

$$d'_w = d_w/(1-\beta/2\nu). \tag{3.47}$$

Note that $d'_w > d_w$ since the finite clusters slow down the motion of the random walker. The probability of being at the origin can be calculated in essentially the same way, starting from the expression for finite s clusters and performing the average over all clusters. This procedure is also described in Sect. 3.4 where the fracton density of states is discussed, which is closely related to $\langle P(0,t)\rangle$. Due to the presence of the finite clusters, $\langle P(0,t)\rangle$ tends to a constant at large times. The leading time-dependent correction is

$$\langle P(0,t)\rangle - \langle P(0,\infty)\rangle \sim t^{-d'_s/2}, \tag{3.48}$$

where $d'_s = 2d/d_w$.

The above results obtained at p_c can be generalized to $p > p_c$. As in (3.40) we assume that $\langle r^2(t)\rangle$ can be written as

$$\langle r^2(t)\rangle \sim t^{2/d'_w}g(t/t_\xi), \tag{3.49}$$

where $g(x) \sim x^0$ for $x \ll 1$ and $g(x) \sim x^{1-2/d'_w}$ for $x \gg 1$. The first relation trivially satisfies (3.46); the second relation yields $D' \sim \lim_{t\to\infty}\langle r^2(t)\rangle/t \sim t_\xi^{2/d'_w-1} \sim (p-p_c)^\mu$, in agreement with the result for the dc conductivity, (3.1).

It is important to note, however, that according to the derivation, d'_w represents the exponent characterizing the second moment of the distribution function. Other moments, $\langle r^k(t)\rangle \sim t^{k/d'_w(k)}$, can be calculated in the same way. It is easy to verify that

$$\frac{k}{d'_w(k)} = \frac{k + d_f(2-\tau)}{d_w}. \tag{3.50}$$

Hence different moments are characterized by different exponents d'_w. This is in contrast to diffusion in the infinite percolation cluster where d_w does not depend on the moment considered.

3.3.3 Conductivity in the Percolation System

Next we consider the total conductance $\Sigma(p,L) \equiv \rho^{-1}$ of a random insulator/conductor mixture of size L^d and with a concentration p of conductors above the critical concentration p_c. On length scales larger than the correlation length ξ the system is homogeneous and (3.1) and (3.2) hold, whereas on length scales smaller than ξ the clusters are fractals and (3.11) holds. Again, since ξ is the only length scale here, we can bridge both regimes by the scaling ansatz [3.39]

$$\Sigma(p, L) = L^{-\tilde{\zeta}} F(L/\xi), \qquad (3.51)$$

where $F(x) \sim x^0$ for $x \ll 1$ trivially satisfies (3.11). To satisfy (3.1) and (3.2) we must require that $F(x) \sim x^{\tilde{\mu}}$ for $x \gg 1$ *and* the relation

$$\tilde{\zeta} = 2 - d + \mu/\nu \qquad (3.52)$$

between the exponents. This identity can be used to determine μ from measurements of the total conductivity as a function of L close to p_c.

3.3.4 Transport in Two-Component Systems

Next we consider a random mixture of two conductors A (with concentration p) and B (with concentration $1-p$) with conductivities σ_A and σ_B, and assume $\sigma_A \gg \sigma_B$ [3.40]. To describe the underlying diffusion process let us imagine a random walker who jumps from site to site over potential barriers. In A the jump frequency is $f_A \sim \sigma_A$, in B the jump frequency is $f_A \sim \sigma_B$. When the walker reaches the boundary between A and B, it cannot simply leave A for B but is reflected with a probability $\sim f_A/(f_A + f_B)$ (see Fig. 3.5).

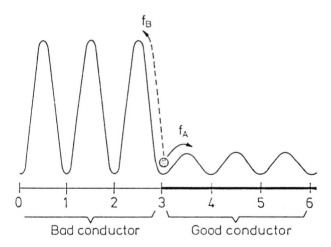

Fig. 3.5. Sketch of potential barriers representing a system with two types of conductivity

In the limit $f_A = 1$, $f_B = 0$ the walker is reflected with probability 1, and the model reduces to the random conductor/insulator mixture. In the limit $f_A \to \infty$, $f_B = 1$ the model describes a random superconductor mixture. Above p_c, the dc conductivity is infinite; below p_c there exist only finite superconducting clusters and the conductivity is finite. When approaching p_c, the size of the superconducting clusters increases and the conductivity diverges as [3.41,42,79]

$$\sigma \sim (p_c - p)^{-s}, \qquad p < p_c, \tag{3.53}$$

which defines the exponent s.

In the corresponding diffusion model, a random walker can visit any site in the superconducting cluster it started on, practically without loosing time, before it gets the chance to leave the cluster from any perimeter site. In the limit $f_A \to \infty$ this exit site is practically chosen randomly. The walker then continuously tries to escape from this cluster but, due to the fractal structure of the cluster surface, continually fails because most of the perimeter sites are "screened": the walker may temporarily leave the cluster but if it leaves in the screened region it will again stumble on the same cluster. Hence we expect a plateau in $\langle r^2(t) \rangle$ at time scales below some crossover time $t_\times \sim (p_c - p)^{-z'}$, which will be specified later. The height of the plateau is determined by the mean-square cluster radius, which scales as $R^2 \sim (p_c - p)^{-2\nu+\beta}$, see (2.52). At $t \sim t_\times$, the walker can completely escape the cluster by exiting from one of the unscreened "tip like" portions of the hull; $t_\times$ is therefore determined by the fractal structure of the unscreened surface sites. Above $t_\times$, we require $\langle r^2(t) \rangle \sim (p_c - p)^{-s}t$. Since here $t_\times$ is the characteristic time scale, we can assume that the short- time and long-time regimes are bridged by the ansatz

$$\langle r^2(t) \rangle = t(p_c - p)^{-s}T(t/(p_c - p)^{-z'}). \tag{3.54}$$

In order to satisfy (3.53) we require that $T(x) \sim x^0 = \text{const}$ for $x \gg 1$. In order to describe correctly the plateau at short-times we require that $T(x) \sim x^{-1}$ for $x \ll 1$, which determines z',

$$z' = -s + 2\nu - \beta. \tag{3.55}$$

Note the formal analogy between the crossover time t_ξ in random conductor/insulator mixtures and the crossover time $t_\times$ here: $t_\xi \sim \xi^{d_w}$ can be written as $t_\xi \sim |p - p_c|^{-\mu-2\nu+\beta}$. Hence the crossover exponent in random superconducting mixtures is obtained from the crossover exponent in random conductor/insulator mixtures by simply interchanging the corresponding transport exponents μ and $-s$. Using duality arguments it can be shown [3.43] that $\mu = s$ in two dimensions.

In the case that both conductivities $\sigma_A \sim f_A$ and $\sigma_B \sim f_B$ are finite and nonzero one can observe both types of anomalous behavior and both crossover times, provided that $t_\times \gg t_\xi$. A schematic picture is drawn in Fig. 3.6.

Consider $f_B = 1$ and $f_A \gg f_B$. Now the walker does not instantaneously reach the surface of the A cluster. First the walker explores the fractal interior of the cluster it starts on, and $\langle r^2 \rangle \sim f_A t^{2/d'_w}$ below $t_\xi \sim f_A \xi^{d_w}$. For $t_\xi < t < t_\times = f_B \xi^{z'/\nu}$, the walker explores the fractal surface of the cluster until it finds an unscreened surface site. For $t \gg t_\times$ we have normal diffusion.

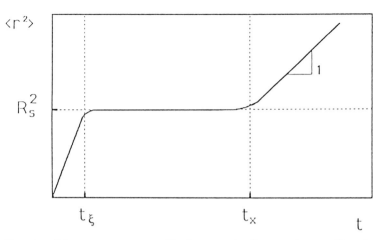

Fig. 3.6. Sketch of the mean square displacement of a random walker in a two-component random network, for p below p_c and $t_\times \gg t_\xi$

3.3.5 Elasticity in Two-Component Systems

Consider a mixture of a rigid and a soft material, e.g., a rigid material such as alumina powder in a gel. If the concentration p of the rigid material is increased, the elastic constant increases and diverges as [3.44–48,57]

$$E \sim (p - p_c)^{-s'}, \tag{3.56}$$

when p approaches the critical concentration p_c.

The value of the superelastic exponent s' has been discussed controversially (for a recent review see [3.44]). Although Bergman [3.57] argued that s' is equal to the superconductivity exponent s, numerical simulations [3.45–47] and experiments [3.48] suggest that $s' < s$ (see Table 3.1 in Sect. 3.6.5). Recent measurements of the Young modulus in gel-alumina and gel-zirconia mixtures by Benguigui and Ron [3.48] suggest that $s' = 0.67 \pm 0.05$ in 3d systems.

3.4 Fractons

In the previous section we considered transport properties of random networks. Next we consider the elastic properties of a random network of (harmonic) springs. For simplicity we will mostly assume scalar force constants (as in Sect. 3.2), and consider only briefly the case of vectorial force constants where different components of displacements of neighboring masses are also coupled. For a recent review on fractons see [3.16]. The scalar force constants are either $k_A = 1$ (with probability p) or $k_B = 0$ (with probability $1 - p$). At the end of this section, we shall also discuss briefly the related problem of electrons in a percolation system. We begin with the elastic modulus of the percolation network.

3.4.1 Elasticity

Consider a small constant stress field applied at opposite faces of the elastic network. Due to the stress field, the network is expanded; the relative expansion $\Delta L/L$ is proportional to the strength of the field. The proportionality constant is the elastic coefficient e, which characterizes the response of the network to the external stress field. The inverse elastic coefficient is the elastic modulus κ [3.47]. The behavior of κ is similar to the behavior of the dc conductivity.

Below p_c, the network is disconnected and the elastic modulus is zero. Above p_c, opposite faces become increasingly connected by springs. If p is increased, the network becomes stiffer and the macroscopic elastic modulus κ increases,

$$\kappa \sim (p - p_c)^{\mu_e}. \tag{3.57}$$

For scalar force constants, the elasticity exponent μ_e is identical to the conductivity exponent μ [3.50].

For vectorial force constants Webman and Kantor [3.51] showed that μ_e is considerably larger than μ. As with the conductivity exponent, the elasticity exponent μ_e cannot be derived exactly and only close upper and lower bounds can be derived, see (3.90d). The bounds yield $3.67 \leq \mu_e \leq 4.17$ in $d = 2$ and $3.625 \leq \mu_e \leq 3.795$ in $d = 3$. These values are considerably larger than those for scalar spring constants, $\mu_e \cong 1.3$ in $d = 2$ and $\mu_e \cong 2.0$ in $d = 3$. Experimentally, elastic modula between 3 and 3.6 have been observed in gels [3.52], showing the importance of *vectorial* force constants in real materials.

Next we consider the vibrations of the network. As with diffusion, it is convenient to discuss the vibrations of the infinite cluster and the vibrations of the whole system separately.

3.4.2 Vibrations of the Infinite Cluster

In Sect. 3.2 we have shown that the fracton density of states is proportional to the Laplace transform of the probability of a return to the origin in the corresponding diffusion problem. At the critical concentration we obtained $z(\omega) \sim \omega^{d_s - 1}$, where [3.1]

$$d_s = 2d_f/d_w \tag{3.58}$$

is the spectral dimension of the percolation cluster. Above p_c, the infinite cluster is self-similar on length scales below ξ and homogeneous on length scales above ξ. Accordingly, the time $t_\xi \sim \xi^{d_w}$ a random walker needs to explore the fractal labyrinth is the relevant time scale in diffusion (see Sect. 3.3), and $\epsilon_\xi = 1/t_\xi$ is the relevant frequency scale. Since the frequency scales for diffusion and vibrations are related by $\epsilon = \omega^2$ (see Sect. 3.2.2), the characteristic frequency ω_ξ for vibrations scales as

$$\omega_\xi \sim t_\xi^{-1/2} \sim \xi^{-d_w/2} \sim (p - p_c)^{\nu d_w/2}. \tag{3.59}$$

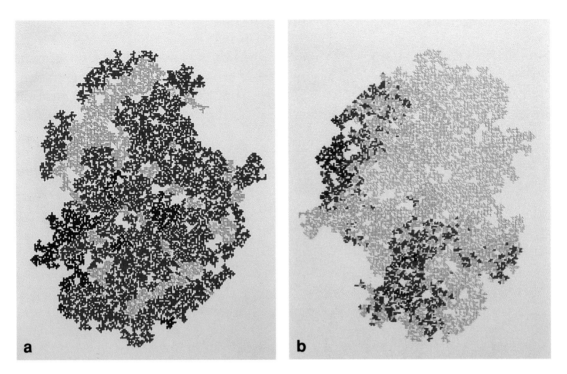

Fig. 3.7a,b. Vibrational amplitudes in the infinite percolation cluster on the square lattice above p_c $(p = 0.64)$, for (a) $\omega = 0.03$ and (b) $\omega = 0.35$. Different colors correspond to different orders of magnitude of the amplitudes. Courtesy of S. Ruß

Below ω_ξ, corresponding to large time and length scales, $z(\omega)$ shows normal phonon behavior, $z(\omega) \sim \omega^{d-1}$, whereas above ω_ξ, $z(\omega)$ shows anomalous fracton behavior, $z(\omega) \sim \omega^{d_s-1}$. Both frequency regimes can be bridged by the scaling ansatz [3.53]

$$z(\omega) = \omega^{d_s-1} n(\omega/\omega_\xi), \tag{3.60a}$$

where we require that $n(x) \sim x^0$ for $x \gg 1$ and $n(x) \sim x^{d-d_s}$ for $x \ll 1$. The second relation yields a prediction of the dependence of $z(\omega)$ on $(p - p_c)$ in the phonon regime:

$$z(\omega) \sim \omega_\xi^{d_s-d} \omega^{d-1} \sim (p - p_c)^{\frac{\nu d_w}{2}(d_s-d)} \omega^{d-1}. \tag{3.60b}$$

The different behavior of phonons and fractons can be observed directly by looking at the vibrational amplitudes in the different frequency regimes. Figure 3.7 shows the vibrational amplitudes in the infinite percolation cluster above p_c for two different frequencies: (a) $\omega < \omega_\xi$, phonon regime; and (b) $\omega \gg \omega_\xi$, fracton regime. One can clearly see the different features in the two regimes. In the phonon regime (a), even close to ω_ξ, large regions of the cluster vibrate, i.e., the phonons represent extended vibrations. In the fracton regime (b), only small portions of the cluster vibrate, i.e., the vibrations are localized.

3.4.3 Vibrations in the Percolation System

Similarly to diffusion, the vibrational spectrum is characterized by different exponents d_s and d'_s on the infinite cluster and in the percolation system, respectively. To see this, we consider first $z(\omega)$ at the critical concentration and utilize the relation between $z(\omega)$ and the probability of a return to the origin, $\langle P(0, t) \rangle$.

As in Sect. 3.3.2, we consider first the restricted ensemble of clusters of fixed size s. The mean probability of being at the origin on these clusters can be written as

$$\langle P_s(0, t) \rangle \sim \begin{cases} s^{-1} & s < S_\times(t) \sim t^{d_f/d_w}, \\ t^{-d_f/2} & s > S_\times(t), \end{cases} \tag{3.61}$$

and the average of $\langle P_s(0, t) \rangle$ over all clusters gives $\langle P(0, t) \rangle$,

$$\langle P(0, t) \rangle \sim \sum_{s=1}^{\infty} s^{1-\tau} \langle P_s(0, t) \rangle. \tag{3.62}$$

The sum can be performed as in (3.43–46). For large times, $\langle P(0, t) \rangle$ approaches a constant. The leading time-dependent correction is

$$\langle P(0, t) \rangle - \langle P(0, \infty) \rangle \sim t^{-d'_s/2}, \quad d'_s = 2d/d_w, \tag{3.63}$$

which immediately yields

$$z(\omega) \sim \omega^{d'_s - 1}, \quad \omega > 0. \tag{3.64}$$

Note that d'_s is obtained from d_s by exchanging the fractal dimension d_f, describing the infinite percolation cluster, with the space dimension d that describes the percolation system as a whole.

Following the scaling approach from Sect. 3.4.1, one can easily extend the discussion to $p > p_c$. Since ω_ξ is the only characteristic frequency scale, we expect that

$$z(\omega) = \omega^{d'_s - 1} m(\omega/\omega_\xi), \quad \omega > 0, \tag{3.65}$$

where $m(x) \sim$ const for $x \gg 1$ and $m(x) \sim x^{d-d'_s}$ for $x \ll 1$. As above, the scaling ansatz allows a prediction of the p dependence of $z(\omega)$ in the phonon regime,

$$z(\omega) \sim (p - p_c)^{\frac{d\nu}{2}(2-d_w)} \omega^{d-1}. \tag{3.66}$$

The existence of both frequency regimes and the validity of the scaling assumptions (3.60) and (3.65) have been confirmed by numerical calculations [3.54,55]. It has been found that the scaling functions n and m are smooth and monotonic functions of the scaling variable x.

By employing the analogy between the diffusion equation and the vibrational equation the more general case of a nonzero second force constant f_B can also be treated [3.56]. We mention here only the behavior close to p_c, where diffusion is characterized by the time scales $t'_\times = (f_A/f_B)$ and $t''_\times = (f_A/f_B)^{\nu d_w/(\mu+s)}$ (see Sect. 3.3). Below $t'_\times$, diffusion is anomalous and characterized by d'_w; above $t''_\times$ diffusion is normal. The crossover times correspond to the vibrational crossover frequencies $\omega'_\times = (t'_\times)^{-1/2}$ and $\omega''_\times = (t''_\times)^{-1/2}$. In close analogy to the above treatment one obtains $z(\omega) \sim \omega^{d-1}$ for $\omega \ll \omega''_\times$ and $z(\omega) \sim \omega^{d'_s-1}$ for $\omega \gg \omega'_\times$. Now, however, the phonon and the fracton regimes do not fit smoothly to each other but are separated by a frequency gap. In the frequency gap between $\omega''_\times$ and $\omega'_\times$, there occurs a pronounced maximum of $z(\omega)$, which is reminiscent of the Van Hove singularity in an ordered lattice consisting only of B springs. Thus, in contrast to the crossover discussed above, a broad maximum in $z(\omega)$ separates the phonon regime from the fracton regime.

To some extent, this situation is similar to the situation with vectorial force constants, where, according to Feng [3.57], a second crossover length l_c occurs, which depends on the relative strength of the bond-stretching and bond-bending force constants. For $\xi \ll l_c$, the fracton properties are the same as those for scalar spring constants, whereas for $\xi \gg l_c$ the density of states shows a crossover from an effective spectral dimension $\tilde{d}_s \cong 0.8$ at intermediate frequencies to $d_s = 2d_f/d_w$ at larger frequencies. Numerical simulations [3.58] confirm this behavior.

Support of the fracton concept comes from neutron and light scattering experiments in several amorphous structures and in aerogels [3.57–67] (for a detailed discussion see Chap. 8). By combining neutron and Raman spectroscopies, Vacher *et al.* [3.63] have recently observed two crossovers in the vibrational density of states of aerogels, where $z(\omega)$ changes from ω^2 at very low frequencies to $\omega^{0.3}$ at intermediate and $\omega^{1.2}$ at large frequencies. This has been interpreted as the first experimental indication of the crossovers from a phonon regime to a bending and a stretching regime as predicted by [3.57].

Since the behavior of $z(\omega)$ at small frequencies determines the behavior of the specific heat C at low temperatures, we also expect the corresponding crossover in the specific heat as a function of temperature (see Chap. 8). Small frequencies correspond to low temperatures, large frequencies correspond to high temperatures. Accordingly, below a crossover temperature T_ξ determined by ω_ξ, we expect "phonon behavior", $C \sim T^d$, whereas above T_ξ we expect "fracton behavior", $C \sim T^{d_s}$ [3.64]. For vectorial force constants, a third crossover should eventually occur. A crossover behavior of the specific heat at low temperatures has indeed been observed in a large number of amorphous systems, but the results so far do not unambiguously support the fracton interpretation (see Chap. 8 and references therein).

3.4.4 Quantum Percolation

A problem related to fractons is quantum percolation, which is based on the one-band tight-binding Hamiltonian

$$H = \sum_n |n\rangle \epsilon_n \langle n| + \sum_{n \neq m} |n\rangle V_{nm} \langle m|. \tag{3.67}$$

Here, $|n\rangle$ represents a localized wave function centered at site n, and $V_{nm} = 0$ unless n and m are nearest-neighbor sites on the same percolation cluster, for which $V_{nm} = 1$. Usually, one assumes $\epsilon_n = 0$ for simplicity.

It is generally accepted that in $d = 2$ all wave functions are localized even for $p > p_c$, i.e., the quantum percolation threshold $p_q(E)$ is equal to 1 for all energy eigenvalues E [3.65]. In $d = 3$, numerical simulations indicate the existence of a quantum percolation threshold $p_c < p_q(E) < 1$, above which extended states occur. This is analogous to the Anderson model in 3d disordered systems, where $V_{nm} = 1$ between nearest-neighbor sites (no geometrical constraints as in percolation), but ϵ_n are uncorrelated random numbers.

It has recently been claimed that quantum percolation belongs to a different universality class than the Anderson model, with the exponent ν_q describing the critical behavior of the localization length, $\lambda \sim (p - p_q(E))^{-\nu_q}$, considerably smaller than the analogous Anderson exponent ν_A. The actual value of ν_q has been discussed controversially. Based on series expansions, Chang et al. [3.66] claim that $\nu_q < 2/3$, violating the theoretical lower bound, $2/d$, obtained by Chays et al. [3.67].

3.5 ac Transport

Next we consider the frequency-dependent transport quantities [ac conductivity $\sigma(\omega)$ and dielectric constant $\epsilon(\omega)$] of a percolation system. We discuss two models: (I) the (microscopic) "lattice-gas" model, where N mobile particles perform hops between available nearest-neighbor sites, and (II) the "equivalent-circuit" model, where the percolation system is treated as a random mixture of resistors and condensors.

3.5.1 Lattice-Gas Model

a) **General treatment.** In the lattice-gas model, the ac conductivity can be measured in the following way. We apply a (small) time-dependent field of frequency ω along one axis (which we call the x axis), $E_x(t) = E_0 \sin(\omega t)$, and determine the current density $j_x(t)$ of the mobile particles. In a computer experiment, $j_x(t)$ is simply determined by the mean numbers of particles N^+ and N^- that jump between t and $t + \Delta t$ across an interface perpendicular to

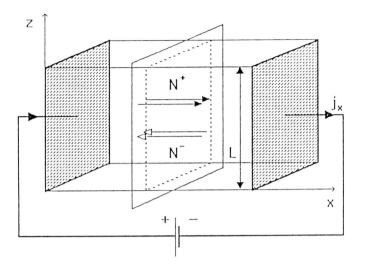

Fig. 3.8. Determination of the current density $j_x(t)$ on a cubic lattice. N^+ and N^- denote the number of particles that jump between time t and $t + \Delta t$ across the interface, with the field $(+)$ and against the field $(-)$, respectively

the direction of the applied field, here along the positive or negative direction of the x axis (see Fig. 3.8),

$$j_x(t) = \frac{N^+ - N^-}{L^{d-1}\Delta t}. \qquad (3.68)$$

For small fields, $j_x(t)$ is also sinusoidal and in general is described by $j_x(t) = j_0 \sin(\omega t - \gamma)$ where γ is a phase shift with respect to the external field. In general, both the amplitude j_0 of the current and the phase shift γ depend on the frequency ω of the external field. The exceptions include noninteracting particles in homogeneous systems, where the current follows the field instantaneously, i.e., $\gamma = 0$, with a frequency-independent amplitude, $j_0 = \text{const}$. The frequency-dependent conductivity connects the Fourier transform

$$j_x(\omega) = \frac{1}{2\pi} \int_{-\infty}^{+\infty} dt \exp(i\omega t) j_x(t) \qquad (3.69a)$$

of the current density with the Fourier transform $E_x(\omega)$ of the external field,

$$j_x(\omega) = \sigma_{xx}(\omega) E_x(\omega). \qquad (3.69b)$$

It is easy to show that $\sigma_{xx}(\omega) \equiv \sigma'_{xx}(\omega) + i\sigma''_{xx}(\omega)$ is related to j_0, E_0, and γ by

$$\sigma'_{xx}(\omega) = \frac{j_0}{E_0} \cos\gamma, \qquad \sigma''_{xx}(\omega) = \frac{j_0}{E_0} \sin\gamma. \qquad (3.70)$$

Accordingly, σ is real for $\gamma = 0$.

From $\sigma(\omega)$ we can deduce immediately the behavior of the (complex) dielectric constant $\epsilon(\omega)$. Both quantities are related by (see, e.g., [3.70])

$$\epsilon(\omega) = \epsilon_\infty + 4\pi i \frac{\sigma(\omega)}{\omega}. \tag{3.71}$$

The frequency-dependent conductivity can be approximately related to the mean square displacement, as we will show now. The treatment is exact for non-interacting particles. According to the fluctuation dissipation theorem, $\sigma_{xx}(\omega)$ is related to the Laplace transform of the current correlation function in the absence of an external field [3.71]:

$$\sigma_{xx}(\omega) = \lim_{\epsilon \to 0} \frac{\beta}{V} \int_0^\infty dt \langle J_x(t)J_x(0)\rangle e^{i\omega t - \epsilon t}. \tag{3.72}$$

As usual, $\beta = 1/k_B T$ is the inverse temperature and V is the volume of the d-dimensional lattice. The angled brackets now also include a thermal average involving the energy of the lattice- gas on the random structure. The thermal average is needed if the mobile particles interact with each other. By definition, the total current $J_x(t)$ is identical to the x component of the total velocity of all particles multiplied with their charge e,

$$J_x(t) = e \sum_{i=1}^N v_{ix}. \tag{3.73}$$

If $\rho = N/V$ denotes the concentration of mobile particles, (3.72) becomes

$$\sigma_{xx}(\omega) = \lim_{\epsilon \to 0} \beta \rho e^2 \int_0^\infty dt \frac{1}{N} \sum_{i,j=1}^N \langle v_{ix}(t)v_{jx}(0)\rangle e^{i\omega t - \epsilon t}. \tag{3.74}$$

The formula simplifies considerably if cross correlations between different particles $i \neq j$ can be neglected, i.e., $\langle v_{ix}(t)v_{jx}(0)\rangle = \langle v_{ix}(t)v_{ix}(0)\rangle \delta_{i,j}$, as is the case for noninteracting particles. Since the system is isotropic, the conductivity is independent of the direction of the field, $\sigma_{xx} \equiv \sigma$ and $\langle v_{ix}(t)v_{ix}(0)\rangle = (1/d)\langle \mathbf{v}_i(t) \cdot \mathbf{v}_i(0)\rangle$, and we simply have

$$\sigma(\omega) = \lim_{\epsilon \to 0} \frac{\beta \rho e^2}{d} \int_0^\infty dt \langle \mathbf{v}(t) \cdot \mathbf{v}(0)\rangle e^{i\omega t - \epsilon t}. \tag{3.75}$$

Here, $\langle \mathbf{v}(t) \cdot \mathbf{v}(0)\rangle \equiv \frac{1}{N}\sum_{i=1}^N \langle \mathbf{v}_i(t) \cdot \mathbf{v}_i(0)\rangle$ is the mean velocity correlation function of a mobile particle, which is related to the mean square displacement of a tagged particle by

$$\langle r^2(t)\rangle \equiv \frac{1}{N}\sum_{i=1}^N \langle [\mathbf{r}_i(t) - \mathbf{r}_i(0)]^2\rangle = \int_0^t dt' \int_0^t dt'' \langle \mathbf{v}(t') \cdot \mathbf{v}(t'')\rangle. \tag{3.76}$$

Due to time-reversal symmetry and the homogeneity of time, the correlation function depends on only the absolute value of $(t' - t'')$, and we obtain

$$\frac{d}{dt}\langle r^2(t)\rangle = 2\int_0^t dt'\langle \mathbf{v}(t')\cdot\mathbf{v}(0)\rangle. \tag{3.77}$$

In the limit of small t we require $\langle \mathbf{v}(t)\cdot\mathbf{v}(0)\rangle \cong \langle v^2(0)\rangle$. Using this result we obtain from (3.64–66), after a partial integration,

$$\sigma(\omega) = -\omega^2\frac{\beta\rho e^2}{2d}\lim_{\epsilon\to 0}\int_0^\infty dt\langle r^2(t)\rangle e^{i\omega t-\epsilon t}, \tag{3.78}$$

which generalizes the Einstein equation (3.15) to nonzero frequencies [3.71].

b) Noninteracting particles. For noninteracting particles, (3.78) is exact and we can obtain $\sigma(\omega)$ directly from the time derivative of the mean square displacement. In the percolation system we have (see Sect. 3.3)

$$\frac{d}{dt}\langle r^2(t)\rangle \sim \begin{cases} t^{-n} & t \ll t_\xi, \\ D' & t \gg t_\xi, \end{cases} \tag{3.79}$$

where $n = 1 - 2/d'_w$. Equation (3.78) implies

$$\sigma(\omega) \sim \begin{cases} (-i\omega)^n \equiv \exp(-i\frac{\pi}{2}n)\omega^n & \omega \gg \tilde\omega_\xi \equiv 1/t_\xi, \\ D' & \omega \ll \tilde\omega_\xi. \end{cases} \tag{3.80}$$

Accordingly for $\omega \gg \tilde\omega_\xi$ both σ' and σ'' are proportional to ω^n, and the phase shift γ is equal to $-\pi n/2$.

The negative phase shift in the power-law regime $\omega \gg \tilde\omega_\xi$ can be understood as follows. Due to the bias field, the particles are driven into dangling ends and backbends of the clusters where they cannot continue to move along the direction of the applied field even if the strength of the field is enhanced, but instead move backwards against the field. Accordingly, the current density reaches its maximum value *before* the field is at its maximum. Due to the self-similarity of the clusters on length scales below ξ, the phase shift is constant and determined by the exponent n in the frequency-dependent conductivity [3.37,72,73].

At small frequencies $\omega \ll \tilde\omega_\xi$, $\sigma'(\omega)$ is proportional to the (real) diffusion constant D' and σ'' is zero. Since by definition σ' is even and σ'' odd in ω, the leading term to σ'' at small frequencies is proportional to ω. The proportionality constant contains ϵ'. To find the dielectric constant ϵ', we perform a scaling analysis (see also [3.37]). Since $\tilde\omega_\xi \equiv 1/t_\xi$ is the only frequency scale here, we expect

$$\sigma(\omega) = \omega^n[f_1(\omega/\tilde\omega_\xi) + if_2(\omega/\tilde\omega_\xi)], \tag{3.81}$$

where the functional form of the scaling functions can be different below p_c and above p_c. For $\omega \gg \tilde\omega_\xi$ we require, according to (3.80), $f_1 + if_2 \sim \exp(-i\frac{\pi}{2}n)$. To satisfy $\sigma''(\omega) \sim \omega$ at small frequencies both below and above p_c, we require that $f_2(x) \sim x^{1-n}$ for small x, which yields

$$\sigma''(\omega) \sim \tilde\omega_\xi^{n-1}\omega \sim (p-p_c)^{-(2\nu-\beta)}\omega, \tag{3.82}$$

both below and above p_c. Accordingly, $\epsilon'(0)$ diverges [3.37] as p_c is approached,

$$\epsilon'(0) \sim |p - p_c|^{-2\nu+\beta}, \tag{3.83}$$

with the same exponent below and above p_c. Note that ϵ' diverges in the same way as the mean square cluster radius R^2 diverges (see Sect. 2.5).

c) Charged particles. The picture becomes more complex if we consider N charged particles coupled by Coulomb interactions. Now the probability that a particle jumps to an empty nearest-neighbor site on the same cluster is not constant but depends on the energy difference between the configurations before and after the attempted jump. In a Monte Carlo simulation, a randomly chosen particle jumps with a probability of 1 to an attempted empty cluster site if the energy U' of the attempted configuration is lower than the energy U of the present configuration. If $U' > U$, the particle jumps with probability $W = \exp[-\beta(U'-U)]$. After each trial, the time is enhanced by $1/N$, such that after 1 time step, on average, all particles in the system have attempted to jump.

Due to the combination of structural disorder and long-range interaction, pronounced backward correlations occur [3.74–76]. When a particle has the chance to jump to a neighbor site, it is out of equilibrium and has an enhanced probability of jumping back immediately. These backward correlations occur at length scales of the order of a unit jump distance. Accordingly, they do not change the asymptotic behavior of the mean square displacement of an ion but rather lead to a new power-law behavior at intermediate time scales $t \ll t_1$:

$$\langle r^2(t) \rangle \sim t^{k'}, \quad k' < 1. \tag{3.84}$$

Both the exponent k' and the new crossover time t_1 depend on the effective charge q of the particles and on the temperature, and increase with decreasing temperature and increasing charge.

It has been found numerically that (3.78) is also a good approximation for $\sigma(\omega)$ for charged particles where the cross correlations cannot be neglected in (3.74). Within the corresponding range of frequencies, the ac conductivity scales as $\sigma(\omega) \sim \omega^{n'}$, where n' is again simply related to k' by $n' = 1 - k'$. The value of n' can be considerably larger than n, see above, which is not changed by the interaction. A representative example of $\sigma'(\omega)$, above the percolation threshold, is shown in Fig. 3.9. The same effects occur also for screened interactions [3.74].

Experimentally, temperature dependent exponents have been observed, and the dependence of n' (as well as of the size of the dispersive regime) on temperature is in substantial agreement with the experimental situation [3.77].

3.5.2 Equivalent Circuit Model

In an alternative (macroscopic) approach, each bond in the random two-component network represents a circuit consisting of a resistor with resistivity $1/\sigma_A^0$ (or $1/\sigma_B^0$) and a capacitor with capacitance C_A (or C_B), and the

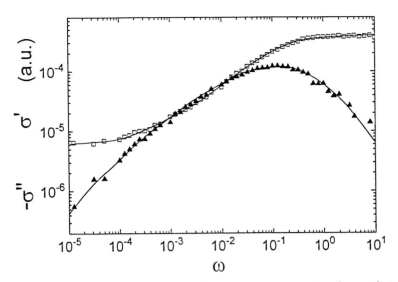

Fig. 3.9. Real and imaginary parts of $\sigma(\omega)$ for charged particles ($\rho = 1/100$) on the infinite percolation cluster well above p_c, as a function of ω for $\Gamma = 40$. The cluster was generated on a simple cubic lattice, the concentration p of available sites is $p = 0.40$. After [3.76]

(complex) conductivity of each bond is therefore either $\sigma_A = \sigma_A^0 - i\omega C_A$ or $\sigma_B = \sigma_B^0 - i\omega C_B$. At the percolation threshold, according to Dykhne [3.43], Webman *et al.* [3.78], and Straley [3.79], the total conductivity follows the power-law

$$\sigma = \sigma_A(\sigma_A/\sigma_B)^{-u}, \tag{3.85a}$$

where $u = \mu/(\mu+s)$ is related to the exponents μ and s from Sect. 3.3. Equation (3.85a) yields, in the limit of $\sigma_B^0 \to 0$, $C_A \to 0$ (conductor/capacitor limit),

$$\sigma(\omega) = \sigma_A^0(C_B/\sigma_A^0)^u(-i\omega)^u, \tag{3.85b}$$

similar to (3.80), where now the exponent n is replaced by the exponent u. The values of u are larger than those of n: $u = 0.5$ in $d = 2$ and $u \cong 0.71$ in $d = 3$.

Below and above p_c, standard scaling theory (see Sect. 2.5) gives [3.42,80]

$$\sigma = \sigma_A(\sigma_A/\sigma_B)^{-u}S[|p - p_c|(\sigma_A/\sigma_B)^\Phi], \tag{3.86}$$

where the complex scaling function $S(z)$ can be different above and below p_c. The exponent Φ, as well as the asymptotic behavior of the scaling function, is determined by the asymptotic behavior of σ in the limit $\omega \to 0$ and $(\sigma_A/\sigma_B) \to \infty$, yielding $S(z) \sim z^\mu$ above p_c, $S(z) \sim z^{-s}$ below p_c, and $\Phi = 1/(\mu + s)$. This immediately gives $\sigma'(0) \sim (p - p_c)^\mu$ above p_c, as in the diffusion model.

In the conductor/capacitor limit, the scaling variable z is proportional to $|p - p_c|[\sigma_A^0/(C_B\omega)]^\Phi \sim (\tau\omega)^{-\Phi}$, where $\tau = |p - p_c|^{-1/\Phi}C_B/\sigma_A^0$ defines the characteristic time scale in the short-circuit model [3.42,80]. This time scale is related to the polarizab ility of the medium between the clusters and scales differently from t_ξ, which is the time scale for diffusion inside the clusters. The exponent $1/\Phi = \mu+s$, which characterizes τ, is smaller than the exponent νd_w, which characterizes t_ξ. Hence the characteristic frequency $\epsilon_\tau \equiv 1/\tau$ is smaller than $\tilde{\omega}_\xi \equiv 1/t_\xi$, and the dispersion regime in the short-circuit model must start at considerably smaller frequencies than in the lattice-gas model [3.37].

From the scaling form (3.86) the real part of the dielectric constant can be determined as in Sect. 3.5.1 [3.42,80],

$$\epsilon'(0) \sim |p - p_c|^{-s}, \tag{3.87}$$

with the same exponent s below and above p_c. The divergency of ϵ' at p_c has a simple physical interpretation: each pair of neighbored clusters forms a capacitor. The effective surface increases when p_c is approached and tends to infinity at p_c. Accordingly, the effective capacitance of the system also diverges.

The divergence of the dielectric constant is described by different exponents in the two models, by the static exponent $2\nu - \beta$ in the lattice-gas model and by the dynamic exponent s in the equivalent circuit model. Again, this discrepancy is not surprising: in the lattice-gas model, the capacitances on bonds connecting the different clusters are not taken into account, and the polarization of the medium within which these clusters are embedded is taken to be zero. In the equivalent circuit model, on the other hand, this polarization effect has been considered, but not the effect of anomalous diffusion inside the clusters, and not effects of interaction among the mobile particles.

Accordingly, the equivalent circuit model is useful when the medium connecting the clusters is polarizable. This might be the case, for example, for Teflon systems [3.81] or for thin gold films near their percolation threshold [3.82], and possibly also for dispersed ionic conductors [3.83]. But it should be noticed that the equivalent circuit model is also not fully sufficient to describe the experimental situation here since exponents n considerably greater than the theoretical values have been observed [3.81,82]. It seems that a microscopic theory including both the polarizability of the medium and interaction effects between charges is needed to describe correctly the experimental situation.

3.6 Dynamical Exponents

In the previous sections we have defined several exponents which characterize transport in percolation systems. We applied scaling theory to derive relations between the dynamical exponents, such as $d_w' = d_w/(1 - \beta/2\nu)$ and $\tilde{\mu} = \mu/\nu$, and relations between the dynamical and the static exponents, such as $d_w = d_f + \tilde{\zeta}$, and $d_s = 2d_f/d_w$. However, although some exact values are known for static exponents, the exact values for d_w, $\tilde{\zeta}$, s, μ_e, and d_s are not yet known, except for the trivial cases $d = 1$ and $d = 6$. Determining the dynamical exponents has been a challenge in the past decades and many conjectures have been proposed (see e.g., [3.39]). Perhaps the most successful is the Alexander-Orbach conjecture [3.1] $d_w = \frac{3}{2}d_f$ or $d_s = 4/3$, which was found to be quite accurate (not exact) for percolation in $d \geq 2$.

The reason for the difficulty in determining exactly the dynamical exponents is the complex structure of the percolation clusters, which consist of branches, loops, and dangling ends in all length scales. Thus, due to the lack of an exact theory, the values for the dynamical exponents are usually estimated from numerical methods such as the Monte Carlo technique, the exact enumeration method, or the transfer matrix technique, and from analytical approximations such as the ϵ expansion or series expansions. In this section we describe some of these techniques and present the updated values of the dynamical exponents.

We begin by deriving rigorous bounds for the dynamical exponents, which are very useful and easy to derive, and shed some light on the complicated physics of the system.

3.6.1 Rigorous Bounds

The resistance between any two sites on the infinite percolation cluster at criticality is given by the resistance of the backbone connecting these sites. This is because the backbone is the only part of the cluster on which the current flows. The backbone can be viewed as a chain consisting of blobs connected by red bonds [3.84] (see Fig. 3.10). An upper bound for the resistance can be obtained by assuming that the effect of loops can be neglected and each blob is replaced by a single shortest path. A lower bound can be obtained by assuming that the resistance of the blobs in the backbone can be neglected. The reason for this is that cutting the loops increases the resistance, and taking the blob resistance as zero decreases the resistance. Thus the values derived for $\tilde{\zeta}$ and d_w when neglecting loops or blobs can serve as upper or lower bounds respectively [3.66,67].

In a loopless cluster, there exists only one path of length ℓ between two sites, and the resistance ρ between these sites is proportional to the chemical distance ℓ between them. Since the chemical distance ℓ scales with the Euclidean distance r as $r \sim \ell^{\tilde{\nu}}$ (see (2.10) in Chap. 2) we obtain

$$\rho \sim r^{1/\tilde{\nu}} \equiv r^{d_{\min}}, \tag{3.88}$$

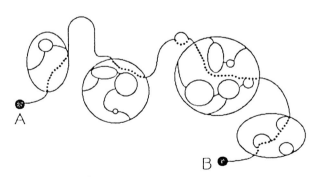

Fig. 3.10. Sketch of a percolation backbone consisting of blobs and red bonds, which is the only relevant part contributing to the resistance between A and B. The dashed lines represent the shortest paths within the blobs

where $d_{\min}$ is the fractal dimension of the minimum path. Hence the static exponent $\tilde{\nu}$ characterizes the dynamical properties when loops can be neglected.

Assuming that the blobs have zero resistance, the total resistance is simply proportional to the number of red bonds, n_{red}, between both sites. Since $n_{\mathrm{red}} \sim r^{1/\nu}$ [3.86,87] (see (2.81) in Chap. 2), one has

$$\rho \sim r^{1/\nu}. \tag{3.89}$$

Thus, the bounds for $\tilde{\zeta}$ are

$$\frac{1}{\nu} \le \tilde{\zeta} \le \frac{1}{\tilde{\nu}}. \tag{3.90a}$$

From (3.90a), (3.16) and (3.58) follow bounds for d_w and d_s,

$$d_f + \frac{1}{\nu} \le d_w \le d_f + \frac{1}{\tilde{\nu}}, \tag{3.90b}$$

$$\frac{2d_f}{d_f + 1/\tilde{\nu}} = \frac{2d_\ell}{d_\ell + 1} \le d_s \le \frac{2d_f}{d_f + 1/\nu}. \tag{3.90c}$$

Similarly we obtain for the elastic modulus (see Sect. 3.4)

$$d\nu + 1 \le \mu_e \le d\nu + \frac{\nu}{\tilde{\nu}}. \tag{3.90d}$$

The lower bound assumes springs with nonzero force constant only along the shortest path. The upper bound assumes elastic springs only along the singly connected bonds, while the other bonds are rigid (see Fig. 3.11).

The lower and upper bounds in (3.90) become identical in $d \ge 6$ dimensions where loops can be neglected and $\nu = \tilde{\nu} = 1/2$, yielding $\tilde{\zeta} = 2, d_w = 6, d_s = 4/3$, and $\mu_e = 4$. For $d < 6$, the two bounds can be estimated from the known static exponents ν, $\tilde{\nu}$, d_ℓ, and d_f (see Tables 2.2 and 2.3). The bounds are particularly good for $d \ge 3$. For $d = 3$ for example, where accurate numerical estimates

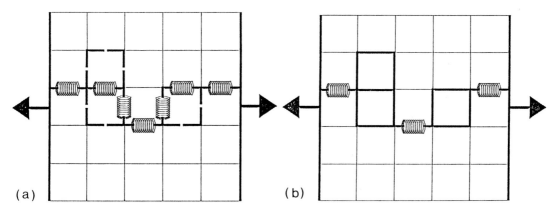

Fig. 3.11a,b. Sketch of a percolation backbone consisting of blobs and red bonds which is the only relevant part contributing to the elasticity between A and B. (a) Springs are taken only along the shortest path between A and B, the other bonds are disconnected. (b) Springs are taken only along the singly connected bonds, the other bonds are rigid

are difficult to obtain, the bounds give $1.13 \leq \zeta \leq 1.35$, $3.65 \leq d_w \leq 3.87$, $1.30 \leq d_s \leq 1.38$, and $3.62 \leq \mu_e \leq 3.80$. The importance of these bounds has been demonstrated by several numerical estimates, which are clearly *outside* these bounds. The best numerical estimates for the transport exponents that are consistent with the bounds, are listed in Table 3.1 in Sect. 3.6.5.

It appears from the table that the numerical values for $\tilde{\zeta}$ are *larger* than the lower bound $1/\nu$. Since the total resistance of the backbone is the sum of the resistance of the blobs and the resistance of the red bonds, it follows that for $d < 6$ the dominant contribution to the resistance is from the blobs.

3.6.2 Numerical Methods

a) Monte Carlo method of random walks. In this method random walks on percolation clusters are simulated by the Monte Carlo technique, and the diffusion exponent d'_w is determined from the mean square displacement. First a large percolation system is generated and an occupied site is chosen randomly as the origin of the random walk. Then one of its neighbor sites is chosen randomly and the random walker attempts to move to that site. If it is blocked, the move is rejected, otherwise the walker moves. In both cases, the time is enhanced by one unit ("blind ant"). This procedure is repeated until the desired number of time steps has been performed. At each time step, the square displacement of the random walker r^2 from the origin is recorded. Averages are performed over many starting points, which can be on either the infinite or the finite clusters. From the asymptotic behavior of $\langle r^2(t) \rangle$, (3.46), the exponent d'_w can be determined. This yields $d'_w = 3.01 \pm 0.02$ [3.88] for $d = 2$ and $d'_w = 5.0 \pm 0.24$ [3.88] and 5.7 ± 0.05 [3.89] in $d = 3$. These values correspond to $d_w = 3.75 \pm 0.19$ and 4.30 ± 0.05 for $d = 3$. Note that the last value is well above the upper bound of 3.87 from (3.90b). Recent results by Duering and Roman [3.90] yield $d'_w = 5.13 \pm 0.15$, corresponding to $d_w = 4.00 \pm 0.15$, where the error bars include the upper bound.

In an alternative numerical method, which saves computing time and is particularly suitable for parallel computers, one generates clusters and walks simultaneously [3.91]. A site is chosen both as the seed of the cluster and as the origin of the walk. Then the nearest-neighbors of the seed are either occupied with probability p or blocked with probability $1 - p$. The random walker now jumps randomly to one of the nearest-neighbor occupied sites. Next those nearest-neighbors of this site which were not yet specified are randomly occupied or blocked as before and the random walk continues until the desired number of time steps is reached. The advantage of this method is that one does not need to store the whole percolation system but only the neighboring sites of the trail of the walker. Thus many random walks can be performed in parallel. This method yields $d'_w = 2.95 \pm 0.05$ for $d = 2$ and $d'_w = 4.90 \pm 0.10$ for $d = 3$ [3.91]. These values correspond to $d_w = 2.80 \pm 0.10$ for $d = 2$ and $d_w = 3.73 \pm 0.10$ for $d = 3$.

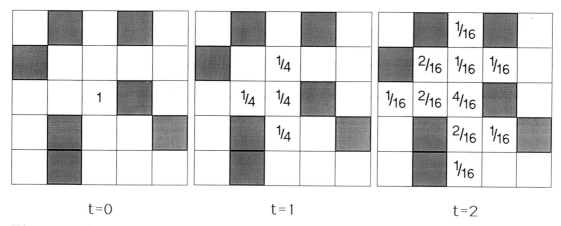

Fig. 3.12. The evolution of the probability of a random walker for three successive time steps

The superconductivity exponent can be derived by performing random walks on conductor–superconductor random mixtures [3.92].

b) Exact enumeration of random walks. In this technique [3.6,36,9 3], the discretized version of the diffusion equation is solved numerically by iteration in time, using the fact that the probability of the random walker being at site $\mathbf{r}$ at time $t+1$ is determined by the probabilities of it being at its nearest-neighbor sites at time t:

$$P(\mathbf{r}, t+1) = P(\mathbf{r}, t) + \sum_{\delta} w_{\mathbf{r},\mathbf{r}+\delta}[P(\mathbf{r}+\delta, t) - P(\mathbf{r}, t)]. \qquad (3.91)$$

Here, the sum over δ is over the z nearest-neighbor sites $\mathbf{r}+\delta$ of $\mathbf{r}$, and the transition probability $w_{\mathbf{r},\mathbf{r}+\delta}$ is $1/z$ when $\mathbf{r}+\delta$ is a cluster site and 0 otherwise.

The iteration starts at time $t=0$, where the random walker starts at the origin, and $P(\mathbf{r},0) = \delta_{\mathbf{r},0}$, and continues until the desired number of time steps is reached. The method is illustrated in Fig. 3.12.

To obtain the average probability density $\langle P(\mathbf{r}, t)\rangle$ one has to average $P(\mathbf{r}, t)$ over many starting points or different cluster configurations. The mean square displacement is obtained from

$$\langle r^2(t)\rangle = \int \langle P(r, t)\rangle r^2 r^{d_f - 1} dr \sim t^{2/d_w}. \qquad (3.92)$$

Fluctuations of $P(r, t)$ can also be calculated (see Sect. 3.7). This way, d_w and the other transport exponents are obtained. Using this method in $d=2$ yields $d_w = 2.87 \pm 0.02$ [3.93].

A more efficient way to calculate these exponents is by applying the exact enumeration method to diffusion on the percolation backbone [3.94]. The diffusion exponent of the percolation backbone d_w^B, defined by $\langle r^2\rangle \sim t^{2/d_w^B}$, is related to the fractal dimension of the backbone d_B and the conductivity exponent $\tilde{\zeta}$ by $d_w^B = d_B + \tilde{\zeta}$ (see (3.16)). Since the dangling ends do not contribute to the conductivity, $\tilde{\zeta}$ is the same as for percolation clusters, and thus by calculating d_w^B and d_B one obtains $\tilde{\zeta}$ for percolation. This procedure consumes less computer time than diffusion on percolation, since the backbone contains much fewer sites than the whole cluster and thus correspondingly fewer sites have to be analyzed.

c) **Finite-size scaling of conductivity.** A numerical method for solving Kirchhoff's equations and directly calculating the conductivity exponent $\tilde{\mu}$ and the superconductivity exponent $\tilde{s} = s/\nu$ is the *transfer matrix* technique [3.95,96]. Random resistor networks and random superconducting networks are studied at criticality, and the conductance of a strip of size $n \times L$ in $d = 2$ and of a bar of size $n \times n \times L$ in $d = 3$ is calculated as a function of n in the limit of $L \to \infty$. The resistance of a strip of size n scales as

$$R_n \sim \begin{cases} n^{\tilde{\mu}} & \text{[resistor network]}, \\ n^{-\tilde{s}} & \text{[superconductor network]}. \end{cases} \tag{3.93}$$

Note again that in $d = 2$, because of duality, $\tilde{\mu} = \tilde{s}$. This method yields: $\tilde{\mu} = \tilde{s} = 0.9745 \pm 0.0015$ [3.97] in $d = 2$ and $\tilde{s} = 0.85 \pm 0.04$, $\tilde{\mu} = 2.2 \pm 0.1$ in $d = 3$ [3.96].

An alternative finite-size technique was recently used [3.98,99] to calculate $\tilde{\mu}$. In this method, a voltage difference is applied at two remote sites of a percolation cluster at p_c, and the number of bonds carrying the current between the two sites is successively reduced by a transformation that leaves the total conductivity between the two points unchanged. Finally, the conductance is reduced to the conductivity of a single bond. By varying the distance L between the two sites the exponent $\tilde{\mu}$ (defined in (3.10)), has been obtained: $\tilde{\mu} = 0.975 \pm 0.001$ in $d = 2$ [3.98] and $\tilde{\mu} = 2.276 \pm 0.012$ in $d = 3$ [3.99]. Note that the $d = 3$ result is very close to the upper bound from (3.79a), $\tilde{\mu} \leq 2.35$.

3.6.3 Series Expansion and Renormalization Methods

The series expansion method for dynamical exponents [3.100] is similar to the method for the static exponents, which was discussed in Sect. 2.8. To calculate the conductivity exponent ζ [3.101,102] we consider the moments of the "resistance susceptibility"

$$\chi_k = \left\langle \sum_j v_{ij} R_{ij}^k \right\rangle, \tag{3.94}$$

where R_{ij} is the resistance between sites i and j, and v_{ij} is 1 if sites i and j are connected, and 0 otherwise. As usual, the brackets $\langle \ \rangle$ denote the configurational average over all clusters. For small clusters, one can calculate the resistances R_{ij} between all pairs and the configurational average exactly. As a result, one obtains χ_k as a power series in p, similar to the power series in Sect.

2.8. The susceptibility χ_k diverges as $\chi_k \sim (p_c - p)^{-\gamma_k}$ when p_c is approached from below. Using scaling arguments one can show that $\gamma_k = \gamma + k\zeta$.

For a square lattice, 16 terms of the resistance susceptibility have been derived [3.100]. From these one obtains, using elaborate extrapolation methods, $\mu = 1.28 \pm 0.01$ (or $\tilde{\mu} = 0.9600 \pm 0.0075$). For $d = 3$, 13 terms have been determined [3.102], yielding $\mu = 2.02 \pm 0.05$ ($\tilde{\mu} = 2.31 \pm 0.06$).

The small-cell real-space renormalization-group technique has been used by Bernasconi [3.103] to calculate μ. He obtained $\mu = 1.33 \pm 0.02$ ($\tilde{\mu} = 0.992 \pm 0.01$) in $d = 2$ and $\mu = 2.14 \pm 0.02$ ($\tilde{\mu} = 2.45 \pm 0.02$) in $d = 3$.

The resistance exponent ζ was calculated by Harris, Kim, and Lubensky [3.104] and by Wang and Lubensky [3.105] using the renormalization-group ϵ expansion method. The method is based on mapping the random resistor network problem onto the dilute q-state Potts model, in the limit $q \to 0$, using the replica assumption (see also Sect. 2.8). For $d = 6 - \epsilon$ dimensions they find

$$\zeta = 1 + \frac{1}{42}\epsilon + \frac{4}{3087}\epsilon^2 + O(\epsilon^3). \tag{3.95a}$$

Using the relation $d_w = d_f + \tilde{\zeta}$ and the ϵ expansion for d_f and ν from (2.89) one obtains,

$$\tilde{\mu} = (d - 2) + \tilde{\zeta} = 6 - \frac{25}{21}\epsilon - \frac{355}{9261}\epsilon^2 + O(\epsilon^3), \tag{3.95b}$$

$$d_w = 6 - \frac{2}{3}\epsilon - \frac{4}{147}\epsilon^2 + O(\epsilon^3), \tag{3.95c}$$

and

$$d_s = \frac{4}{3} - \frac{2}{189}\epsilon - \frac{97}{27783}\epsilon^2 + O(\epsilon^3). \tag{3.95d}$$

From inspection of these equations one might conclude that the expansion of ζ shows the best convergence. However, in $d = 3$ one obtains $\zeta \cong 1.089$ compared with $\zeta = 1.03 \pm 0.05$ obtained from series expansions and $\zeta = 0.98 \pm 0.05$ obtained by numerical simulations. The value $\tilde{\mu} \cong 2.084$ from (3.95b) deviates even more from the numerical estimate 2.28 ± 0.01, and is also smaller than the rigorous lower bound $\tilde{\mu} \geq 2.13$. This deviation is probably due to the bad convergency of the expansion for ν. In contrast, the diffusion exponent from (3.95c) yields $d_w \cong 3.775$, in very good agreement with numerical data and just below the value 3.87 following from the upper bound of (3.90b).

3.6.4 Continuum Percolation

The dynamical exponents discussed above have been derived for lattice percolation and do not depend on the type of lattice. Next we consider the dynamical exponents for continuum percolation systems. The static exponents, such as β and ν, are the same as for lattice percolation [3.106]. The natural question arises whether this universality holds also for the dynamical exponents.

Halperin, Feng, and Sen [3.107] have shown that this is *not* the case. They studied transport properties of a class of "Swiss-cheese" or random-void systems consisting of a uniform conducting media in which spherical holes are randomly placed (see Fig. 2.3). They find that the continuum system can be mapped onto a random percolation network with a power-law distribution of bond conductivities $\{\sigma\}$,

$$p(\sigma) = (1-p)\delta(\sigma) + p\phi(\sigma), \tag{3.96a}$$

where

$$\phi(\sigma) \sim \sigma^{-\alpha}, \qquad 0 < \sigma \leq 1, \tag{3.96b}$$

and $\alpha = (2d-5)/(2d-3)$. In $d = 2$, $\alpha = -1$ and the distribution is nonsingular, while for $d = 3$ the distribution is singular, with $\alpha = 1/3$. This type of power-law distribution arises natually also in fluid-flow permeability in porous media [3.107].

   The problem of percolation networks with a power-law distribution of conductivities has been studied by several authors [3.108,109]. The conclusions [3.107] from these studies are that $\mu(\alpha)$ can be bounded as follows:

$$\mu_1(\alpha) \equiv 1 + (d-2)\nu + \frac{\alpha}{1-\alpha} \leq \mu(\alpha) \leq \mu + \frac{\alpha}{1-\alpha}, \qquad 0 \leq \alpha \leq 1. \tag{3.97}$$

The lower bound $\mu_1(\alpha)$ comes from the singly connected bonds, while the upper bound is due to the effect of the blobs. For $\alpha < 0$, $\mu(\alpha) = \mu$, i.e., the same as for the discrete lattice. Moreover, Straley [3.109] argued that $\mu(\alpha)$ is exactly given by

$$\mu(\alpha) = \min(\mu, \mu_1(\alpha)), \tag{3.98}$$

where μ is the lattice conductivity exponent. Thus for

$$\alpha \geq \alpha_c = \frac{\mu - \mu_1(0)}{1 + (\mu - \mu_1(0))}, \tag{3.99}$$

μ has the value of the lower bound, $\mu = \mu_1(\alpha)$, and the resistance is solely determined by the singly connected red bonds. For $\alpha < \alpha_c$ ($\alpha_c \cong 0.23$ in $d = 2$ and $\alpha_c \cong 0.13$ in $d = 3$), the resistance is dominated by the blobs and is the same as for lattice percolation. Thus for continuum percolation in $d = 3$ where $\alpha = 1/3 > \alpha_c$ we expect that $\mu = \mu_1(1/3) \cong 2.38$, which is different from the lattice value, $\mu \cong 2.03$. Indeed, the experimental result, shown in Fig. 3.1, is consistent with the value of μ for continuum percolation.

   Equation (3.98) can be easily understood [3.109]. For $\alpha > 0$, many bonds can have very large resistances. When one of these bonds is a red bond, it might dominate the total resistance. On the other hand, if a very large resistance occurs inside a blob, it might have comparatively little effect since there exists at least one other path where the current could pass more easily. For $\alpha > \alpha_c$ the resistance of the red bonds is, on the average, larger than the total resistance of the blobs and dominates the resistivity.

Equation (3.98) was supported [3.110] by renormalization-group analysis using the ϵ expansion. Experimental support for (3.97) was obtained by Benguigui [3.111] and Lobb and Forester [3.112].

In the analogous diffusion problem, the transition rates w between nearest-neighbor cluster sites are chosen from the power-law distribution

$$\phi(w) \sim w^{-\alpha}, \qquad \alpha < 1, \quad \omega \leq 1, \tag{3.100}$$

with the same α as for the conductivity problem. The corresponding results for the diffusion exponent are [3.113],

$$d_w(\alpha) = d_f + \frac{1}{(1-\alpha)\nu}, \tag{3.101}$$

for $\alpha > \alpha_c$, and $d_w(\alpha) = d_w(0)$ for $\alpha < \alpha_c$. Numerical simulations [3.113] support (3.98) and (3.101).

In continuum percolation, however, the transport exponents depend not only on the way the percolation system is generated, but also on the way the random walk in the system is performed. This has been demonstrated by Petersen *et al.* [3.114], who studied the mean square displacement of random walks with random orientation but *fixed* step length in the Swiss-cheese model. They found that asymptotically this system belongs to the universality class of the Lorenz-gas model [3.115].

The above treatment can be extended to continuum models of elastic percolation systems, which can be mapped onto elastic networks with a power-law distribution of bond-bending and bond-stretching force constants, with the same exponent α as before. Above the critical concentration, the elastic modulus decreases as $\kappa \sim (p - p_c)^{\mu_e}$ when approaching p_c [see (3.47)]. The elasticity exponent μ_e can be bounded in analogy to the conductivity exponent by

$$d\nu + \frac{1}{1-\alpha} \leq \mu_e(\alpha) \leq \mu_e + \frac{\alpha}{1-\alpha}. \tag{3.102}$$

As with (3.98) we expect that for $\alpha \geq \alpha_c$

$$\mu_e(\alpha) = d\nu + \frac{1}{1-\alpha}. \tag{3.103}$$

Thus the elastic exponent for the $d = 3$ continuum percolation model is $\mu_e = \mu_e(1/3) \cong 4.15$ compared to μ_e between 3.6 and 3.8 in the lattice model [3.51].

3.6.5 Summary of Transport Exponents

In this subsection, the best estimates for the transport exponents in percolation are listed, for $d = 2$, 3, and 6. From the different methods described above it seems that numerical simulations yield, to date, the best estimates for the

Table 3.1. Transport exponents in percolation for two- and three-dimensional lattices and the Cayley tree: [a]numerical finite size scaling of conductivity [3.95–99], [b]Monte-Carlo method (random walks) [3.88,90,94,116], [c]series expansion [3.100,102], [d]epsilon expansion [3.104,105], [e]exact enumeration of random walks [3.91,93], [f]exact, [g]numerical simulaions [3.117], [h]numerical simulations [3.47]

Transport exponents	Space dimension		
	d=2	d=3	d≥ 6
$\tilde{\mu}$	0.9745 ± 0.0015^a	2.2 ± 0.1^a	6^f
	0.97 ± 0.01^b	2.276 ± 0.012^a	
	0.9600 ± 0.075^c	2.31 ± 0.06^c	
		2.084^d	
$\tilde{s}$	0.9745 ± 0.0015^a	0.85 ± 0.04^a	
s'	1.24 ± 0.03^h	0.65 ± 0.03^h	
d_w	2.871 ± 0.001^a	3.80 ± 0.02^a	6^f
	2.87 ± 0.02^e	3.755^d	
d'_w	2.95 ± 0.02^b	5.00 ± 0.03^b	∞
d_s	1.321 ± 0.001^a	1.328 ± 0.01^a	$4/3^f$
	1.30 ± 0.02^b	1.31 ± 0.02^b	
	1.31 ± 0.02^e		
d_w^B	2.60 ± 0.02^g	3.05 ± 0.15^g	4

critical exponents. The rigorous bounds (3.90), however, are found useful in particular in $d \geq 3$, where the bounds are close to each other and sometimes within the error bars of the numerical data.

3.7 Multifractals

In recent years, it has been established that several physical quantities describing random systems do not obey the conventional scaling laws. Examples are the growth probabilities of diffusion limited aggregation [3.118] (see also Chaps. 1, 4, and 10), the voltage distribution in random resistor networks [3.119], and the probability density of random walks on random fractals [3.26]. These quantities have a very broad distribution, and their moments cannot be described by a single exponent. Instead, an infinite hierarchy of exponents is needed to characterize them. This phenomenon is called multifractality and was first found by Mandelbrot [3.120] in the context of turbulence. For more detailed definitions and further examples see Chaps. 1, 4, and 10. In this section we will focus on multifractality related to transport properties of percolation.

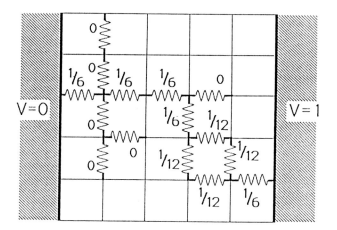

Fig. 3.13. Voltage drops in a small percolation cluster. The external voltage is $V = 1$

3.7.1 Voltage Distribution

Consider the infinite percolation cluster at p_c where each bond has unit conductivity. When one applies a voltage on two opposite sites of the cluster separated by a distance L, a voltage drop V_i results on each bond i (see Fig. 3.13).

We are interested in the distribution $n(V)$ of these voltage drops. Few features can be easily understood. Some bonds carry a small fraction of the total current while others carry a large fraction. The maximum voltage drop, $V_{\max}$, will be found on singly connected bonds, which carry the total current I. If, for simplicity, the external voltage is chosen as one unit, we have $V_{\max} = I = \rho^{-1}$, where ρ is the total resistance of the cluster. Since ρ scales with the linear cluster size L as $\rho \sim L^{\tilde{\zeta}}$, (3.11), the maximum voltage drop scales as

$$V_{\max} \sim L^{-\tilde{\zeta}}. \tag{3.104}$$

In contrast, the minimum voltage drop $V_{\min}$ occurs on nearly ballanced bonds, which are embedded inside blobs. The scaling of $V_{\min}$ is not yet clear and is a matter of current studies (see, e.g., [3.121]).

The range between $V_{\max}$ and $V_{\min}$ can be of many orders of magnitude, which indicates that the distribution of the voltage drops $\{V_i\}$ is very broad. The distribution is characterized by a long tail at small voltage drops (Fig. 3.14), which was obtained numerically by Arcangelis, Redner, and Coniglio [3.119]. The notion multifractality describes the fact that different ranges in the histogram can be viewed as different fractal sets. Each set is characterized by its own fractal dimension.

In general, the voltage distribution can be characterized by the moments

$$Z_q \equiv \langle V^q \rangle = \sum_V V^q n(V), \tag{3.105}$$

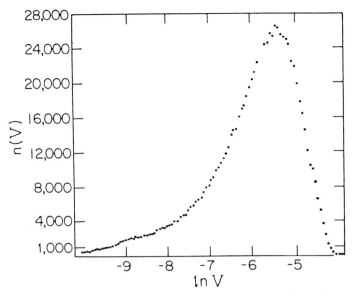

Fig. 3.14. Voltage drop distribution on the infinite percolation cluster at criticality. After [3.119]

where $n(V)dV$ is the probability of finding a voltage drop between V and $V + dV$. The hierarchy of exponents $\tau(q)$ is defined by (see also Chaps. 1, 4, and 10)

$$Z_q \sim L^{-\tau(q)}. \tag{3.106}$$

It is clear from the definition (3.105) that different moments are effected by different regimes of the distribution function. Large negative moments are governed by $V_{\min}$, while large positive moments are governed by $V_{\max}$. If the distribution is described by a single exponent Δ, $\tau(q)$ is linear in q, $\tau(q) = q\Delta + \tau_0$. To see that $\tau(q)$ is not linear, consider the moments Z_0, Z_2, and Z_∞, which can be related to known exponents. In the zeroth moment $Z_0 = \sum_V n(V) \sim L^{-\tau(0)}$, the summation is over all bonds containing nonzero voltage. These bonds constitute the percolation backbone and we can identify $\tau(0) = -d_B$. The second moment Z_2 is proportional to the resistance of the cluster and thus $\tau(2) = -\tilde{\zeta}$. The infinite moment Z_∞, finally, contains only the largest voltage drops, i.e., the voltages of the red bonds and thus $\tau(q \to \infty) = q\tilde{\zeta} - 1/\nu$. These three exponents do not lie on a straight line, showing that the distribution cannot be characterized by a single exponent. Numerical results for moments with small q values [3.119] suggest that indeed an infinite hierarchy of exponents $\tau(q)$ is needed to characterize the voltage distribution.

The scaling behavior of $n(V)$ can be obtained using the multifractal formalism described in the Appendix of Chap. 1. The dominant contribution of the distribution for a given q is assumed to scale as $n(V) \sim L^{f(\alpha)}$, where $f(\alpha)$ is the Legendre tranformation of $\tau(q)$. As in other multifractal systems, $f(\alpha)$ has a bell-shaped form.

A different infinite set of exponents was obtained for a nonlinear random resistor network, where the voltage drop V_i and local current I_i are related in a nonlinear fashion [3.122],

$$V_i = r_i |I_i|^\alpha \text{sign}(I). \qquad (3.107)$$

It can be shown [3.122] that the total resistance depends on L via

$$\rho_\alpha(L) \sim L^{\tilde{\zeta}(\alpha)}, \qquad (3.108)$$

where here $\tilde{\zeta}(\infty) = d_{\text{red}}$, $\tilde{\zeta}(1) = \tilde{\zeta}$, $\tilde{\zeta}(0^+) = d_{\min}$, and $\tilde{\zeta}(-1) = d_B$. Thus $\zeta(\alpha)$ is a hierarchy of exponents interpolating between several known exponents. Note, however, that these exponents do not represent different moments of a distribution function as in the previous or the following examples.

3.7.2 Random Walks on Percolation

Consider a random walk starting from a given site on a percolation cluster. We are interested in the probability density $P(r,t)$ of finding the random walker after t time steps on a site a distance r from its starting point. For large clusters, there will be many sites at distance r (see Figs. 3.0 and 3.15), some of them with very different chemical distances from the origin, and thus the

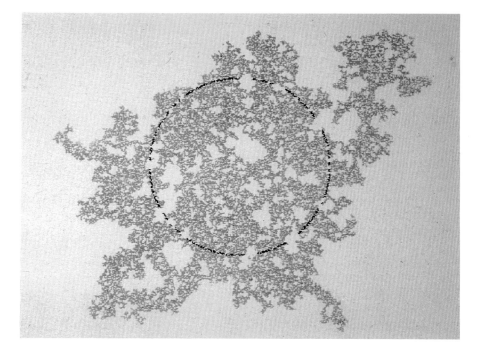

Fig. 3.15. Large percolation cluster at p_c. The random walker starts at the origin of the circle and after t steps can be found at distance r [×], with very different probabilities at different sites

probability of being on different sites may be very different. We will show, using an analytical approach, that the histogram of these probabilities is very broad and has multifractal features [3.26]. On the other hand, one can also consider the probability $P(\ell, t)$ of finding the walker on sites at a fixed *chemical distance* ℓ from the starting point (see Fig. 3.0). We will show that this histogram is relatively narrow and can be represented by a single exponent.

The derivation is similar to that of random walks on linear fractals presented in Chap. 1 (Sect. 1.14). Here we show that the multifractal behavior is a more general feature and is expected to occur in many random fractals and in particular on the infinite percolation cluster.

We consider the qth moment $\langle P^q(r,t) \rangle$ of the probability density. By definition, we have

$$\langle P^q(r,t) \rangle = \left\langle \frac{1}{N(r)} \sum_{i=1}^{N(r)} P_i^q(r,t) \right\rangle, \tag{3.109}$$

where the sum is over all $N(r)$ sites located a distance r from the origin and $\langle \cdots \rangle$ denotes the average over N configurations. For evaluating $\langle P^q(r,t) \rangle$ we anticipate that the length ℓ of the shortest path connecting a site with the origin is the relevant physical length in the diffusion problem, such that the fluctuations of the probability $P_i(\ell, t)$ of finding the random walker after t time steps on a site i at chemical distance ℓ from the origin, are small. For simplicity, we assume that $P_i(\ell, t)$ only depends on ℓ and t, for all configurations considered and all sites i at fixed ℓ, and decays as

$$P_i(\ell, t) = P(\ell, t) = \langle P(\ell, t) \rangle \sim \exp\left(-(\ell/\langle \ell(t) \rangle)^v\right), \tag{3.110}$$

where $\langle \ell(t) \rangle \sim t^{d_{\min}/d_w} \equiv t^{1/d_w^\ell}$ is the mean chemical distance travelled by the random walker. For $\ell \gg \langle \ell(t) \rangle$, we have [3.6]

$$v = \frac{d_w^\ell}{d_w^\ell - 1} = \frac{d_w}{d_w - d_{\min}}. \tag{3.111}$$

The ansatz (3.110) with (3.111) is rigorous for linear fractal structures such as the random-walk chain or other self-avoiding walks, where $v = 2$. For percolation structures, its validity has been shown by Monte-Carlo simulations [3.6,26].

Assuming that among the $N(r)$ sites there are $N(\ell, r)$ sites at chemical distance ℓ from the origin (see Sect. 2.3) and employing (3.110), we can write (3.110) in the form

$$\langle P^q(r,t) \rangle = \left\langle \frac{1}{N(r)} \sum_{\ell \geq \ell_{\min}(r)} N(\ell, r) P^q(\ell, t) \right\rangle \cong \sum_{\ell \geq \ell_{\min}(r)} \langle N(\ell, r)/N(r) \rangle \langle P(\ell, t) \rangle^q. \tag{3.112}$$

According to the definition (Sect. 2.3), $\langle N(\ell, r)/N(r) \rangle$ can be identified with the probability $\phi(\ell \mid r)$ that two sites separated by a distance r are at chemical distance ℓ. Thus, the moments in r space and ℓ space are related by

$$\langle P^q(r,t)\rangle = \int_{\ell_{\min}(r,N)}^{\infty} \phi(\ell \mid r)\langle P(\ell,t)\rangle^q \, d\ell. \tag{3.113}$$

Here, $\ell_{\min}(r,N)$ denotes the smallest ℓ value one can find at distance r from the origin in N configurations (see (2.17)). Equation (3.113) enables us to calculate the probability density $\langle P(r,t)\rangle$ and the moments $\langle P^q(r,t)\rangle$ [3.26,28–30]. If one averages over *all* configurations ($N \to \infty$, $\ell_{\min} = r$), one can reproduce (with (2.15) and (3.110,111)) Eqs. (3.17a,b) from Sect. 3.2.2. If we assume the validity of (3.110) it is easy to show that the moments satisfy the general scaling relation [3.26]

$$\langle P^q(r,t)\rangle \sim \langle P(rq^{1/(vd_{\min})},t)\rangle, \tag{3.114}$$

which together with (3.17) yields in the asymptotic regime

$$\langle P^q(r,t)\rangle \sim \langle P(r,t)\rangle^{\tau(q)}, \qquad \tau(q) \sim q^{\gamma}, \qquad q > 0 \tag{3.115}$$

with

$$\gamma = (1 + v(d_{\min} - 1))^{-1} < 1. \tag{3.116}$$

Since $\tau(q)$ is nonlinear, the average moments cannot be described by a finite number of exponents, showing the multifractal nature of $P(r,t)$.

Next we calculate the histogram $H(\ln P)$ giving the number of sites with values P between $\ln P$ and $\ln P + d \ln P$ [$P \equiv P(r,t)$] [3.26]. We use the relation

$$\langle P^q\rangle \sim \frac{1}{r^{d_f-1}} \int_0^{\infty} P^q H(\ln P) d \ln P, \tag{3.117}$$

and compare (3.113) with (3.117). By changing the variables in (3.113) from ℓ to $\ln P$ and using relation (3.110) we obtain, with $b(r) \propto \left(r/\langle r^2(t)\rangle^{1/2}\right)^{d_{\min}v}$ and $P_0 \equiv P(0,t)$,

$$H(\ln P) \sim r^{d_f-1} \frac{1}{|\ln(P/P_0)|} \left(\frac{b(r)}{|\ln(P/P_0)|}\right)^{\alpha} \exp\left[-\left(\frac{b(r)}{|\ln(P/P_0)|}\right)^{\beta}\right], \tag{3.118}$$

with

$$\alpha = \frac{g}{vd_{\min}} \tag{3.119}$$

and

$$\beta = (v(d_{\min} - 1))^{-1}. \tag{3.120}$$

From the moments one can deduce easily the relative width $\Delta \equiv (\langle P^2\rangle - \langle P\rangle^2)/\langle P\rangle^2$ of the histogram $H(\ln P)$ in ℓ and r space. In ℓ space, Δ is constant (similar to $\Delta = 0$ in linear fractals), while in r space Δ increases exponentially

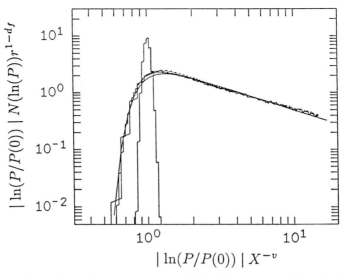

Fig. 3.16. Scaling plot of the histogram $H(\ln P)$ for $r = 35$ and 70 and $t = 1000$ with $X = (r/\langle r^2(t)\rangle^{1/2})^{d_{\min}}$ (*broad curve*). The smooth line represents the theoretical prediction (3.118). The narrow curve represents the histogram in ℓ space, for $\ell = 100$ and $t = 1000$, with $X = \ell/\langle\ell\rangle$. After [3.31]

with r. Numerical results (in scaled form for different r values) for $H(\ln P)$ in $d = 2$ are shown in Fig. 3.16. The theoretical results for the histogram (dashed line), (3.118), are in agreement with the numerical results.

The large fluctuations in r space are in marked contrast to the very narrow fluctuations in ℓ space, which a posteriori justifies the ansatz (3.110) for the probability distribution in the chemical ℓ space, and $\langle P^q(\ell, t)\rangle \sim \langle P(\ell, t)\rangle^q$ used before.

The origin of the multifractality can be understood here. The broad logarithmic distribution (3.118) leading to the multifractal behavior originates from the convolution of two relatively narrow distributions in ℓ space, $\phi(\ell \mid r)$ and $\langle P^q(\ell, t)\rangle$. Although $\phi(\ell \mid r)$ is rather narrow in ℓ, the resulting histogram of $P(r, t)$ is logarithmically broad since ℓ scales *logarithmically* with P through (3.104).

Finally, we would like to note that the situation becomes more complex if averages are performed only over a finite number N of configurations. In this case in $\langle P(r, t)\rangle_N$ a second crossover distance $r_2(N)$ occurs (see (3.21,22)), which increases logarithmically with N. It can be shown [3.31] that the relation $\langle P^q(r, t)\rangle \sim \langle P(r, t)\rangle^{\tau(q)}$ is only valid below some critical value $q_c(r, N) \sim (r_2(N)/r)^{vd_{\min}}$. Above q_c, one has $\tau(q) = q$, i.e., the multifractality vanishes.

3.8 Related Transport Problems

In the last section we deal with dynamical properties of percolation clusters in the presence of additional physical constraints. Because of the limited space available we present only a few examples from the many existing ones. First we discuss the influence of an external bias field on diffusion in a percolation system, both at p_c and above p_c, and show how drastically the bias field changes the transport properties. Next we study the so-called dynamical percolation, where the percolation system is not static as considered before but undergoes configurational changes on time scales comparable with experimental observation times, and describe the effect of the fluctuating environment on the diffusion properties. Finally, we discuss several models for diffusion limited reactions and show how the reaction rates and the concentration of "surviving" particles that have not yet reacted are modified by the fractal nature of the percolation system.

3.8.1 Biased Diffusion

Consider a random walker on the infinite cluster under the influence of a bias field. The bias field is modeled by giving the random walker a higher probability P_+ of moving along the direction of the field and a lower probability P_- of moving against the field,

$$P_\pm \propto (1 \pm E), \tag{3.121}$$

where $0 \le E \le 1$ is the strength of the field. The field can be either uniform in space ("Euclidean" bias) or directed in topological space [3.124]. In a topological bias (see Fig. 3.17) every bond between two neighbored cluster sites experiences a bias, which drives the walker away in chemical space (see Sect. 2.3) from a point source.

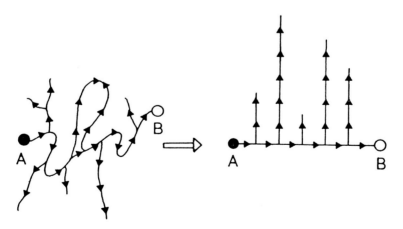

Fig. 3.17. Illustration of a percolation cluster under the influence of a topological bias field and its mapping to a random comb model

For convenience, let us start with the topological bias field. If we apply such a field in a uniform system, then the mean distance R of the walker from the "source" A is increased linearly in time, giving the walker a radial velocity. In a Euclidean bias field, the walker gets a velocity along the direction of the field. The question is how this behavior is changed in the infinite percolation cluster [3.124–129]. First we discuss $p = p_c$, where the cluster is self-similar, having loops and dangling ends on all length scales. We consider a walker travelling from a site A to another site B on the cluster. On his way the walker is driven into the loops and dangling ends, which emanate from the shortest path between A and B.

In a topological bias field the walker can get "stuck" in loops, as he can get stuck in dangling ends. Therefore, both loops and dangling ends act as random delays on the motion of the walker, and the percolation cluster can be imagined as a random comb where the teeth in the comb act as the random delays on the motion of the walker (see Fig. 3.17). The distribution of the length of the loops and dangling ends in the fractal structure determines the biased diffusion.

At the critical concentration, due to self-similarity on all length scales, the lengths L of the teeth are expected to follow a power-law distribution,

$$P(L) \sim L^{-(\alpha+1)}, \quad \alpha > 0. \tag{3.122}$$

The time τ spent in a tooth increases exponentially with its length L, $\tau \sim [(1+E)/(1-E)]^L$ [3.124]. Since the lengths of the teeth are distributed according to (3.108), it is easy to show that the waiting times τ follow the singular *waiting time distribution* [3.124]

$$\phi(\tau) \sim 1/(\tau \ln^\alpha \tau), \tag{3.123}$$

and the system can be mapped onto a linear chain (the backbone of the comb) where each site i is assigned to a waiting time τ_i according to (3.123). A random walker has to wait, on average, τ_i time steps before he can jump from site i to one of the neighboring sites.

The singular waiting time distribution changes the asymptotic laws of diffusion drastically, from the power-law (3.14) to the *logarithmic* form [3.124]

$$\ell \sim \left(\frac{\ln t}{A(E)} \right)^\alpha, \tag{3.124}$$

where

$$A(E) \sim \ln[(1+E)/(1-E)], \tag{3.125}$$

and ℓ is the mean distance the walker has traveled along the backbone of the comb. Equation (3.124) is rigorous for the random comb, but is also in agreement with numerical data for the infinite percolation cluster at p_c, with $\alpha \cong 1$ [3.124]. Accordingly, we have the paradoxical situation that on the fractal structure of the percolation cluster the motion of a random walker is dramatically *slowed down* by a bias field: the larger the bias field E the stronger the effect.

A logarithmic dependence of the mean displacement $\langle x \rangle$ along the direction of the field was found also for the Euclidean bias,

$$\langle x \rangle \sim \frac{\ln t}{A(E)}, \tag{3.126}$$

with $A(E)$ from (3.125) for not too large values of E, $E \leq 0.6$ [3.126]. In an Euclidean bias field, a random walker can also get stuck in backbends of the shortest path between two points, as also happens in linear fractal structures such as self-avoiding random walks or random walks. One can show rigorously [3.130] that in these structures (3.126) holds, and it is an open question how general the simple logarithmic behavior (3.126) and (3.125) is for diffusion in random fractals in the presence of an external bias field (see also [3.11]).

Finally, let us consider $p > p_c$. Above p_c, the percolation cluster under the influence of the topological bias can be approximately mapped on a random comb with an exponential distribution of dangling ends [3.124,127],

$$P(L) \sim \lambda^L \equiv \exp[-L/L_0(p)], \tag{3.127}$$

where the mean length $L_0(p)$ of the dangling ends decreases monotonically with increasing p. In this case, diffusion is characterized by power-law relations and a dynamical phase transition. Below a critical bias field E_c,

$$E_c = (1 - \lambda)/(1 + \lambda), \tag{3.128}$$

biased diffusion is classical and $\ell \sim t$. Above E_c, diffusion is anomalous, $\ell \sim t^{1/d_w(E,\lambda)}$, where now [3.124]

$$d_w(E, \lambda) = \ln[(1 - E)/(1 + E)]/\ln \lambda \tag{3.129}$$

depends *continuously* on the bias field E and on the concentration p, via λ. Figure 3.18 shows that these results derived for a random comb actually also hold for diffusion on the infinite cluster with a topological bias field.

Thus, due to the bias field, diffusion is much more complex above p_c than without a bias field at p_c. Now, for each concentration p above p_c there exists a dynamical phase transition at a critical bias field E_c that decreases with increasing p. Below E_c, diffusion is normal; above E_c, diffusion is anomalous and the exponent depends continuously on the field strength E and p. Here we have again the paradoxical situation that the motion of the walker is slowed down when the bias field is enhanced.

3.8.2 Dynamic Percolation

In the conventional percolation model, the occupied sites represent structural pathways for diffusing particles. Since the occupied sites and the blocked ones do not change with time, the structural pathways do not depend on time t. This assumption is not always realistic, for example, in highly viscous liquids such as

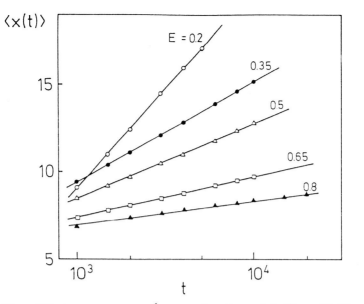

Fig. 3.18. The diffusion exponent d_w^ℓ as a function of $\ln[(1 + E)/(1 - E)]$ on a square lattice, for a topological bias field E and several values of p: $p = 0.623(\circ)$, $p = 0.653(\square)$ and $p = 0.703(\triangle)$. After [3.124]

polymeric ionic conductors above the glass transition temperature, transport of charge carriers is dominated by microscopic segmental motions of the host medium.

Two relatively simple percolation models have been proposed, which take into account both the host dynamics and the carrier hopping between unblocked sites. In the first model, Druger, Ratner, and Nitzan [3.131] described the host dynamics by a series of instantaneous "renewal" events. These events occur at random times governed by a renewal time distribution $f(t)$, whose first moment $\tau_{\mathrm{ren}} = \int_0^\infty t f(t) dt$ is the average renewal time. In the renewal process, at each time chosen randomly from the distribution, the position of all unblocked bonds (or sites) is reassigned, and the carrier motion proceeds on the newly defined network of unblocked bonds (or sites). In a second percolation model proposed by Harrison and Zwanzig [3.132], individual bonds rather than the whole lattice are changed at each time step. In both models, the host dynamics is spatially uncorrelated, since changes in one location occur independently of those in another location, and is also uncorrelated with the position of the hopping particles.

Both models have some features in common with the more complicated two-component lattice-gas model where the pathways of tracer ions hopping with high jump rates f_A between empty nearest-neighbor sites are blocked by a second species of particles, which move with a considerably lower jump frequency f_B (see e.g., [3.133]). In the limit $f_B \to 0$ we recover ordinary percolation.

The first model seems to be simpler, and one can obtain some exact solutions of quite general nature. For uncorrelated renewals following a Poisson process, the waiting time distribution $f(t)$ for the time between renewals is exponential,

$$f(t) = \tau_{\text{ren}}^{-1} \exp(-t/\tau_{\text{ren}}). \tag{3.130}$$

In this case, Druger *et al.* [3.130] obtained the remarkable result that the frequency-dependent conductivity $\sigma(\omega)$ is simply obtained by analytical continuation from the ac conductivity $\sigma_0(\omega)$ for static disorder,

$$\sigma(\omega) = \sigma_0(\omega - i\tau_{\text{ren}}^{-1}). \tag{3.131}$$

This relation provides a convenient frame for analysing, for example, the temperature and pressure dependence of the conductivity in cases where τ_{ren} is a known function of intrinsic properties of the polymer network [3.134].

Using the generalized Einstein equation (3.78), which relates $\sigma(\omega)$ to $\langle r^2(t) \rangle$, one can then use (3.130) to express $\langle r^2 \rangle$ by the mean square displacement $\langle r^2(t) \rangle_0$ for static disorder. This yields [3.130]

$$\frac{d^2}{dt^2} \langle r^2(t) \rangle = \exp(-t/\tau_{\text{ren}}) \frac{d^2}{dt^2} \langle r^2(t) \rangle_0, \qquad t > 0, \tag{3.132}$$

from which $\langle r^2(t) \rangle$ is easily obtained after the initial conditions $\langle r^2(0) \rangle = \langle r^2(0) \rangle_0$ and $d\langle r^2(t) \rangle/dt|_{t=0} = d\langle r^2(t) \rangle_0/dt|_{t=0}$ have been specified.

To some extent, dynamic percolation might also be useful [3.134] to mimic some aspects of the complicated physics of diffusion of interacting particles, which has been discussed in Sect. 3.5.

3.8.3 The Dynamic Structure Model of Ionic Glasses

Recently, to describe ionic transport in glasses a new type of dynamic percolation called the *dynamic structure model* has been proposed [3.135,136] by which important anomalies in glasses (mixed alkali effect, decoupling of structural and electrical relaxation, drastic increase of the ionic conductivity as a function of mobile ion concentration (for a review see, e.g., [2.19])) can be explained. In the model, the sites have a memory. A mobile ion (we call it an A-ion) prefers to jump to an empty nearest-neighbor "A-site" which has been left by an A-ion not more than τ time steps ago and therefore is adapted to the A-ion. The ion does not like to jump to an unadapted "C-site" which cannot accommodate easily the arrival of the ion. In this way, the distinction between the two types of ionic environments can lead to "mismatch" effects (between ions and sites) when a particular ion tries to diffuse through a glass, and this strongly influences the process of ion migration. A direct consequence is the existence of fluctuating pathways for the mobile ions which tend to freeze in at lower temperatures, and an element of percolation is introduced which determines the ion transport. The numerical exploration of this model led to an effective power-law relationship between the ionic conductivity σ and the concentration c of the A-ions, $\sigma \sim c^{\text{const}/k_B T}$, which is now confirmed in the literature [3.136,137].

If two types of mobile ions, A-ions and B-ions, can move in the glass, both ions tend to maintain their characteristic environments. A-ions prefer to jump to nearest-neighbor "A-sites" while B-ions prefer B-sites. Due to this memory effect, both types of ions create their own fluctuating pathways and have the tendency to stay on them. If, for example, A-ions are progressively substituted by B-ions, the diffusion coefficient of the A-ions decreases since more and more A-pathways become blocked by B-paths, whereas the diffusion coefficient of the B-ions increases. There is an intersection of the diffusion curves of both types of ions at some ratio of concentration where both A- and B-ions have the same mobility. As a consequence, the total ionic conductivity goes through a deep minimum when A-ions are progressively substituted by B-ions. The minimum is close to the intersection of the diffusion curves and becomes more pronounced if the temperature is lowered.

This effect indeed occurs in glasses, where it is known as "mixed alkali effect". It occurs in *all* vitreous alkali conductors of the general formula xX_2O-$(1-x)Y_2O$-$(SiO_2$ or B_2O_3 or $GeO_2)$ where X_2O and Y_2O are different network modifying alkali oxides, and in related sulphide systems. Mixed "alkali" effects are also seen in glasses containing protons, Ag^+ and Tl^+ ions, and there is a corresponding effect in mixed F^-/Cl^- glasses where the anions are the principal current carriers. The dynamic structure model can explain the effect in a natural way as a percolation phenomenon.

3.8.4 Trapping and Diffusion Controlled Reactions

Diffusion-controlled reactions play an important role in various branches of biology, chemistry, and physics (see, e.g., [3.135–138]). In a simple model one considers two types of particles A and B, which diffuse on a given lattice. The concentration of A and B particles is small, so both types of particles can diffuse freely for a long time before meeting each other. If they meet at the same site they immediately react and the product is C. So the reaction can be described as microscopic events $A + B \to C$. Special cases are $C = B$ where particle A is annihilated or $C = 0$ where both particles are annihilated.

Applications of such an elementary picture include processes in which elementary excitations are trapped, quenched, or annihilated by another kind of excitation or particle. Physical examples include electron trapping and recombination, exciton trapping, quenching, or fusion, and soliton–antisoliton recombination. In many realizations such reactions occur in disordered media.

We begin by considering diffusion in percolation clusters in the presence of trapping centers which absorb every particle entering it. This process is schematically described by $A + B \to B$, which was originally studied by Smoluchowski [3.142] in ordinary lattices.

a) Single trap. First we assume a single static trap B on the infinite percolation cluster at p_c. The trap is located at $\mathbf{r} = \mathbf{0}$. We want to consider two cases.

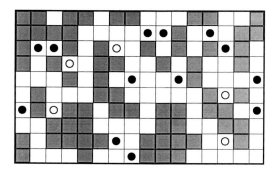

Fig. 3.19. Percolation cluster with A particles (*full circles*) and trapping centers B (*open circles*)

(a) If a single particle A starts to diffuse at time $t = 0$ from a site at $\mathbf{r} = \mathbf{r_0}$, what is the asymptotic survival probability, $S(r_0, t)$, of this particle?

(b) If independent A particles are distributed initially randomly on the infinite fractal with concentration $c(0)$, what will be the concentration of the particles (profile), $c(r, t)$, near the trap B after t time steps (see Fig. 3.19)?

We can show that for isotropic systems the quantities $S(r_0, t)$ and $c(r, t)$, when r is identified with r_0, are the same.

Let $P(r, r_0, t)$ be the probability density of a single particle that started at $r = r_0$ in the presence of a trap at $r = 0$. The survival probability $S(r_0, t)$ of the single particle in (a) is given by

$$S(r_0, t) = \int P(r, r_0, t)dr. \tag{3.133}$$

In (b), one starts with a uniform distribution of particles with concentration $c(r, 0) = c_0$, and $c(r, t)$ is obtained by integrating P over all initial positions r_0,

$$c(r, t) = c_0 \int P(r, r_0, t)dr_0. \tag{3.134}$$

For isotropic systems we can expect that on interchanging r and r_0, $P(r, r_0, t)$ will remain the same. Thus $S(r_0, t)$ depends on r_0 and t in the same way that $c(r, t)$ depends on r and t.

Next we consider the flux F of A particles into the trap. In (a), the flux is simply $F_a(t) = -dS(r_0, t)/dt$, while in (b) $F_b(t) = dc(r, t)/dr|_{r=0}$. We begin with case (b).

The total number of particles trapped in (b) is equal to the number of distinct sites $N(t)$ visited by a random walker on the cluster [3.142–144]. As has been mentioned in Sect. 3.2, $N(t)$ scales as [3.1]

$$N(t) \sim t^{d_s/2}, \tag{3.135}$$

where $d_s \cong 4/3$ (see Sect. 3.6). The relation (3.135) is quite general and holds for all fractals where $d_s \leq 2$. For $d_s \geq 2$ (which is not relevant here) one has $N(t) \sim t$. From (3.135) we have the rate of flow into the trap,

$$F_b(t) = \frac{dN(t)}{dt} \sim t^{-(1-d_s/2)}. \tag{3.136}$$

From $F_b(t)$ we can calculate the profile $c(r,t)$ of A particles around the trap. For length scales much smaller than the typical diffusion distance, $r \ll t^{1/d_w}$, the concentration of particles at distance r from the origin can be found from the identity $F_b(t) = dc(r,t)/dr|_{r=0}$ and the scaling assumption $c(r,t) = G(r/t^{d_w})$. This yields [3.144]

$$c(r,t) \sim \left(\frac{r}{t^{1/d_w}}\right)^{(1-d_s/2)d_w}. \tag{3.137}$$

For $r \gg t^{1/d_w}$, $c(r,t)$ is constant. Note that (3.137) is valid also for homogeneous systems: for $d=1$, $d_s = 1$ and $d_w = 2$ yield $c(x,t) \sim x/t^{1/2}$ [3.145]. For $d=2$, the exponent $(1 - d_s/2)d_w$ is zero, which is consistent with the known result $c(r,t) \sim \ln r / \ln t$ [3.146,147].

To derive $F_a(t)$, we again use the identity between $c(r,t)$ and $S(r_0,t)$ discussed above. We obtain from (3.137) [3.148]

$$S(r_0,t) \sim t^{-(1-d_s/2)}, \tag{3.138}$$

and therefore

$$F_a(t) \equiv -\frac{dS}{dt} = t^{-(2-d_s/2)}. \tag{3.139}$$

Next we consider percolation with a random distribution of traps.

b) Random distribution of traps. We denote the concentration of traps by c_T. In d-dimensional Euclidean space, the survival probability of a random walker in the presence of *randomly* distributed traps is asymptotically [3.149],

$$S(t) \sim \exp\left(-ac_T^{2/(d+2)} t^{d/(d+2)}\right). \tag{3.140}$$

As has been shown by Grassberger and Procaccia [3.150], this relation can be derived as follows.

The asymptotic behavior of $S(t)$ is dominated by those particles that move in large trap-free regions of the cluster. The probability of finding a trap-free region of volume V in an infinite lattice follows a Poisson distribution, $P_0(V) \sim \exp(-c_T V)$.

Consider now the probability $P(V,t)$ that a given particle survives for time t in a trap-free region of volume V, which is enclosed by a trapping boundary. This probability is asymptotically

$$P(V,t) \sim \exp\left(-\text{const} \times \frac{t}{V^{2/d}}\right). \tag{3.141}$$

For randomly distributed traps, the probability of surviving is dominated by the large but rare trap-free regimes. Hence, for $t \to \infty$,

$$S(t) \sim \max_{\{V\}}[P_0(V)P(V,t)], \qquad (3.142)$$

from which (3.140) follows.

The above derivation can be easily extended to find the survival probability on a random fractal such as the infinite percolation cluster at criticality. The only difference is that, because of anomalous diffusion, the scaling relation between time and volume covered by the random walker is $t \sim V^{d_w/d_f} = V^{2/d_s}$ (see also Sect 3.4). Thus (3.141) is modified to

$$P(V,t) \sim \exp\left(-\mathrm{const} \times \frac{t}{V^{2/d_s}}\right), \qquad (3.143)$$

which immediately yields [3.151],

$$S(t) \sim \exp\left(-\mathrm{const} \times c^{2/(d_s+2)}t^{d_s/(d_s+2)}\right). \qquad (3.144)$$

From (3.144) we obtain the Euclidean result (3.140) by simply replacing d_s by d. Finally we consider the annihilation reaction.

c) Annihilation. The annihilation reaction is schematically described by $A + B \to 0$. When both A and B particles are initially randomly distributed with equal concentrations, $c_A(0) = c_B(0) \equiv c(0)$, their concentration after time t (which is identical to the survival probability) is [3.152]

$$c(t) = c(0)t^{-d_s/4}. \qquad (3.145)$$

This result can be understood by considering spatial fluctuations in the initial concentrations. In a domain of linear size L there are $c(0)L^{d_f} \pm [c(0)L^{d_f}]^{1/2}$ A and B particles. After a time on the order of $t \sim L^{d_w}$ all pairs will be annihilated and the number of either A or B particles left in the regime will be of the order of $[c(0)L^{d_f}]^{1/2}$. The concentration is therefore $c(t) \sim [c(0)L^{d_f}]^{1/2}/L^{d_f}$, which together with $t \sim L^{d_w}$ leads to (3.145).

If the particles A and B are identical, i.e., $A + A \to 0$, one has

$$c(t) = c(0)t^{-d_s/2}. \qquad (3.146)$$

For more general reviews on kinetic reactions on fractals we refer to [3.14] and [3.153]. Again, the results for homogeneous lattices can be obtained simply from (3.145) and (3.146) by replacing d_s by d [3.154]. See, however, a recent work by Lindenberg *et al.* [3.155], questioning the generality of (3.145).

The first systematic experimental study of chemical reaction in porous media which is of fractal nature was performed by Kopelman [3.153] using the reaction

$$N^* + N^* \to N_2^{**} \to N + N + \text{photon}.$$

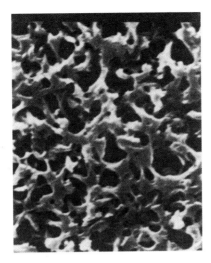

Fig. 3.20. Porous nylon membrane. Magnification about $\times 10^4$. After [3.153]

Here N^* is a naphthalene molecule excited to its first triplet state and N_2^{**} is the transitory dimer in the first excited singlet state. The experiment was performed in a solution embedded in various porous membranes, such as the nylon membrane shown in Fig. 3.20.

The naphthalene molecules diffuse inside the pores. The concentration $c(t)$ behaves as $t^{-0.66}$ in good agreement with (3.146) and the known value of $d_s \cong$ 1.32 for three-dimensional lattices.

d) Initially separated reactants. Recently it was realized that diffusion-reaction systems in which the reactants are initially separated [3.157] can be studied experimentally [3.158,159]. These systems are characterized by the presence of a dynamical interface or a front that separates the reactants and their dynamics has many surprising features [3.159–162]. The reaction front appears in many biological, chemical and physical processes [3.163,164]. We first discuss the diffusion reaction on Euclidean d-dimensional lattices and later on fractals.

Gàlfi and Ràcz [3.157] studied the diffusion-controlled reaction system with initially separated reactants. They studied the kinetics of the diffusion reaction process using a set of mean-field (MF) type equations,

$$\frac{\partial c_A}{\partial t} = D_A \frac{\partial^2 c_A}{\partial x^2} - k c_A c_B, \qquad \frac{\partial c_B}{\partial t} = D_B \frac{\partial^2 c_B}{\partial x^2} - k c_A c_B. \qquad (3.147)$$

Here $c_A \equiv c_A(x,t)$ and $c_B \equiv c_B(x,t)$ are the concentrations of A and B particles at position x at time t respectively, D_A and D_B are the diffusion constants and k is the reaction constant. The rate of production of the C particles at site x and time t, which we call the reaction-front profile, is given by $R(x,t) \equiv k c_A c_B$ and the total production rate of C particles is $R(t) \equiv \int_{-\infty}^{\infty} R(x,t)dx$. Initially, at $t = 0$, the A species are uniformly distributed on one side of $x = 0$ and the B species on the other side.

Using scaling arguments, Gàlfi and Ràcz [3.157] find that the total rate varies with time t as $R(t) \sim t^{-1/2}$. The width w of the reaction front scales as $w \sim t^{\alpha}$ with $\alpha = 1/6$ and the reaction rate at the center of the front scales as $h \sim t^{-\beta}$ with $\beta = 2/3$. For a detailed disscusion see Chap. 7 in [3.165].

Next we discuss, how the total production rate $R(t)$ of C particles changes on fractal systems [3.166]. Here, a straight line on the fractal separates the initial locations of the A and B particles. Since the A and B particles must meet in order to react, only those particles that are at distances smaller than their root-mean-square displacement can react. Thus, the total number of reactants up to time t scales as the root-mean-square displacement of a random walker on a fractal, i.e., $\langle r^2 \rangle^{1/2} \sim t^{1/d_w}$, where d_w is the fractal dimension of the random walk (see Sect. 3.2). From this it follows that the rate $R(t)$ is the time-derivative of t^{1/d_w},

$$R(t) \equiv \int_{-\infty}^{\infty} R(x,t)dx \sim t^{-\gamma}, \qquad \gamma = 1 - 1/d_w. \qquad (3.148)$$

Finally, let us consider the whole percolation system at the critical concentration. As for diffusion, one has to distinguish between diffusion reaction on the infinite cluster and in the percolation system, also containing small clusters (see Sect. 3.3). Equation (3.148) is valid only for the infinite percolation cluster. The reaction rate on the infinite cluster is smaller and decrease slower compared with the system containing clusters of all sizes. This can be understood as follows. At any finite time we can divide all clusters into two groups according to their sizes: *active* clusters of mass $s > s^*$, in which at time $t < t^* \sim s^{* d_w/d_f}$ particles are not aware of the finitness of their cluster (this group contains the infinite cluster); and *inactive* clusters of mass $s < s^*$ on which at least one of the reactants has vanished and the reaction rate is zero. According to this picture, in the full percolation system at any time there are active clusters of finite size that can contribute to the reaction rate. Therefore, the rate of reaction in the percolation system is always higher than that on the infinite cluster. Also, at any time there are some finite clusters that become inactive, causing an additional (compared with the infinite cluster network) decrease in the rate of reaction in the percolation system. Since the system is self-similar, one expects a change in the reaction-rate exponent.

To quantify the above considerations one can look on each cluster of mass s and linear size $r \sim s^{1/d_f}$ as a reservoir of particles divided by the *front line* into A and B parts. We introduce an *active front* of a cluster as the sites belonging both to the cluster and to the front line. The length ℓ_s of the *active front* of a single cluster of size s is expected to be

$$\ell_s \sim r^{d_f - 1} \sim s^{\frac{d_f - 1}{d_f}}. \qquad (3.149)$$

Next we assume that the rate of reaction on a cluster of mass s per unit length of active front is

$$R_0(t) \sim \begin{cases} t^{-\gamma}, & t < t^*, \\ 0, & t > t^*, \end{cases} \qquad (3.150)$$

where $t^* = s^{d_w/d_f}$. Therefore, the total contribution of active clusters of size s to the reaction rate is

$$R_s(t) \sim \varphi_s s^{\frac{d_f - 1}{d_f}} t^{-\gamma}, \tag{3.151}$$

where φ_s is number of clusters of size s that intersect the front line, $\varphi_s \sim s^{1/d_f} n_s$. Substituting this in (3.151) we get

$$R_s(t) \sim s^{1/d_f} n_s s^{\frac{d_f - 1}{d_f}} t^{-\gamma} = t^{-\gamma} s n_s. \tag{3.152}$$

Thus, the reaction rate in the percolation system is

$$R(t) = \sum_{s=s^*}^{\infty} R_s(t) = t^{-\gamma}(s^*)^{2-\tau} = t^{-\gamma} t^{-\delta} = t^{-\gamma'}, \tag{3.153a}$$

where

$$\delta = \gamma' - \gamma = \frac{d_f}{d_w}(\tau - 2). \tag{3.153b}$$

Acknowledgements. We wish to thank our cooperators in this field, in particular D. Ben-Avraham, I. Dayan, W. Dieterich, J. Dräger, K. Funke, M.D. Ingram, J. Kantelhardt, J.E. Kiefer, R. Kopelman, P. Maaß, M. Meyer, K.L. Ngai, R. Nossal, M. Porto, S. Rabinovich, H.E. Roman, S. Ruß, S. Schwarzer, A. Shehter, H.E. Stanley, H. Taitelbaum, B.L. Trus, B. Vilensky, G.H. Weiss, and H. Weissman for many valuable discussions.

References

3.1   S. Alexander, R. Orbach: J. Phys. Lett. **43**, L625 (1982)
3.2   B.J. Last, D.J. Thouless: Phys. Rev. Lett. **27**, 1719 (1971)
3.3   U. Lauer, J. Maier: Ber. Bunsenges. Phys. Chem. **96**, 111 (1992)
3.4   P.G. de Gennes: La Recherche **7**, 919 (1976)
3.5   R. Kopelman: in *Topics in Applied Physics,* Vol 15, ed. by F.K. Fong (Springer Verlag, Heidelberg 1976)
3.6   S. Havlin, D. Ben-Avraham: Adv. in Phys. **36**, 695 (1987)
3.7   S. Alexander, J. Bernasconi, W.R. Schneider, R. Orbach: Rev. Mod. Phys. **53**, 175 (1981);
       S. Alexander: in *Percolation Structures and Processes,* ed. by G. Deutscher, R. Zallen and J. Adler (Adam Hilger, Bristol, 1983), p. 149
3.8   J.W. Haus, K.W. Kehr: Phys. Rep. **150**, 263 (1987)
3.9   A. Aharony: in *Directions in Condensed Matter Physics,* ed. by G. Grinstein, G. Mazenko (World Scientific, Singapore 1986)
       D. Stauffer, A. Aharony: *Introduction to Percolation Theory,* 2nd edition (Taylor and Francis, London 1992)
3.10   D. Avnir, ed.: *The Fractal Approach to Heterogeneos Chemistry* (John Wiley, Chichester 1989)
3.11   J.P. Bouchaud, A. Georges: Phys. Rep. **195**, 127 (1990)
3.12   J.P. Clerc, G. Giraud, J.M. Laugier, J.M. Luck: Adv. in Phys. **39**, 191 (1990)
3.13   E.W. Montroll, M.F. Shlesinger: in *Nonequilibrium Phenomena II: From Stochastics*

to *Hydrodynamics,* Studies in Statistical Mechanics, Vol. 2, ed. by J.L. Lebowitz, E.W. Montroll (North-Holland, Amsterdam 1984)

3.14 A. Blumen, J. Klafter, G. Zumofen: in *Optical Spectroscopy of Glasses,* ed. by I. Zschokke (Dordrecht: Reidel), p. 199

3.15 G.H. Weiss, R.J. Rubin: Adv. Chem. Phys. **52**, 363 (1983); G.H. Weiss: *Aspects and Applications of the Random Walk* (North Holland, Amsterdam 1994)

3.16 T. Nakayama, K. Yakubo, L. Orbach: Rev. Mod. Phys. **66**, 381 (1994)

3.17 W. Feller: *Introduction to Probability Theory and its Applications* Vol. 1 (John Willey, New York 1973)

3.18 J.M. Ziman: *Models of Disorder. The theoretical physics of homogeneously disordered systems* (Cambridge University Press, London 1979)

3.19 Y. Gefen, A. Aharony, B.B. Mandelbrot, S. Kirkpatrick: Phys. Rev. Lett. **47**, 1771 (1981)

3.20 B.B. Mandelbrot, J.A. Given: Phys. Rev. Lett. **52**, 1853 (1984)

3.21 A.B. Harris, A, Aharony: Europhys. Lett. **4**, 1355 (1987)

3.22 M.E. Fisher: J. Chem. Phys. **44**, 616 (1966); M.E. Fisher, F.R.J. Burford: Phys. Rev. **156**, 583 (1967)

3.23 C. Domb: Adv. Chem. Phys. **15**, 229 (1969)

3.24 P.G. de Gennes: *Scaling Concepts in Polymer Physics* (Cornell University, Ithaka 1979)

3.25 For random walks on random walks, see K.W. Kehr, R. Kutner: Physica A **110**, 535 (1982)

3.26 A. Bunde, S. Havlin, H.E. Roman: Phys. Rev. A **42** 6274 (1990); S. Havlin, A. Bunde: Physica D **38**, 184 (1989)

3.27 S. Havlin, R. Nossal, B.L. Trus, G.H. Weiss: Phys. Rev. A **45**, 7511 (1992)

3.28 J. Dräger, St. Russ, A. Bunde: Europhys. Lett. **31**, 425 (1995)

3.29 A. Bunde, J. Dräger: Phil. Mag. B **71**, 721 (1995); Phys. Rev. E **52**, 53 (1995)

3.30 S. Rabinovich, H.E. Roman, S. Havlin, A. Bunde: preprint (1995)

3.31 J. Dräger, A. Bunde: preprint (1995)

3.32 R.A. Guyer: Phys. Rev. A **32**, 2324 (1984)

3.33 R. Rammal, G. Toulouse: J. Phys. Lett. **44**, L13 (1983)

3.34 H.E. Roman, S. Ruß, A. Bunde: Phys. Rev. Lett. **66**, 1643 (1991); A. Bunde, H.E. Roman, S. Ruß, A. Aharony, A.B. Harris: Phys. Rev. Lett. **69**, 3189 (1992)

3.35 Y.A. Levy, B. Souillard: Europhys. Lett. **4**, 233 (1987); A. Aharony, A.B. Harris: J. Stat. Phys. **54**, 1091 (1989); A. Aharony, A.B. Harris: Physica A**163**, 38 (1990)

3.36 D. Ben-Avraham, S. Havlin: J. Phys. A **15**, L691 (1982); S. Havlin, D. Ben-Avraham, H. Sompolinsky: Phys. Rev. A **27**, 1730 (1983)

3.37 Y. Gefen, A. Aharony, S. Alexander: Phys. Rev. Lett. **50**, 77 (1983)

3.38 I. Webman: Phys. Rev. Lett. **47**, 1496 (1991)

3.39 D. Stauffer: *Introduction to Percolation Theory* (Taylor and Francis, London 1985); D. Stauffer, A. Aharony: *Introduction to Percolation Theory*, 2nd edition (Taylor and Francis, London 1992)

3.40 A. Bunde, A. Coniglio, D.C. Hong, H.E. Stanley: J. Phys. A **18**, L137 (1985); D.C. Hong, H.E. Stanley, A. Coniglio, A. Bunde: Phys. Rev. B **33**, 4565 (1986); H. Taitelbaum, S. Havlin: J. Phys. A **21**, 2265 (1988)

3.41 I. Webman, M.H. Cohen, J. Jortner: Phys. Rev. B **11**, 2885 (1975); B **16**, 2593 (1977)

3.42 A.L. Efros, B.I. Shklovskii: Phys. Stat. Sol. B **76**, 475 (1976)

3.43 A.M. Dykne: Zh. Eksper. Theor. Fiz. **59**, 111 (1970)

3.44 M. Sahimi: *Application of Percolation Theory* (Taylor and Francis, London 1993)

3.45 G.W. Milton, in: *Physics and Chemistry of Porous Media* Ed. by D.L. Johnson and P.N. Sen (AIP, NY 1984)

3.46 L. Limat: Phys. Rev. B**38**, 7219 (1988)

3.47 S. Arabi, M. Sahimi: Phys. Rev. Lett. **65**, 725 (1990)

3.48 L. Benguigui, P. Ron: Phys. Rev. Lett. **70**, 2423 (1993); J. Phys. I France **5**, 451 (1995)

3.49 M. Born, K. Huang: *Dynamical Theory of Crystal Lattices,* Chapter 5, (Oxford University Press, New York 1954)

3.50   P.G. de Gennes: J. Phys. Lett. **37**, L1 (1976)

3.51   I. Webman, Y. Kantor: in *Kinetics of Aggregation and Gelation,* ed. by F. Family, D. P. Landau (North Holland, New York 1984)

3.52   M. Adam, M. Delsanti, D. Durand, C. Hild, J.P. Munch: Pure and Appl. Chem. **53**, 489 (1981);
       M. Adam, M. Delsanti, D. Durand: Macromolecules **18**, 2285 (1985);
       L. Benguigui: Phys. Rev. Lett. **53**, 2028 (1984)

3.53   A. Aharony, S. Alexander, O. Entin-Wohlman, R. Orbach: Phys. Rev. B **31**, 2565 (1985)

3.54   G.S. Grest, I. Webman: J. Physique **45**, L1155 (1984)

3.55   K. Yakubo, T. Nakayama: Phys. Rev. B **40**, 517 (1989)

3.56   S. Ruß, H.E. Roman, A. Bunde: J. Phys. C **3**, 4797 (1991)

3.57   S. Feng: Phys. Rev. B **32**, 5793 (1985)

3.58   K. Yakubo, T. Nakayama: in *Phonons 89,* ed. by W. Ludwig, S. Hunklinger (World Scientific, Singapore 1990), p. 682

3.59   A. Fontana, F. Rocca, M.P. Fontana: Phys. Rev. Lett. **58**, 503 (1987)

3.60   T. Freltoft, J.K. Kjems, D. Richter: Phys. Rev. Lett. **59**, 1212 (1987)

3.61   For references on aerogels: E. Courtens, R. Vacher, E. Stoll: Physica D **38**, 41 (1989)

3.62   Y. Tsujimi, E. Courtens, J. Pelaus, R. Vacher: Phys. Rev. Lett. **60**, 2757 (1988)

3.63   R. Vacher, E. Courtens, G. Coddens, A. Heidemann, Y. Tsujimi, J. Pelous, M. Foret: Phys. Rev. Lett. **65**, 1008 (1990)

3.64   R. Calemczuk, A.M. de Goer, B. Sulce, R. Maynard, A. Zarembowitch: Europhys. Lett. **3**, 1205 (1987)

3.65   C.M. Soukulis, Q. Li, G.S. Grest: Phys. Rev. **77**, 24 (1992)

3.66   I. Chang, Z. Lev, A.B. Harris, J. Adler, A. Aharony: Phys. Rev. Lett. **74**, 2094 (1995)

3.67   J.T. Chayes, D.S. Fisher, T. Spenser: Phys. Rev. Lett. **57** 2999 (1986)

3.68   J.K. Kjems, D. Posselt: in *Random Fluctuations and Pattern Growth,* ed. by H.E. Stanley, N. Ostrowsky (North-Holland, Amsterdam 1988), p. 7

3.69   J.E. Oliveira, J.N. Page, H.M. Rosenberg: Phys. Rev. Lett. **62**, 780 (1989)

3.70   J.D. Jackson: *Classical Electrodynamics* (Wiley, New York 1975)

3.71   H. Scher, M. Lax: Phys. Rev. B **7**, 4491 (1973)

3.72   H. Harder, A. Bunde, S. Havlin: J. Phys. A **19**, L927 (1986)

3.73   J.P. Bouchaud, A. Georges: J. Phys. A **23**, L1003 (1990)

3.74   A. Bunde, P. Maass: J. Non-Cryst. Solids **131-133**, 1022 (1991)

3.75   P. Maass, J. Petersen, A. Bunde, W. Dieterich, H.E. Roman: Phys. Rev. Lett. **66**, 52 (1991);
       D. Knödler, W. Dieterich: Physica A **191**, 426 (1992);
       M. Meyer, P. Maass, A. Bunde: Phys. Rev. Lett. **71**, 573 (1993)

3.76   P. Maass, M. Meyer, A. Bunde: Phys. Rev. B **51**, 8164 (1995)

3.77   A.K. Jonscher: Nature **267**, 673 (1977); *Dielectric Relaxation in Solids* (Chelsea Dielectrics Press, London 1983);
       H. Böttger, V.V. Bryskin: *Hopping Conduction in Solids* (Verlag Chemie, Weinheim 1985);
       K.L. Ngai: Comments on Solid State Phys. **9**, 127 (1979); **9**, 141 (1980);
       K.L. Ngai, A.K. Rajagopal, S. Teitler: J. Chem. Phys. **88**, 5086 (1988);
       K. Funke: Z. Phys. Chem. Neue Folge **154**, 251 (1987);
       K. Funke, W. Wilmer: Europhys. Lett. **12**, 363 (1990);
       G. Niklasson: Physica D **38**, 260 (1989)

2.78   I. Webman, M.H. Cohen, J. Jortner: Phys. Rev. B **16**, 2593 (1977)

3.79   J.P. Straley: Phys. Rev. B 15, 5733 (1977)

3.80   D. Stroud, D.J. Bergman: Phys. Rev. B **25**, 2061 (1982)

3.81   Y. Song, T.W. Noh, S.-I. Lee, J.R. Gaines: Phys. Rev. B **33**, 904 (1986)

3.82   R.B. Laibowitz, Y. Gefen: Phys. Rev. Lett. **53**, 380 (1984)

3.83   R. Blender, W. Dieterich: J. Phys. C **20**, 6113 (1987)

3.84   H.J. Herrmann, H.E. Stanley: Phys. Rev. Lett. **53**, 1121 (1984)

3.85   S. Havlin, D. Djordjevic, I. Majid, H.E. Stanley, G.H. Weiss: Phys. Rev. Lett. **53**, 178 (1984)

3.86   A. Coniglio: Phys. Rev. Lett. **46**, 250 (1981)

3.87   A. Coniglio: J. Phys. A **15**, 3829 (1982)

3.88   R.B. Pandey, D. Stauffer, A. Margolina, J.G. Zabolitzky: J. Stat. Phys. **34**, 427 (1984)

3.89   R.B. Pandey, D. Stauffer, J.G. Zabolitzky: J. Stat. Phys. **49**, 849 (1987)

3.90   E. Duering, H.E. Roman: J. Stat. Phys. **64**, 851 (1991)

3.91   S. Havlin, D. Ben-Avraham: J. Phys. A **16**, L483 (1983)

3.92   P.G. de Gennes: J. Physique Coll. **41**, C3 (1980);
       A. Bunde, A Coniglio, D.C. Hong, H.E. Stanley: J. Phys. **18**, L137 (1985);
       J.Adler, A. Aharony, D. Stauffer: J. Phys. **18**, L129 (1985)

3.93   I. Majid, D. Ben-Avraham, S. Havlin, H.E. Stanley: Phys. Rev. B **30**, 1626 (1984)

3.94   D.C. Hong, S. Havlin, H.J. Herrmann, H.E. Stanley: Phys. Rev. B **30**, 4083 (1984)

3.95   B. Derrida, J. Vannimenus: J. Phys. A **15**, L557 (1982)

3.96   H.J. Herrmann, B. Derrida, J. Vannimenus: Phys. Rev. B **30**, 4080 (1984)

3.97   J. M. Normand, H.J. Herrmann, M. Hajjar: J. Stat. Phys. **52**, 441 (1988); Int. J. Mod. Phys. C **1** (1990)

3.98   R. Fogelholm: J. Phys. C **13**, L571 (1980)

3.99   D.G. Gingold, C.J. Lobb: Phys Rev. B **42**, 8220 (1990)

3.100 R. Fisch, A.B. Harris: Phys. Rev. B **18**, 416 (1978)

3.101 J.W. Essam, F. Bhatti: J. Phys. A **18**, 3577 (1985)

3.102 J. Adler, Y. Meir, A. Aharony, A.B. Harris, L. Klein: J. Stat. Phys. **58**, 511 (1990)

3.103 J. Bernasconi: Phys. Rev. B **18**, 2185 (1978)

3.104 A.B. Harris, T.C. Lubensky: J. Phys. A **17**, L609 (1984);
       A.B. Harris, S. Kim, T.C. Lubensky: Phys. Rev. Lett. **53**, 743 (1984)

3.105 J. Wang, T.C. Lubensky: Phys. Rev. B **33**, 4998 (1986)

3.106 E.T. Gawlinski, H.E. Stanley: J. Phys. A **14**, L291 (1981)

3.107 B.I. Halperin, S. Feng, P.N. Sen: Phys. Rev. Lett. **54**, 2391 (1985);
       S. Feng, B.I. Halperin, P. Sen: Phys. Rev. B **35**, 197 (1987); B.I. Halperin: Physica D **38**, 179 (1989)

3.108 P.M. Kogut, J.P. Straley: J. Phys. C **12**, 2151 (1979);
       A. ben-Mizrachi, D.J. Bergman: J. Phys. C **14**, 909 (1981)

3.109 J.P. Straley: J. Phys. C **15**, 2333, 2343 (1982)

3.110 T.C. Lubensky, A.M.S. Tremblay: Phys. Rev. B **37**, 7894 (1988);
       J. Machta: Phys. Rev. B **37**, 7892 (1988)

3.111 L. Benguigui: Phys. Rev. B **34**, 8177 (1986)

3.112 C.J. Lobb, M.G. Forrester: Phys. Rev. B **35**, 1899 (1987)

3.113 A. Bunde, H. Harder, S. Havlin: Phys. Rev. B **34**, 3540 (1986)

3.114 J. Petersen, H.E. Roman, A. Bunde, W. Dieterich: Phys. Rev. B **39**, 893 (1989)

3.115 J. Machta, S.M. Moore: Phys. Rev. A **32**, 3164 (1985)

3.116 H.E. Stanley, I. Majid, A Margolina, A. Bunde: Phys. Rev. Lett. **53**, 1706 (1984)

3.117 M. Porto, A. Bunde, S. Havlin, H.E. Roman: preprint (1996)

3.118 T.C. Halsey, P. Meakin, I. Procaccia: Phys. Rev. Lett. **56**, 854 (1986);
       P. Meakin, H.E. Stanley, A. Coniglio, T.A. Witten: Phys. Rev. A **32**, 2364 (1985);
       P. Meakin, A. Coniglio, H.E. Stanley, T.A. Witten: Phys. Rev. A **34**, 3325 (1986);
       For a review see e.g., H.E. Stanley, P. Meakin: Nature **335**, 405 (1988)

3.119 L. de Arcangelis, S. Redner, A. Coniglio: Phys. Rev. B **31**, 4725 (1985); B **34**, 4656 (1986);
       R. Rammal, C. Tannous, A.M.S. Tremblay: Phys. Rev. A **31**, 2662 (1985)

3.120 B.B. Mandelbrot: J. Fluid Mech. **62**, 331 (1974)

3.121 B. Kahng: Phys. Rev. Lett. **64**, 914 (1990)

3.122 S.W. Kenkel, J.P. Straley: Phys. Rev. Lett. **49**, 767 (1982)

3.123 R. Blumenfeld, Y. Meir, A. Aharony, A.B. Harris: Phys. Rev. B **35**, 3524 (1987);
       R. Rammal, A.M.S. Trembley: Phys. Rev. Lett. **58**, 415 (1987)

3.124 A. Bunde, S. Havlin, H.E. Stanley, B. Trus, G.H. Weiss: Phys. Rev. B **34**, 8129 (1986);
       S. Havlin, A. Bunde, H.E. Stanley, D. Movshovitz: J. Phys. A **19** L693 (1986)

3.125 S. Havlin, A. Bunde, Y. Glaser, H.E. Stanley: Phys. Rev. A **34**, 3492 (1986)

3.126 A. Bunde, H. Harder, S. Havlin, H.E. Roman: J. Phys. A **20**, L865 (1987)

3.127 M. Barma, D. Dhar: J. Phys. C **16**, 1451 (1983)

3.128 S.R. White, M. Barma: J. Phys. A **17**, 2995 (1984)

3.129 R.B. Pandey: Phys. Rev. B **30**, 489 (1984);

D. Dhar: J. Phys. A **17**, L257 (1984);

Y. Gefen, I. Goldhirsch: J. Phys. A **18**, L1037 (1985)

3.130 H.E. Roman, M. Schwartz, A. Bunde, S. Havlin: Europhysics Lett. **7**, 389 (1988)

3.131 S.D. Druger, A. Nitzan, M.A. Ratner: J. Chem. Phys. **79**, 3133 (1983);

S.D. Druger, M.A. Ratner, A. Nitzan: Phys. Rev. **31** 3939 (1985);

3.132 A.K. Harrison, R. Zwanzig: Phys. Rev. A **32**, 1072 (1985)

3.133 R. Granek, A. Nitzan: J. Chem. Phys. **92**, 1329 (1989)

3.134 G.S. Grest, I. Webman, S.A. Safran, A.L.R. Bug: Phys. Rev. A **33**, 2842 (1986)

R. Kutner, K.W. Kehr: Phil. Mag. A **48**, 199 (1983);

3.135 A. Bunde, M.D. Ingram, P. Maass, K.L. Ngai: J. Phys. A **24**, L881 (1991);

P. Maass, A. Bunde, M.D. Ingram: Phys. Rev. Lett. **68**, 3064 (1992)

3.136 A. Bunde, M.D. Ingram, P. Maass: J. Non-Cryst. Solids **172-174**, 1222 (1994)

3.137 A. Pradel, M. Ribes: J. Non-Cryst. Solids **172-174**, 1315 (1994)

3.138 N.G. van Kampen: *Stochastic Processes in Physics and Chemistry* (North Holland, Amsterdam 1981)

3.139 S.A. Rice: *Diffusion-Limited Reactions* (Elsevier, Amsterdam 1985)

3.140 A.A. Ovchinnikov, Y.B. Zeldovich: Chem. Phys. **28**, 215 (1978);

V. Kuzovkov, E. Kotomin: Rep. Prog. Phys. **51**, 1479 (1988);

3.141 M. Bramson, J.L. Lebowitz: Physica A **168**, 88 (1990)

3.142 M. von Smoluchowski: Phys. Zeit. **16**, 321 (1915); **17**, 557, 585 (1917)

3.143 R. Kopelman, P. Argyrakis: J. Chem. Phys. **72**, 3053 (1980)

3.144 S. Havlin, R. Kopelman, R. Schoonover, G.H. Weiss: Phys. Rev. A **43**, 5228 (1991)

3.145 G.H. Weiss, R. Kopelman, S. Havlin: Phys. Rev. A **39**, 466 (1989)

3.146 S. Havlin, H. Larralde, R. Kopelman, G.H. Weiss: Physica A **169**, 337 (1990)

3.147 S. Redner, D. Ben-Avraham: J. Phys. A **23**, L1169 (1990);

H. Taitelbaum: Phys. Rev. A **43** (1991)

3.148 S. Havlin, R. Nossal, B.L. Trus, G.H. Weiss: Phys. Rev. A **43** 5228 (1991)

3.149 N.D. Donsker, S.R.S. Varadhan: Commun. Pure Appl. Math. **32**, 721 (1979)

3.150 P. Grassberger, I. Procaccia: J. Chem. Phys. **77**, 6281 (1982)

3.151 J. Klafter, G. Zumofen, A. Blumen: J. Phys. Lett. **45**, L49 (1984);

I. Webman: Phys. Rev. Lett. **52**, 220 (1984); J. Stat. Phys. **36**, 603 (1984)

3.152 P. Meakin, H.E. Stanley: J. Phys. A **17**, L173 (1984);

K. Kang, S. Redner: Phys. Rev. Lett. **52**, 955 (1984)

3.153 R. Kopelman: Science **241**, 1620 (1988)

3.154 D. Toussaint, F. Wilczek: J. Chem. Phys. **78**, 2642 (1983);

M. Bramson, J.L. Lebowitz: Physica A **168**, 88 (1990)

3.155 K. Lindenberg, W.S. Sheu, R. Kopelman: Phys. Rev. A **43** 7070 (1991); J. Stat. Phys. **65**, 1285 (1991);

W.S. Sheu, K. Lindenberg, R. Kopelman: Phys. Rev. A **42**, 2279 (1990)

3.156 G. Zumofen, J. Klafter, A. Blumen: Phys. Rev. A **43**, 7068 (1991); J. Stat. Phys. **65**, 1015 (1991)

3.157 L. Gálfi, Z. Rácz: Phys. Rev. A **38**, 3151 (1988)

3.158 Y. E. Koo, L. Li, R. Kopelman: Mol. Cryst. Liq. Cryst. **183**, 187 (1990);

Y. E. Koo, R. Kopelman: J. Stat. Phys. **65**, 893 (1991)

3.159 H. Taitelbaum, Y. E. Koo, S. Havlin, R. Kopelman, G. H. Weiss: Phys. Rev. A **46** 2151 (1992)

3.160 Z. Jiang and C. Ebner: Phys. Rev. A **42**, 7483 (1990)

3.161 H. Taitelbaum, S. Havlin, J. Kiefer, B. L. Trus, G. H. Weiss: J. Stat. Phys. **65**, 873 (1991)

3.162 S. Cornell, M. Droz, B. Chopard: Phys. Rev. A **44**, 4826 (1991);

M. Araujo, S. Havlin, H. Larralde, H. E. Stanley: Phys. Rev. Lett. **68**, 1791 (1992);ibid **71**, 3592 (1993);

E. Ben-Naim, S. Redner: J. Phys. A **25**, L575 (1992);

H. Larralde, M. Araujo, S. Havlin, H. E. Stanley: Phys. Rev. A **46**, 855 (1992)

3.163 D. Avnir, M. Kagan: Nature, **307**, 717 (1984);

G. T. Dee: Phys. Rev. Lett. **57**, 275 (1986);

B. Heidel, C. M. Knobler, R. Hilfer, R. Bruinsma: Phys. Rev. Lett. **60**, 2492 (1986)

3.164 R. E. Liesegang: Naturwiss. Wochensch. **11**, 353 (1896);
    T. Witten, L. M. Sander: Phys. Rev. Lett. **47**, 1400 (1981);
    R. A. Ball: Ausr. Gemmol. **12**, 89 (1984);
    K. F. Mueller: Science **255**, 1021 (1984)
3.165 A. Bunde, S. Havlin, eds: *Fractals in Science* (Springer, Heidelberg 1994)
3.166 S. Havlin, A. M. Araujo, H. Larralde, A. Shehter, H. E. Stanley: Chaos, Solitons and Fractals **6**, 33 (1995)

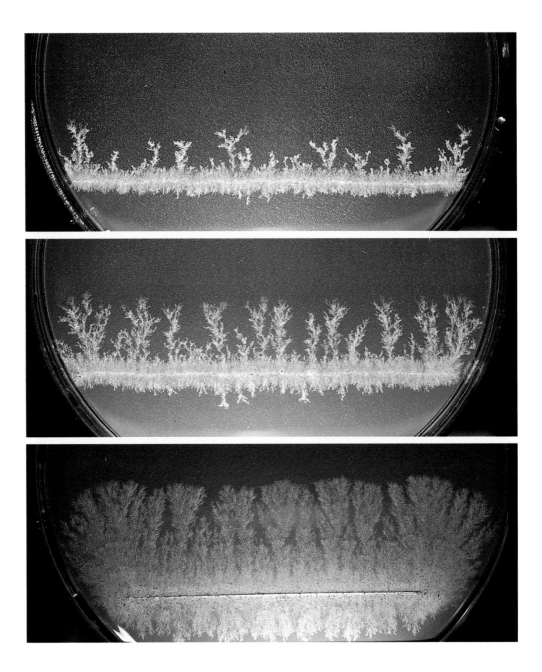

4 Fractal Growth

Amnon Aharony

4.1 Introduction

Aggregation of particles often produces fractal clusters. Many experiments (see also Chap. 8) and numerical simulations have recently explored the properties of aggregation kinetics, gelation, and sedimentation, with emphasis on their fractal structure. Displacement of a fluid in porous media usually results in a displacement front. If the fluid is driven by a fluid of lower viscosity, the displacement front is unstable and its motion is mathematically analogous to the kinetics of some aggregation processes, again yielding fractal structures. This chapter aims to review some of these phenomena. Lack of space and personal preferences naturally limit the scope of this review. Much more information can be found in recent books, reviews, and conference proceedings on the subject [4.1–10].

A typical aggregate, of the kind we wish to discuss here, is shown in Fig. 10.0. This aggregate was generated in a computer simulation of *diffusion limited aggregation* (DLA) [4.11], which will be explained below. It turns out that the shape shown in Fig. 10.0 looks very similar to those which arise in many natural aggregation processes, including diffusion limited electrodeposition [4.12], growth in aqueous solutions [4.13], dielectric breakdown [4.14], viscous fingers in porous media [4.15], and fungi and bacterial growth [4.16–18]. Figure 4.0 shows an example of fungi growth on a strip. In what follows, we shall give more details of some of these systems and we shall discuss the reasons for their similarities. We shall also discuss more compact aggregates, specifically those arising in the Eden model [4.19] and in invasion percolation [4.20,21].

In order to discuss such aggregates, one must first characterize their structure quantitatively. Since many of these structures seem to be self-similar, i.e., they

◄ **Fig. 4.0.** Fungi colonies grown in a glucose-rich medium in a strip geometry (for details see [4.17]). Courtesy of S. Miyazima

look the same under different magnifications, it has become common to describe them as *fractals* [4.22] and characterize them with a *fractal dimensionality, d_f*. It turns out that the fractal dimensionality alone is not sufficient for a unique description of the structures. We have thus been led to identify many additional fractal dimensionalities, or exponents, each of which characterizes a subset of the points on the structure which determines one of its physical (or geometrical) properties [4.23]. One infinite family of such exponents is related to *multifractals* [4.24,25]. Since these concepts are also discussed in Chaps. 1, 3, and 10, we shall give only a brief summary of them, in Sect. 4.2. At the moment, it is not at all clear that we have identified all the relevant exponents which are necessary to describe the aggregates uniquely.

Assuming we have a clear identification of a structure, our next aim is to understand the relations between the microscopic interactions which are responsible for its growth and the specific complex macroscopic shapes that result. This is done using *growth models*. In such a model, one sets a few simple "microscopic" growth rules, by which particles are added to the aggregate. Repeated iteration of these rules then gives rise to the "macroscopic" cluster.

At present (this chapter was finalized in January 1990 and updated by the editors in January 1995), we have some empirical numerical experience with a variety of growth models (Sects. 4.3 and 4.4). In some cases, it is possible (within some approximation) to translate the physical process into simple algorithmic growth rules, and then to follow these rules in a computer simulation. This yields an *algorithmic description* of the resulting complex aggregate. There has been much recent work on variations of the growth rules, and on listing their effects on the resulting structures. However, unlike the situation in critical phenomena [4.26], there does not yet exist a full theory that classifies the structures into universality classes according to a few characteristics and that is able to calculate the properties of the macroscopic structure (e.g. its fractal dimensionality) analytically from the microscopic growth rules.

Following our description of several growth models, we end with three more specific topics. In Sect. 4.5 we review some simulations, experiments and analytical model calculations of growth on dilute networks, at and near the percolation threshold [4.27–30] (see also Chaps. 2 and 3). Section 4.6 contains an application of the multifractal spectrum to studying a crossover from one growth behavior to another, as a result of competing growth rules [4.31]. Finally, Sect. 4.7 contains some critical remarks on the use of multifractality for negative moments [4.32].

4.2 Fractals and Multifractals

Ideal fractals are *self-similar* on all length scales, and therefore all their properties follow power laws in their linear scale, L; there exists no basic length which can be used as a unit of length. In physical problems, there always exists a smallest length, e.g., the linear size of one particle in the aggregate, a. Self-similarity then arises only *asymptotically*, for $L \gg a$. When we study finite clusters, the results should also depend on their linear size, L_m (which can be defined, for example, as the largest linear size of a box containing the cluster in the Euclidean d-dimensional space in which the cluster is embedded). The statement that the cluster is fractal is usually equivalent to the statement that the number of particles on the cluster, within a box of size L which covers part of the cluster, with $a \ll L \ll L_m$, scales asymptotically as

$$M(L) = A(L/a)^{d_f}, \tag{4.1}$$

where d_f is the fractal dimensionality. (In all other publications from this author, d_f is denoted by D.) Since (4.1) is asymptotic, it may require corrections, with smaller powers of L, for small values of L. In many of the cases of interest here, (4.1) also applies for the total mass of the cluster,

$$M_t = M(L_m) \sim (L_m/a)^{d_f}. \tag{4.2}$$

However, sometimes one needs to worry about boundary effects. Clearly, (4.2) is useful only if we have a distribution of cluster sizes, or if we consider different stages in the growth of the same cluster.

An alternative measure of the cluster's linear size is its radius of gyration, R_g, defined via

$$R_g^2 = \frac{1}{M_t} \sum_i (\boldsymbol{r}_i - \boldsymbol{r}_c)^2, \tag{4.3}$$

where $\boldsymbol{r}_c = (1/M_t) \sum_i \boldsymbol{r}_i$ is the cluster's center of mass. Usually, one also expects that

$$M_t \sim (R_g/a)^{d_f}. \tag{4.4}$$

However, both (4.2) and (4.4) are *asymptotic*, i.e., they apply for L_m and R_g much larger than a. For some aggregates, the corrections to (4.4) are found to be much smaller than those to (4.2) (see [4.33]).

An alternative way to define d_f is called *box counting*: divide the d-dimensional hypercubic Euclidean box, of linear size L_m, which contains the cluster, into $(L_m/\delta)^d$ smaller boxes, each of linear size δ. (As usual, d is the Euclidean dimension of the space in which the cluster is embedded.) The number of small boxes which contain pieces of the cluster, N, scales as

$$N(L_m, \delta) \sim (L_m/\delta)^{d_f}. \tag{4.5}$$

For fixed L_m, d_f is found as the limit of $-\ln N/\ln \delta$ for small δ. For aggregates, δ is never smaller than a.

One definition of *multifractals* (see also Chaps. 1, 3, and 10) is related to box counting. Within each box i, we define a *measure* p_i, such that $\sum_i p_i = 1$. In the example to be used below, p_i will be the probability that the next particle to be added to the aggregate will be added within the box i. We now consider the moments

$$M_q(L_m, \delta) = \sum_i p_i^q. \tag{4.6}$$

The distribution of the measures p_i is called *multifractal* if all of these moments scale as power laws [4.24,25], with an infinite set of independent exponents $\tau(q)$:

$$M_q(L_m, \delta) = A_q(L_m/\delta)^{-\tau(q)}. \tag{4.7}$$

For $q = 0$, (4.6) reduces to the number of boxes which have a nonzero measure. If every box has some finite measure, M_0 coincides with N, hence $\tau(0) = -d_f$. Also, the normalization $\sum_i p_i = 1$ requires that $\tau(1) = 0$.

If there are $n(p)dp$ boxes with p_i between p and $(p + dp)$, then one may write

$$M_q = \int_0^1 dp\, n(p)p^q. \tag{4.8}$$

Assuming that $n(p)p^q$ has a sharp maximum at p_q^*, such that

$$\frac{\partial \ln n(p)}{\partial \ln p}\Big|_{p_q^*} = -q, \tag{4.9}$$

we approximate M_q by $n(p_q^*)(p_q^*)^q$. If both p_q^* and $n(p_q^*)$ behave as powers of (L_m/δ),

$$p_q^* \sim (L_m/\delta)^{-\alpha(q)}, \quad n(p_q^*) \sim (L_m/\delta)^{f(q)}, \tag{4.10}$$

then

$$\tau(q) = q\alpha(q) - f(q). \tag{4.11}$$

From (4.9)

$$\frac{\partial f}{\partial \alpha} = q, \tag{4.12}$$

and therefore also

$$\frac{\partial \tau}{\partial q} = \alpha. \tag{4.13}$$

Equation (4.11) thus implies that $f(\alpha)$ is the Legendre transform of $\tau(q)$. Assuming the pure power laws (4.10), the function $f(\alpha)$ is expected to be independent of (L_m/δ), and to represent a universal characterization of the cluster.

From (4.12) it follows that $f(\alpha)$ has a maximum at $q = 0$. At that point, $f(0) = -\tau(0) = d_f$. Thus, f has the general shape sketched in Fig. 4.1: it grows for $\alpha < \alpha(0)$, $(q > 0)$, reaches a maximum at $q = 0$, and decreases for $\alpha > \alpha(0)$, $(q < 0)$.

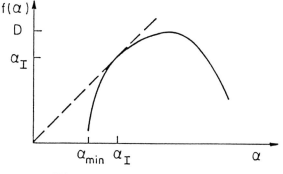

Fig. 4.1. A general sketch of $f(\alpha)$

Since $0 \le p_i \le 1$, it is clear that M_q decreases with q, and therefore $\tau(q)$ increases with q. Since $\tau(0) = -d_f$ and $\tau(1) = 0$, it has become useful to define D_q, via [4.25]

$$\tau(q) = D_q(q - 1). \tag{4.14}$$

For multifractal behavior, D_q exhibits a nontrivial dependence on q. This changes in the special case in which p_i denotes the fraction of the total mass in the box i, $p_i = M_i/M_t$. If $a \ll \delta \ll L_m$, then the mass within each box also scales as $M_i \sim (\delta/a)^{d_f}$. Thus, $p_i \sim (\delta/L_m)^{d_f}$ and $M_q \sim (L_m/\delta)^{d_f(1-q)}$, i.e., $D_q \equiv d_f$, independent of q. Thus, deviations from a constant D_q indicate deviations of the measure p_i from being just proportional to the mass, or deviations from the local fractal behavior $M_i \sim (\delta/a)^{d_f}$. For a mass fractal we thus expect $D_q \equiv f \equiv \alpha \equiv d_f$. Note, however, that transient corrections to the asymptotic behavior may yield transient deviations of D_q from d_f, and thus an apparent nontrivial curve of $f(\alpha)$, which should disappear when $L_m \gg \delta \gg a$.

Even when the mass does not obey the local fractal behavior $M_i \sim (\delta/a)^{d_f}$, it should still be bounded by $1 \le M_i \le (\delta/a)^d$. For $q < 0$, this yields

$$\frac{d_f}{|q| + 1} \le D_q \le \frac{d|q| + d_f}{|q| + 1} < d. \tag{4.15}$$

For $q > 1$, this yields $0 < D_q < (dq - d_f)/(q - 1)$. In both cases, the bound $D_q < d$ should be obeyed. Values in the literature that break this bound [4.31] must arise due to nonasymptotic corrections.

These restrictions do not apply to measures which are not connected to the mass distribution. In particular, we shall see below measures which decay exponentially with (L_m/δ). Such measures clearly indicate a breakdown of the power law (4.7) for $q < 0$, with $\tau(q) \to -\infty$ [4.32].

Since $\sum_i p_i = 1$, we have $M_1 = 1$, or $\tau(1) = 0$. The largest contribution to M_1 will come from the boxes where $n(p)p$ is maximal, i.e., when $\partial f/\partial \alpha = 1$ This defines the value α_I, and we identify

$$f(\alpha_I) = \alpha_I, \quad (\partial f/\partial \alpha)|_{\alpha_I} = 1. \tag{4.16}$$

Since this value of α dominates M_1, we conclude that $f(\alpha) \leq \alpha$, with equality only at α_I. Thus, the curve of $f(\alpha)$ in Fig. 4.1 is below the line $f(\alpha) = \alpha$, and is tangent to it at $\alpha = \alpha_I$ (where the slope is $q = 1$).

In the context of aggregation, p_i will denote the *growth probability* (defined below, in Sect. 4.4). The above discussion shows that for large (L_m/δ), growth practically occurs only in the boxes with $\alpha = \alpha_I$, whose number scales as $(L_m/\delta)^{f(\alpha_I)} = (L_m/\delta)^{\alpha_I}$. The number $D_I = f(\alpha_I) = \alpha_I$ is called the *information dimension*. Differentiating (4.6) easily yields

$$\alpha_I = - \lim_{(L_m/\delta) \to \infty} \frac{\sum_i p_i \ln p_i}{\ln(L_m/\delta)}. \tag{4.17}$$

In the case of growth, it is believed that the largest growth probability, $p_{\max}$, occurs at the tips. If the growth occurs at such a tip, the linear size of the cluster grows by one. Otherwise, it remains unchanged. Thus,

$$\begin{aligned} \Delta L_m / \Delta M_t &= L_m(M_t + 1) - L_m(M_t) \\ &= p_{\max}[L_m(M_t) + 1] + (1 - p_{\max})L_m(M_t) - L_m(M_t) \\ &= p_{\max}. \end{aligned}$$

Using $M_t \sim L_m^{d_f}$ and $p_{\max} \sim L_m^{-\alpha_{\min}}$ now yields [4.35]

$$d_f = 1 + \alpha_{\min}. \tag{4.18}$$

4.3 Growth Models

4.3.1 Eden Model

The simplest growth model is that of Eden [4.19] (see also Sect. 2.6). A site is occupied at the origin. All its empty neighboring sites are then identified as possible growth sites. One of these growth sites is then chosen at random (all having the same growth probability), and is occupied. The empty neighbors of both sites are now labeled as growth sites, and the procedure is iterated. The resulting aggregate is "compact", i.e., its fractal dimension is equal to the Euclidean dimension of the embedding space, d [4.36]. Moreover, in the interior all sites finally become occupied. The surface of the growth zone, however, has a nontrivial behavior [4.37].

In all the growth models, one may use spherical geometry, with a single seed at the origin, or strip geometry, when there are seeds on a $(d-1)$-dimensional "plane" of size L^{d-1}, and growth is studied as a function of the average height. For the growth sites, this height is defined via

$$h = \frac{1}{N_s} \sum_{i=1}^{N_s} h_i, \tag{4.19}$$

where N_s is the total number of growth sites, and h_i is the distance of the ith growth site from the substrate. Defining the width of the growth zone via

$$\sigma(L,h) = \left[\frac{1}{N_s}\sum_{i=1}^{N_s}(h_i - h)^2\right]^{1/2}, \tag{4.20}$$

one can study the scaling of σ as a function of both L and h. This turns out to obey the *self-affine* form [4.22,38]

$$\sigma(L,h) = L^\alpha f(h/L^z), \tag{4.21}$$

with

$$f(x) \sim \begin{cases} x^\beta & \text{if } x \ll 1 \\ \text{const} & \text{if } x \gg 1. \end{cases} \tag{4.22}$$

Self-affinity is reflected by the fact that $\alpha \neq 1$, so the sizes of the growth zones parallel and perpendicular to the substrate are not proportional to each other.

For $L \ll L^z$, $\sigma \sim h^\beta$, independent of L. Therefore $z = \alpha/\beta$. Numerical simulations [4.39] seem to be consistent with $\alpha = 1/d$ and $\alpha + z = 2$. These results seem to be inconsistent, for $d > 2$, with some recent theoretical models [4.40]. For more details and recent developments, see Chap. 7.

4.3.2 Percolation

In the Eden model, all growth sites had the same growth probability. In addition, every growth site could be considered as a candidate for growth many times, until it was eventually occupied. *Percolation* related models modify both of these assumptions.

In the standard percolation model [4.41,42] (see also Chaps. 2 and 3), each site on a lattice is either occupied (with probability p) or empty (with probability $(1 - p)$). Thus, each site is considered only once. This can be translated into a growth model if we start from an occupied origin and then go once through the list of all its neighboring sites, occupying a fraction p of them at random. We next look at the neighbors of all the newly occupied sites, but do not consider those neighbors which have already been declared "unoccupied" [4.43,44]. Repeating this procedure many times, we end up with a distribution of clusters.

Below the percolation threshold concentration, p_c, all of these clusters are finite, with a size distribution function which behaves as $s^{-\tau}$ for $s < \xi^{d_f}$ and decays exponentially for for $s > \xi^{d_f}$. Here, $\xi \sim (p_c - p)^{-\nu}$ is the pair-connectedness correlation length (see Sects. 2.3–5). Numerically, $d_f \cong 1.9$ (2.5), $\nu \cong 1.33$ (0.88) in $d = 2$ (3), and $\tau = 1 + d/d_f$. For linear sizes $L < \xi$, the clusters are fractals, and their mass scales as $s \sim L^{d_f}$. For $L > \xi$, the clusters behave as "lattice animals", with a typically smaller fractal dimensionality d_a [4.41]. Above p_c, one also finds an infinite cluster, which has a fractal dimension d_f

on length scales $L < \xi \sim (p - p_c)^{-\nu}$, and has a uniform density, i.e., $s \sim L^d$, for $L > \xi$ [4.45]. At the percolation threshold, $p = p_c$, $\xi = \infty$ and all large clusters are fractal, with fractal dimensionality d_f.

For $p \neq p_c$, percolation involves the nontrivial length ξ, and therefore all the properties of clusters are functions of ratios like L/ξ. A discussion of such crossover phenomena (from $L \ll \xi$ to $L \gg \xi$) is outside the scope of this paper, and the interested reader is referred to Chaps. 2 and 3 and Refs. [4.41] and [4.42]. At p_c, the static geometry of a specific grown cluster has no length except the cluster's size (e.g. L_m, R_g), hence the power law dependences, like $s \sim R_g^{d_f}$. However, one can look at the cluster at intermediate growth stages, as a function of the "time". In the present model, the "time" t represents the number of neighboring shells, counting from the origin. After t growth steps, t represents the minimal number of steps on the grown cluster necessary to reach the growth sites from the origin. This number is also called the *chemical distance*, or the minimal path (for a discussion see Sect. 2.3) It turns out [4.46] that the average number of points on the cluster within a chemical distance t scales as $t^{d_f/d_{\min}}$, and its average linear size scales as $t^{1/d_{\min}}$, with $d_{\min} \cong 1.13$ (1.36) for $d = 2$ (3).

4.3.3 Invasion Percolation

The invasion percolation model [4.20,21] (see also Sects. 2.6, 7.4, and 8.2) was motivated by the study of the displacement of one fluid by another in a porous medium [4.47,48]. Dynamics of displacement fronts is essential for understanding displacement processes in oil production, and invasion percolation is the simplest model with nontrivial front structure and dynamics.

When water is injected very slowly into a porous medium filled with oil, the capillary forces dominate the viscous forces, and the dynamics is determined by the local pore radius r. Capillary forces are strongest at the narrowest pore necks. It is consistent with both a simple theoretical model and experimental observations to represent the displacement as a series of discrete jumps in which at each time step the water displaces oil from the smallest available pore.

Wilkinson and Willemsen [4.21] simulated the model on a regular lattice (see also Sect. 2.6). Sites and bonds represented pores and throats, and were assigned random "radii". For convenience, one assumes that the easily invaded throats are invaded instantaneously, and one assigns random numbers r in the range $[0, 1]$, representing the pore sizes, to the sites.

Growth sites are identified as the sites that belong to the "defending" fluid and are neighbors of the invading fluid. At every time step the invading fluid is advanced to the growth site that has the lowest random number r.

The invading fluid may trap regions of the defending fluid. As the invader advances it is possible for it to completely surround regions of the defending fluid, i.e., completely disconnect finite clusters of the defending fluid from the exit sites of the sample. If the defending fluid is compressible, then the invader

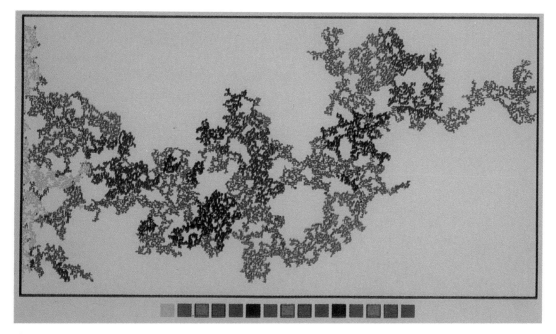

Fig. 4.2. Two-dimensional invasion percolation with trapping (from Ref. [4.49]). The invader (*colored*) enters from sites on the left-hand edge and the defender (*white*) escapes through the right-hand edge. Different colors (*left to right on color scale*) indicate sites added within successive time intervals $\Delta t = 2121$

can still move into the trapped regions. At the point of breakthrough, the resulting invader cluster is equivalent to the spanning infinite cluster of regular percolation (practically all the sites with $r < p_c$ remain untouched) [4.21]. Since oil is incompressible, Wilkinson and Willemsen [4.21] introduced the rule that water cannot invade trapped regions of oil. This rule is implemented by the removal of growth sites in regions completely surrounded by the invading fluid from the list of growth sites.

Figure 4.2 shows the results of a simulation of the invasion process, with trapping, on a square lattice [4.49]. The number of sites, $M(L)$, that belong to the central $L \times L$ part of an $L \times 2L$, with injection from one side, scales as [4.21] $M(L) = AL^{d_f}$, with $d_f \cong 1.82$. This is consistent with experiments by Lenormand and Zarcone [4.50] of air slowly invading a network of ducts filled with glycerol.

Figure 4.2 indicates that the front moves by invading local connected areas, in *bursts*. Once a new site is invaded, the front tends to stay in that vicinity. The growth within a time interval t tends to occur within a *connected* region, and the differently colored regions have similar linear extensions, of order $r_t \sim t^{1/d_f}$. Qualitatively, this can be understood as follows: The invading front exhausts all easily invaded pores, then by forcing the invader through a difficult pore a new region, which may have new easily invaded areas, is made available to the front. The front then moves into this area until it gets stuck again, having exhausted the new easily invaded pores.

To obtain a *quantitative* measurement of these time correlations, we define a *pair correlation function* $N(r,t)drdt$, giving the conditional probability that a site at a distance between r and $(r+dr)$ from a reference site is invaded at a time between t and $(t+dt)$ later than the reference site. During the simulation, the reference site is successively chosen to be the last invaded site, thus averaging over all local surroundings. The correlation function was found [4.49] to obey the dynamic scaling form

$$N(r,t) = r^{-1}f(r^z/t), \tag{4.23}$$

with the dynamic exponent $z = d_f$. The function $f(x)$ was found to be peaked around $x = 1$, and obeyed

$$f(x) \sim \begin{cases} x^a & \text{if } x \ll 1 \\ x^{-b} & \text{if } x \gg 1, \end{cases} \tag{4.24}$$

with $a \cong 1.4$, $b \cong 0.6$.

The factor r^{-1} occuring on the right-hand side of (4.23) results from the normalization $\int N(r,t)dr = 1$. The result $z = d_f$ follows from the requirement that $\int N(r,t)dt/r^{d-1}$ should behave as the pair-connectedness correlation function, $G(r) \sim r^{d_f - d}$. Finally, the values of a and b have recently been discussed by Roux and Guyon [4.51], who gave arguments favoring the relations $a = 1$ and $b = 1/d_f$.

Note that the particular shape of the invasion percolation clusters is a direct consequence of our choice of the distribution of the random numbers, $\{r\}$. Particularly, the connection with percolation is related to the fact that this distribution had a finite width. An extreme alternative limit arises when all the "throats" have the same "radius". In that case, all growth sites have the same growth probability, and the problem reduces to the Eden model. In general, one may expect a crossover between the Eden model and invasion percolation, as a function of the width of the random numbers' distribution.

4.4 Laplacian Growth Model

4.4.1 Diffusion Limited Aggregation

The algorithm for diffusion limited aggregation (DLA), as first suggested by Witten and Sander [4.11], is simple: one places a seed particle at the origin and sends from far away a particle that performs a random walk. When this random walker touches the seed particle, it irreversibly sticks and both particles form a small aggregate. If it returns to the external boundary, it is removed. One proceeds by sending more random walkers (one at a time) and adding them to the aggregate whenever they touch a particle that is already part of the aggregate. Such an aggregate, containing 50 000 particles, is shown in Fig. 4.0 [4.10] (see also Chaps. 1, 8, and 10). One can see that the aggregate is highly branched, with no loops except on very small scales. The open structure is attributed to the observation that the chances that a random walker will find its way deep inside a fjord without touching the aggregate is very small. Thus the tips grow much faster than the screened parts.

The analytic discussion of DLA is based on the relation of the DLA algorithm to the *Laplace equation*. Consider the probability $P(\mathbf{r})$ that the random walker reaches a point $\mathbf{r}$ between the external boundary and the growing aggregate, without having visited the aggregate or the external boundary. This probability must obey the relation

$$P(\mathbf{r}) = \frac{1}{z} \sum_{\boldsymbol{\delta}} P(\mathbf{r} + \boldsymbol{\delta}), \qquad (4.25)$$

where the sum is over the z nearest-neighbor sites $\mathbf{r} + \boldsymbol{\delta}$. Equation (4.25) also applies to neighbors of the external boundary and the aggregate, if we set $P \equiv 1$ on the former and $P \equiv 0$ on the latter.

Equation (4.25) is clearly a discrete version of the Laplace equation, $\nabla^2 P = 0$. Thus $P(\mathbf{r})$ follows the same behavior as the electrical potential between two electrodes connected to the outer boundary and the aggregate. The growth probability at a site on the perimeter of the cluster is simply proportional to the value of $P(\mathbf{r})$ at that site, found from solving the Laplace equation. This value is the same as $|\nabla P|$ on the bond connecting the perimeter site to the aggregate. This analogy explains the relevance of DLA to electrodeposition.

For intermediate sizes (of the order of 10 000 particles), DLA aggregates seem to be self-similar, with fractal dimensions of the order of 1.65–1.75 in two dimensions and of order 2.5 in three dimensions [4.53]. In fact, the results in d dimensions are quite close to the "mean field" approximation [4.54] $d_f = (d^2 + 1)/(d + 1)$. For such sizes, the results seem to be independent of small-scale details, lattice structure, sticking rules, etc. This led to many attemps to formulate a theory for DLA that would yield the universal exponents. Unfortunately, no such theory has yet been able to explain the complex structure.

As indicated above, the growth probability of DLA is very different on the tips and inside the fjords. This opened many discussions of the distribution of growth probabilities, and of its possible multifractal nature. In a DLA simulation, one obtains the growth probabilities p_i by stopping at a given growth stage, and then sending many random walkers. Whenever a walker touches the aggregate, one increases a counter at that perimeter site by one, without letting the walker stick (the walker is then removed). Repeating this with many walkers yields the growth probability distribution. Figure 4.3 shows a two-dimensional off-lattice DLA aggregate, and the location of cluster sites which were contacted more than 1, 10, 100, 1000 and 10000 times (out of 5×10^6 random walkers used to probe the cluster's surface) [4.3].

a.

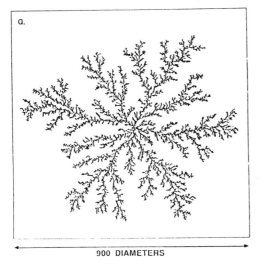

900 DIAMETERS

b.

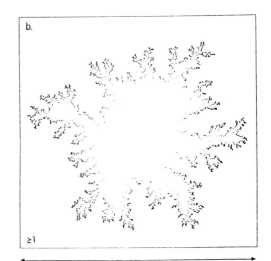

≥1

900 DIAMETERS

c.

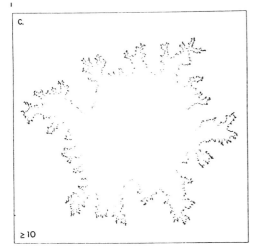

≥ 10

900 DIAMETERS

d.

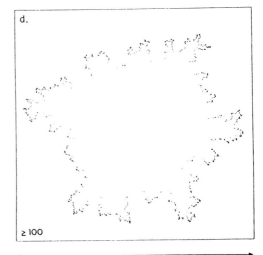

≥ 100

900 DIAMETERS

e.

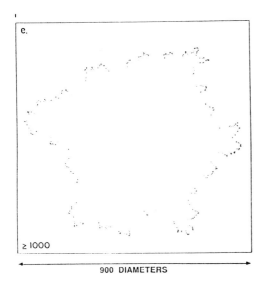

≥ 1000

900 DIAMETERS

f.

≥ 10000

900 DIAMETERS

The results for the moments $M_q = \sum_i p_i^q$ seem to yield multifractality, with a nontrivial $f(\alpha)$ curve, which looks similar to Fig. 4.1. However, the negative moments, related to the small growth probabilities, for which the statistics is poor, fluctuate widely, and seem to yield size-dependent values $f(\alpha)$. More accurate data were obtained using the relation to the Laplace equation: solving this equation directly, or using Green's functions for the solution, one can obtain rather accurate values for $P(\mathbf{r})$, and hence for the growth probabilities p_i [4.55]. The resulting measure is called *the harmonic measure*. The situation with regard to the negative moments is still not clear (see Sect. 4.7 below and Chaps. 1 and 10).

The above technique, of freezing the aggregate and finding its harmonic measure, has also been applied to experiments on viscous fingers, which are expected to be equivalent to DLA clusters (because of the connection to the Laplace equation, see below) [4.56]. Indeed, the results are in qualitative agreement with those from the numerical simulations.

For the harmonic measure in two dimensions, the information dimension α_I is exactly equal to unity [4.57]. Indeed, both simulations [4.58] and experiments [4.59] which looked at the points from which growth actually occurred (by subtracting pictures of the cluster at different times) confirm this value. In fact, Måløy et al. [4.59] used such subtractions to look at the mass distribution of the added pieces, and found indications that it is also multifractal. However, the maximum of $f(\alpha)$ here is equal to $\alpha_I = 1$, the dimension of the subset of sites on which growth actually occurs.

The above picture has become even more complex, since larger-scale simulations have become available. In particular, the shapes of large aggregates grown on lattices turn out to depend on the lattice symmetry. For example, the last 5000 sites added to a 10^5 site cluster on a square lattice have an overall diamond-like shape. This provides support for the work of Turkevich and Scher [4.35], who concentrated on the singularity associated with the largest growth probability, $p_{\max}$, at a *wedge-shaped tip*. As indicated by (4.18), knowledge of the scaling of $p_{\max}$ yields d_f. Turkevich and Scher noted that along a wedge with opening angle θ the potential decays as $P(r) \sim r^{x-1}$, with $x = \pi/(2\pi - \theta)$, where r is the distance from the tip. Integrating up to a, and normalizing by the integral up to the cluster size R, yields $p_{\max} \sim (a/R)^x$, identifying $\alpha_{\min} = x$, hence

$$d_f = (3\pi - \theta)/(2\pi - \theta). \tag{4.26}$$

This yields $d_f = 5/3$ and $7/4$ for square and hexagonal lattices.

◀ **Fig. 4.3a–f.** The results of a simulation in which 5×10^6 random walkers were used to probe the surface of a two-dimensional 50 000-particle off-lattice DLA cluster. After a particle has contacted the cluster, it is removed and a new random walker is launched. (a) The cluster. (b-f) The location of the particles on the cluster which were contacted at least once, ≥ 10, ≥ 100, ≥ 1000, and ≥ 10000 times, respectively (After [4.3], courtesy of P. Meakin)

For off-lattice simulations in two dimensions, or for lattices of high symmetry (more than six distinct arms), the results seem to be universal (at intermediate sizes), with $d_f \cong 1.71$ or $\theta \cong 10°$. For larger sizes, θ seems to decrease towards zero, and d_f approaches 1.5 [4.3]. A detailed analysis of this crossover is still being sought. For studies of the nature of the mass distribution for large sizes see [4.60–63].

In the above discussion, DLA was constructed using particles which approach it in a random walk, whose fractal dimension is equal to 2 (the number of steps t scales as the square of the end to end distance, r^2). This can be generalized by using other rules, e.g., by using particles which have trajectories with $t \sim r^{d_w}$. The resulting fractal dimension d_f will then depend on both d and d_w. One case which attracted some attention is that of *ballistic aggregation*, with $d_w = 1$ (straight lines) [4.65–67]. In that case, it is now clear that the resulting aggregates are compact, with [4.68] $d_f = d$, and probably self-affine.

Another wide family of models, which we have no space to go into here, concerns cluster-cluster aggregation ([4.3,4,69], see also Sect. 8.2).

4.4.2 Dielectric Breakdown Model

The dilectric breakdown model (DBM) was introduced by Niemeyer, Pietronero and Wiesmann [4.14] to simulate, by means of a semi-stochastic model, the formation of complicated discharge patterns with a high level of branching, which occur in charged insulators (see also Sect. 8.2). The fractal dimensionality of such experimental discharge patterns, when grounded via a point electrode in two dimensions, is about 1.7, and they look very similar to DLA aggregates. The DBM algorithm proceeds via the following steps:

(a) Solve the Laplace equation, $\nabla^2 \psi = 0$, in the insulator, with boundary conditions $\psi_{\text{ext}} = 1$ on the external boundary and $\psi_0 = 0$ on the discharge path (assumed to be equipotential).
(b) Identify all the possible perimeter bonds, through which discharge may advance.
(c) On all of these bonds, find the magnitude of the electric field, $E_i = |\boldsymbol{E}_i| = |\nabla \psi_i|$.
(d) Define now a bond probability,

$$p_i = \frac{E_i^{\eta}}{\sum_i E_i^{\eta}}, \qquad (4.27)$$

and choose one of the perimeter bonds, i, with the probability p_i.
(e) Add this bond to the growing discharge pattern by setting the potential at its end equal to zero.
(f) Repeat steps (a) to (e) until the pattern reaches the boundary.

As explained in the previous section, DLA is equivalent to the DBM for $\eta = 1$, apart from small local differences in the boundary conditions and sticking rules near the cluster. For $\eta = 0$, all the perimeter sites have the same growth probability, and the DBM reduces to the Eden model with $d_f = d$. For large η, growth occurs mainly on the tips, and there is much less branching ($d_f \to 1$ for $\eta \to \infty$).

4.4.3 Viscous Fingering

In a porous medium, the velocity of a viscous fluid v is related to the gradient of the pressure P via Darcy's law,

$$v = -k\nabla P, \qquad (4.28)$$

where k is the permeability. The same law applies in a Hele-Shaw cell, when the fluid moves between two parallel plates. When a less viscous fluid is pushed into the viscous one, the interface becomes unstable [4.70], resulting in the so-called *viscous fingers* (see also Sects. 7.4 and 8.2). The shape of the fingers is determined by the capillary number, $N_{\text{ca}} = U\eta/\sigma$, where U is the fingers' tip velocity, η is the viscosity of the displaced fluid and σ is the interfacial tension. N_{ca} is proportional to the ratio of the viscous force to the interfacial one. For low σ, i.e., large N_{ca}, the fingers split many times and have DLA-like shapes.

The analogy to DLA and to the DBM becomes clear when the viscous fluid is incompressible [4.71]. In that case, $\nabla \cdot v = 0$, hence

$$\nabla \cdot k\nabla P = 0. \qquad (4.29)$$

If we ignore the spatial dependence of k, this becomes the Laplace equation. Since the velocity is given by (4.28), the interface moves with a velocity proportional to ∇P, just as the local growth rate of the DLA cluster is proportional to the particle flux.

A priori, the flow equations for the viscous fingers interface are continous and *deterministic*, while those for DLA contain a *stochastic* element (due to the discrete steps of growth). However, porous media also contain an element of randomness, due to the local distribution of values of the permeability. Indeed, Chen and Wilkinson [4.72] demonstrated that a deterministic solution of the flow equations on a square network with a wide distribution of random channel radii yields structures which are very similar to those generated by DLA. Note that in this deterministic solution, the distribution of radii is quenched and fixed throughout the calculation, while in DLA one picks new random numbers at each step of the simulation. Similar DLA-like structures arise in real experiments of air displacing glycerol in a model porous medium produced with random glass spheres [4.15].

4.4.4 Biological Growth Phenomena

In the last decade it was realized that the patterns formed by bacterial and fungi colonies, under appropriate conditions, look very similar to DLA patterns (compare Fig. 4.0 with Fig. 1.12). For details we refer to Chap. 1 and [4.16–18].

4.5 Aggregation in Percolating Systems

In real porous media, the local permeability k (see (4.28)) fluctuates randomly in space. Unlike in the simulations of Chen and Wilkinson [4.72], a finite fraction of the channels is also completely blocked. In addition, the very narrow channels are practically blocked by the capillary forces, as in the invasion percolation limit. To study the effects of such blocking, we have recently studied the various growth models on a dilute network, with only a fraction p_c of the sites present [4.27,28]. In this study, all the open channels were assumed to have the same permeability.

4.5.1 Computer Simulations

We constructed specific realizations of the spanning cluster on a site diluted square lattice, of various sizes, at $p_c \cong 0.593$. On each of these, we simulated DLAs by sending random walkers from the outside and sticking them to the aggregate on touching. We also used the stochastic DBM rule, solving Laplace's equation [i.e., writing (4.25), and requiring that $P(\boldsymbol{r} + \boldsymbol{\delta}) = P(\boldsymbol{r})$ if site $\boldsymbol{r} + \boldsymbol{\delta}$ is blocked, to ensure that $\nabla P = 0$ on the boundary], and then adding the perimeter site i to the aggregate with a probability given by (4.27). For $\eta = 1$ we found [4.27] both types of rules yield practically the same fractal dimensionalities, $d_f \cong 1.3$. However, the amplitude A [mass $= A(\text{length})^{d_f}$] of the DBM was a factor of 1.3 larger than that of the DLA. This difference is related to a local difference in the equations. Equation (4.25), with the boundary condition stated above, is equivalent to

$$P(\boldsymbol{r}) = \sum_{\boldsymbol{\delta}} P(\boldsymbol{r} + \boldsymbol{\delta})/n(\boldsymbol{r}), \qquad (4.30)$$

where $n(\boldsymbol{r})$ is the number of open sites neighboring $\boldsymbol{r}$. In contrast, the DLA algorithm is equivalent to the equation [4.27,73]

$$P(\boldsymbol{r}) = \sum_{\boldsymbol{\delta}} P(\boldsymbol{r} + \boldsymbol{\delta})/n(\boldsymbol{r} + \boldsymbol{\delta}). \qquad (4.31)$$

Since the viscous fluid is incompressible, the invading fluid cannot enter the "dangling ends" of the cluster, which are connected to the other parts of it only via a single bond. Thus, the aggregate which grows on a percolation cluster is limited to its biconnected part, or its *backbone* (see Sect. 2.3). Figure 4.4a shows a specific realization of a spanning cluster on a 147×147 square lattice. The backbone of the cluster is shown in white, and the remaining sites are light blue. In two dimensions, the fractal dimension of the backbone is about 1.6 [4.41,42]. Figure 4.4b shows the simulation of viscous fingering, using the DBM algorithm, on the same cluster. Indeed, the aggregate remains on the backbone

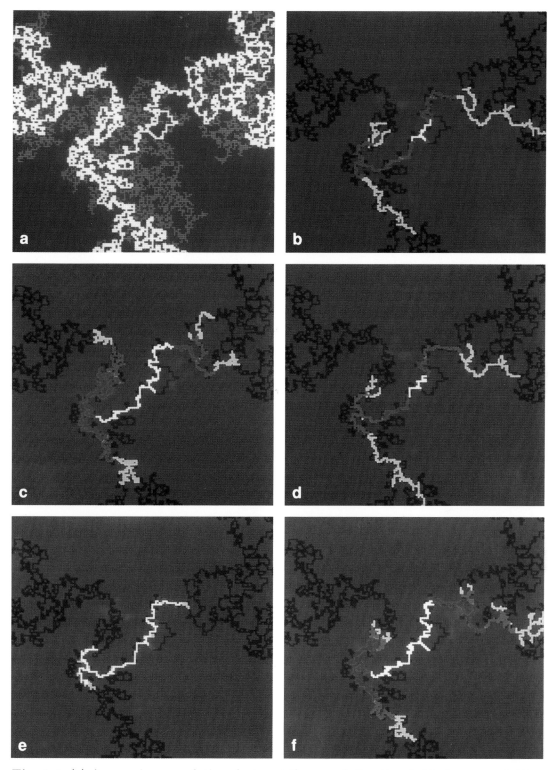

Fig. 4.4. (a) A spanning percolation cluster at the percolation threshold of a square lattice. (b) Simulation of the DBM model with $\eta = 1$ on the same cluster. Colors represent aggregates of mass 30, 86, 213, and 447. (c) Same with $\eta = 0$. (d) Experiment with fast flow. (e) Overlap of fast experiment (*yellow*) and simulation with $\eta = 1$ (*red*). The overlapping points are white. (f) Experiment with slow flow. (From [4.25])

(shown in black). The color code indicates the first 30, 86, 213, and 447 sites [4.28].

A similar simulation, but with $\eta = 0$, is shown in Fig. 4.4c. The fingers now become "thicker", and the fractal dimensionality grows to $d_f = 1.5$. Unlike in the Eden model growth in uniform space, where $d_f = d$, $d_f(\eta = 0)$ is even smaller than the fractal dimensionality of the backbone of the percolating cluster, $d_B \cong 1.6$. Most of the backbone mass, as well as all the "dangling" bonds, remain "trapped".

4.5.2 Viscous Fingers Experiments

An epoxy cast porous model was made, identical to the one shown in Fig. 4.4a and used for the simulations in Figs. 4.4b and 4.4c. The model was filled with glycerol, and air was pushed through the central "seed" site [4.28]. Figure 4.4d shows a digitized picture of the resulting fingers, for a *fast* flow rate (30 pores/sec). These fingers practically overlap those from the computer simulation with $\eta = 1$, Fig. 4.4b, with $d_f \cong 1.3$ (see Fig. 4.4e). Finally, Fig. 4.4f shows experimental results at a *slow* flow rates (0.33 pores/sec). The resulting fingers have $d_f \cong 1.5$, similar to the simulations with $\eta = 0$. Slow flow is thus dominated by *local capillary effects*. Had the pores been *random* in size, one might have ended up with invasion percolation on percolation, having yet another fractal dimensionality. Since our model had more-or-less equal pores, the capillary effects were the same and we ended up with the $\eta = 0$ description (see also Sects. 7.4 and 8.2).

4.5.3 Exact Results on Model Fractals

The above simulations and experiments are special cases of a large variety of *physical phenomena on fractals* [4.23,42]. The simplest geometrical fractal model that imitates the percolation backbone was proposed by Mandelbrot and Given [4.74] (see also Sect. 2.8). In a generalized version of this model, one replaces each bond by L_1 smaller singly connected bonds, in series with a "blob" of L_2 and L_3 bonds in parallel [4.30]. On this structure one can solve the deterministic flow equations exactly, and find the fractal dimensionality of the fingers in terms of those of the backbone, the minimal path (which scales as $L_1 + L_2$, when $L_2 < L_3$), d_{min}, and the singly connected bonds (which scale as L_1), d_{red} (see Sect. 2.3). In the limit of a small length rescale factor $b = 1 + \epsilon$, the result is [4.30]

$$d_f(\eta) = d_B - [(d_B - d_{min})^{\eta+1} - (d_{min} - d_{red})^{\eta+1}]^{1/(\eta+1)} \qquad (4.32)$$

for $\eta > -1$. For two-dimensional percolation, this yields $d_f(1) \cong 1.32$ and $d_f(0) = 2d_{min} - d_{red} \cong 1.51$, in excellent agreement with both computer and real experiments.

4.5.4 Crossover to Homogeneous Behavior

Consider now the percolation structure in which a fraction p_c of the bonds have a high permeability (equal to 1), while the others have a low permeability, equal to R. Flow can now leave the spanning infinite cluster and move to other clusters. Flow is also possible for $p < p_c$.

Two-dimensional simulations of both DLA and DBM, with $\eta = 1$, at $p = p_c$, show aggregates like those discussed above for $R = 0$, with $d_f^{(2)} \cong 1.3$, for small sizes, $L < L_R$, and "homogeneous" DLA-like aggregates, with $d_f^{(1)} \cong 1.7$, for $L > L_R$. The crossover length L_R scales like $L_R \sim R^{-a}$, with $a \cong 0.25$ [4.29]. These results can be interpreted using a scaling theory, in which

$$M(L, R) = L^{d_f^{(2)}} f(L/L_R), \tag{4.33}$$

with

$$f(x) \sim \begin{cases} \text{const} & \text{if } x \ll 1 \\ x^{d_f^{(1)} - d_f^{(2)}} & \text{if } x \gg 1. \end{cases} \tag{4.34}$$

4.6 Crossover in Dielectric Breakdown with Cutoffs

As explained above, the DBM seems to supply a unifying framework, describing many of the aggregation phenomena given in this review. The parameter η uniquely identifies the various *universality classes* of the DBM, in analogy to the number of spin components in critical phenomena [4.27]. As in critical phenomena, one might expect a crossover between different universality classes, whenever there exist competing interactions. Such a competition occurs, for example, between the viscous and capillary forces, with a possible crossover between the DLA pattern ($\eta = 1$) and that of invasion percolation, or the Eden model ($\eta = 0$). Indeed, we have recently studied an example of such a crossover [4.31]. Surprisingly, we found that the details of the crossover can be derived from the knowledge of the information dimension, α_I, and the singularity at the tip, $\alpha_{\min}$.

We studied two models. In Model I, we assumed that the growth probabilities given by (4.27) have a lower cutoff, p_c, so that there is no growth at point i if $p_i \leq p_c$. The rest of the probabilities are renormalized to have a sum of unity. In Model II the cutoff is on the gradients, i.e., there is no growth if $E_i \leq E_c$.

Simulations [4.31] of Model I, with $\eta = 1$, yielded the usual DBM-type growth up to a crossover radius $R_\times$, which behaved as $R_\times \sim p_c^{-1}$. For larger sizes, the aggregates developed spiky fingers, like those arising for $\eta = \infty$. This growth stopped completely at a maximum size $R_{\max} \sim p_c^{-2}$.

To understand these numerical results, we noted [4.31] that the *actual* growth occurs on sites which have the information dimension α_I. At each of these points, the growth probability is $p_I \sim R^{-\alpha_I}$. As long as $p_I > p_c$, the actual growth is not affected by the cutoff. Thus, the effects begin to be felt when $p_I \approx p_c$, hence at $R_\times \sim p_c^{-1/\alpha_I}$. Since for the harmonic measure in two dimensions $\alpha_I = 1$, this explains the result of the simulation.

For $R \geq R_\times$, growth occurs at sites with growth probabilities in the narrow range $p_I \leq p_i \leq p_{\max}$. In practice, growth starts at a few of these sites, which have $p_i \sim p_{\max}(R_\times) \sim R_\times^{-\alpha_{\min}}$, and then proceeds along spikes. Along a spike, $p_{\max} \sim E_{\max}/\sum_i E_i$, and $E_{\max}$ remains of order unity. As sites are added at the tip, the sum in the denominator grows linearly with their number, hence with their radius. Thus [4.31], $p_{\max} \sim 1/aR$, where $a = (\Delta \sum_i E_i/\Delta M)|_{R=R_\times} = R_\times^{\alpha_{\min}-d_f} \sim R_\times^{-1} \sim p_c$ (see (4.18)), i.e., $p_{\max} \sim 1/(p_c R)$. Since growth stops when $p_c > p_{\max}$, we find that $R_{\max} \sim p_c^{-2}$, in agreement with the simulations. A similar analysis of Model II yields [4.31] $R_\times \sim E_c^{-1/(2-d_f)}$.

Thus, α_I, $\alpha_{\min}$, and d_f turn out to be useful tools in identifying crossover phenomena. We expect similar results in many other situations with competition.

4.7 Is Growth Multifractal?

As we discussed in Sect. 4.4, there exist practical difficulties in obtaining accurate values for the negative moments of the growth probabilities. These moments are dominated by the small growth probabilities that occur deep inside screened fjords. Counting DLA random walkers may miss these rare events, and the numerical solutions of the Laplace equations are not very accurate at such small values of the potential. This may be the reason for the large fluctuations observed for $q < 0$. However, the problems with negative moments may have a more fundamental source [4.32,75] (see also Chaps. 1 and 10). The multifractal formalism assumed ((4.10)) that *all* the growth probabilities scale as powers of the size. This may be wrong for the very small growth probabilities: in fact, the electrostatic field (proportional to the growth probability in DLA or in the DBM) at the bottom of a narrow straight slit decays exponentially with its depth. It is therefore conceivable that sites within a tortuous fjord of depth ℓ may be similarly screened, so that the growth probability deep in them obeys

$$\lim_{\ell \to \infty} \ln p_i(\ell)/\ln \ell = -\infty. \tag{4.35}$$

In [4.32], we proposed that (4.35) should apply to some fraction of the growth sites, with ℓ behaving as some power of the aggregates' size L, for *any typical aggregate*. If this hypothesis is true, then it follows that for sufficiently negative q the moment $M_q(L)$ behaves as $p_{\min}^q L^{d_{\min}}$ and therefore

$$\lim_{\ell \to \infty} \tau(q, L) = -\infty, \quad q < 0. \tag{4.36}$$

We then divided the growth sites into two groups, those with exponentially small growth probabilities, and those with power-law dependence on L (multifractal). Using steepest descent for the qth moment within each group, we concluded that

$$M_q = A(p_q^*)^q L^{d_x(q)} + B L^{d_y(q)}, \tag{4.37}$$

where p_q^* is the growth probability dominating the contributions of exponentially small p_is, and the second term comes from the power-law terms. Clearly, for $q \gg 1$ the second term wins, and $d_y(q) = -\tau(q)$. For $q \ll -1$, the first term wins, and $d_x(q) \to d_{\min}$. Thus, there exists a "phase transition" at some finite q, equal to q_c, above which the multifractal formalism applies, and below which $\tau \to -\infty$ and $\alpha \to \infty$ for $L \to \infty$. In [4.32] we assumed that $d_x(q)$ and $d_y(q)$ are analytic near $q = 0$, with the limits $d_x(0) = d_c$ (the fractal dimensionality of all the screened sited) and $d_y(0) = d_u$ (the fractal dimensionality of all the unscreened sites). The value of $q_c(L)$ then follows from equating the two terms in (4.37); for large L it behaves as

$$q_c \cong (d_u - d_c) \frac{\ln L}{\ln p_c(L)}, \tag{4.38}$$

approaching zero for $L \to \infty$.

As explained directly below (4.7), the total number of growth sites can be found from $M_0 \sim L^{d_g}$, where d_g is the fractal dimension of these sites (now believed to be equal to d_f). From (4.37) we conclude that

$$d_g = \lim_{L \to \infty} \lim_{q \to 0} \frac{\ln M_0}{\ln L} = \max(d_c, d_u). \tag{4.39}$$

Thus, if the unscreened sites dominate then $d_g = d_u > d_c$. On the other hand, if the screened sites dominate, then $d_g = d_c > d_u$. In [4.32] we assumed the former, and concluded that q_c approaches zero from below. Recently, Fourcade and Tremblay [4.76] argued that q_c approaches zero from above, using an assumption equivalent to $d_c > d_u$ [4.77].

Although it is clear that some deep fjords in some aggregates will have exponentially small growth probabilities, it is far from clear if they occur sufficiently frequently to be observable. In contrast to Fourcade and Tremblay [4.76], who argue that practically *all* growth sites become screened for sufficiently large aggregates i.e., $d_g = d_c$, it has recently been argued by Harris [4.78] based on a few special configurations, that the probability for a fjord to grow with the aggregate is so low, that its contribution to M_g always becomes negligible. At the moment, the issue of the breakdown of multifractal scaling for negative moments remains open (see also Chaps. 1 and 10 and [4.79,80]).

4.8 Conclusion

In this review, I have attempted to give the reader a feeling for some of the issues which are of current interest in the research on aggregation. Although we have reached a stage in which we can imitate many natural aggregation phenomena with simple algorithmic growth rules, we are still far from having a complete theory.

Acknowledgements. Much of the work described here would not exist without the collaboration with and insights from J. Feder, M. Murat, U. Oxaal, L. Furuberg, R. Blumenfeld, Y. Meir, E. Arian, P. Alstrøm, P. Meakin, T. Jøssang, and H. E. Stanley. I also thank D. Stauffer for a critical reading of the manuscript and constructive suggestions. This reasearch has been supported in part by grants from the Israel Academy of Sciences and Humanities, the U.S.-Israel Binational Science Foundation, VISTA, a research Corporation between the Norwegian Academy of Science and Letters and Den Norske Stats Olje selskap a.s. (STATOIL), and the Norwegian Council of Science and Humanities (NAVF).

References

4.1   J. Feder: *Fractals* (Plenum, New York 1988)

4.2   T. Vicsek: *Fractal Growth Phenomena* (World Scientific, Singapore 1989)

4.3   P. Meakin, in: *Phase Transitions and Critical Phenomena*, Vol.12, ed. by C. Domb and J.L. Lebowitz (Academic Press, New York 1988), p. 335

4.4   F. Family, D.P. Landau, eds.: *Kinetics of Aggregation and Gelation* (Elsevier - North Holland, Amsterdam 1984)

4.5   H.E. Stanley, N. Ostrowsky, eds.: *On Growth and Form: Fractal and Non-Fractal Patterns in Physics* (Martinus Nijhoff, Dordrecht 1986)

4.6   L. Pietronero, E. Tosatti, eds.: *Fractals in Physics* (North Holland, Amsterdam 1986)

4.7   R. Pynn, T. Riste, eds.: *Time Dependent Effects in Disordered Materials* (Plenum, New York 1987)

4.8   H.E. Stanley, N. Ostrowsky, eds.: *Random Fluctuations and Pattern Growth: Experiments and Models* (Kluwer Academic Publishers, Dordrecht 1988)

4.9   A. Aharony, J. Feder, eds.: *Fractals in Physics* (North Holland, Amsterdam 1989)

4.10   H.E. Stanley: Physica D **38**, 330 (1989)

4.11   T.A. Witten, L.M. Sander: Phys. Rev. Lett. **47**, 1400 (1981)

4.12   M. Matsushita, M. Sano, Y. Hayakawa, H. Honjo, Y. Sawada: Phys. Rev. Lett. **53**, 286 (1984)

4.13   Y. Sawada, A. Dougherty, J.P. Gollub: Phys. Rev. Lett. **56**, 1260 (1986)

4.14   L. Niemeyer, L. Pietronero, H.J. Wiesmann: Phys. Rev. Lett. **52**, 1033 (1984)

4.15   K.J. Måløy, J. Feder, T. Jøssang: Phys. Rev Lett. **55**, 2688 (1985)

4.16   M. Matsushita, H. Fujikawa: Physica A **168**, 498 (1990); T. Matsuyama, R.M. Harshey, M. Matsushita: Fractals **1**, 302 (1993)

4.17   S. Matsuura, S. Miyazima: Physica A **191**, 30 (1992); Fractals **1**, 336 (1993)

4.18   E. Ben-Jacob, O. Shochet, A. Tenenbaum, I. Cohen, A. Czirok, T. Vicsek: Nature **368**, 46 (1994); Fractals **2**, 15 (1994)

4.19   M. Eden: Proc. 4th Berkeley Symp. on Math. Stat. and Prob. **4**, 223 (1961)

4.20   R. Chandler, J. Koplik, K. Lerman, J.F. Willemsen: J. Fluid Mech. **119**, 249 (1982)

4.21   D. Wilkinson, J.F. Willemsen: J. Phys. A **16**, 3365 (1983)

4.22   B.B. Mandelbrot: *The Fractal Geometry of Nature* (Freeman, San Francisco 1982)

4.23   A. Aharony, in: *Advances on Phase Transitions and Disorder Phenomena*, ed. by G. Busiello, L. de Cesare, F. Mancini, M. Marinaro (World Scientific, Singapore 1987), p. 185

4.24   B.B. Mandelbrot: J. Fluid Mech. **62**, 331 (1974)

4.25   T.C. Halsey, P. Meakin, I. Proccacia: Phys. Rev. Lett. **56**, 854 (1986) and references therein

4.26   e.g., A. Aharony, in: *Phase Transitions and Critical Phenomena*, Vol. 6, ed. by C. Domb and M.S. Green (Academic Press, New York 1976), p. 357

4.27   M. Murat, A. Aharony: Phys. Rev. Lett. **57**, 1875 (1986)

4.28 U. Oxaal, M. Murat, F. Boger, A. Aharony, J. Feder, T. Jøssang: Nature **329**, 32 (1987)
4.29 P. Meakin, M. Murat, A. Aharony, J. Feder, T. Jøssang: Physica A **155**, 1 (1989)
4.30 Y. Meir, A. Aharony: Physica A **157**, 524 (1989)
4.31 E. Arian, P. Alstrøm, A. Aharony, H.E. Stanley: Phys. Rev. Lett. **63**, 2005 (1989)
4.32 R. Blumenfeld, A. Aharony: Phys. Rev. Lett. **62**, 2977 (1989)
4.33 J. Feder, E.L. Hinrichsen, K.J. Måløy, T. Jøssang: Physica **D38**, 104 (1989)
4.34 J.P. Hansen, J.L. McCauley, J. Muller, A.T. Skjektrop: in [4.8], p. 310
4.35 L.A. Turkevich, H. Scher: Phys. Rev. Lett. **55**, 1026 (1985)
4.36 D. Dhar: Phys. Rev. Lett. **54**, 2058 (1985)
4.37 M. Plischke, Z. Rácz: Phys. Rev. Lett. **53**, 415 (1984)
4.38 B.B. Mandelbrot: in [4.6], p. 3; see also [4.1,2]
4.39 D.E. Wolf, J. Kertesz: Europhys. Lett. **4**, 561 (1987)
4.40 M. Kardar, G. Parisi, Y. C. Zhang: Phys. Rev. Lett. **56**, 889 (1986)
4.41 D. Stauffer and A. Aharony: *Introduction to Percolation Theory* second edition, (Taylor and Francis, London 1992);
4.42 A. Aharony, in: *Directions in Condensed Matter Physics*, ed. by G. Grinstein, G. Mazenko (World Scientific, Singapore 1986), p. 1
4.43 P.L. Leath: Phys. Rev. B **14**, 5046 (1976)
4.44 Z. Alexandrowitz: Phys. Lett. **80A**, 284 (1980)
4.45 A. Kapitulnik, A. Aharony, G. Deutscher, D. Stauffer: J. Phys. A **16**, L269 (1983)
4.46 S. Havlin, R. Nossal: J. Phys. A **17**, L 427 (1984)
4.47 P.G. de Gennes, E. Guyon: J. Mech. **17**, 403 (1978)
4.48 R. Lenormand: C.R. Acad. Sci. Paris B**291**, 279 (1980)
4.49 L. Furuberg, J. Feder, A. Aharony, T. Jøssang: Phys. Rev. Lett. **61**, 2117 (1988)
4.50 R. Lenormand, C. Zarcone: Phys. Rev. Lett. **54**, 2226 (1985)
4.51 S. Roux, E. Guyon: J. Phys. A **22**, 3693 (1989)
4.52 L. Pietronero, H.J. Wiesmann: J. Stat. Phys. **36**, 909 (1984)
4.53 P. Meakin: Phys. Rev. A **27**, 604, 1495 (1983)
4.54 M. Muthukumar: Phys. Rev. Lett. **50**, 839 (1983)
4.55 C. Amitrano, A. Coniglio, F. di Liberto: Phys. Rev. Lett. **57**, 1016 (1986)
4.56 J. Nittman, H.E. Stanley, E. Touboul, G. Daccord: Phys. Rev. Lett. **58**, 619 (1987)
4.57 N.G. Makarov: Proc. London Math. Soc. **51**, 369 (1985)
4.58 P. Meakin, T.A. Witten: Phys. Rev. A **28**, 2985 (1983)
4.59 K.J. Måløy, F. Boger, J. Feder, T. Jøssang: in Ref. [4.7], p. 111
4.60 A. Arneodo, F. Argoul, E. Bacry, J.F. Muzy, M. Tabard: Phys. Rev. Lett. **68**, 3456 (1992)
4.61 B.B. Mandelbrot: Physica A **191**, 95 (1992)
4.62 W. von Bloh, A. Block, H.J. Schellnhuber: Physica A **191**, 108 (1992)
4.63 P. Ossadnik: Phys. Rev. A **45**, 1058 (1992)
4.64 R. Hegger, P. Grassberger: Phys. Rev. Lett. **73**, 1672 (1994)
4.65 M.J. Vold: J. Colloid Sci. **18**, 684 (1963)
4.66 D. Ben Simon, E. Domany, A. Aharony: Phys. Rev. Lett. **51**, 1394 (1983)
4.67 P. Meakin: in Ref. [4.5], p. 111
4.68 R. Ball, T.A. Witten: Phys. Rev. A **29**, 2966 (1984)
4.69 R. Jullien, R. Botet: *Aggregation and Fractal Aggregates* (World Scientific, Singapore 1987)
4.70 P.G. Saffman, G.I. Taylor: Proc. R. Soc. Lond. **245**, 312 (1958)
4.71 L. Paterson: Phys. Rev. Lett. **52**, 1621 (1984)
4.72 J. Chen, D. Wilkinson: Phys. Rev. Lett. **55**, 1892 (1985)
4.73 See also D. Wilkinson: Phys. Rev. Lett. **58**, 2502 (1987);
      M. Murat, A. Aharony: Phys. Rev. Lett. **58**, 2503 (1987)
4.74 B.B. Mandelbrot, J. Given: Phys. Rev. Lett. **52**, 1853 (1984)
4.75 J. Lee, H.E. Stanley: Phys. Rev. Lett. **61**, 2945 (1988)
4.76 B. Fourcade, A.-M. Tremblay: Phys. Rev. Lett. **64**, 1842 (1990)
4.77 R. Blumenfeld, A. Aharony: Phys. Rev. Lett. **64**, 1843 (1990)
4.78 A.B. Harris: Phys. Rev. B **39**, 7292 (1989) and unpublished;
      A.B. Harris, M. Cohen: Phys. Rev. A **41**, 971 (1990)
4.79 S. Schwarzer, J. Lee, A. Bunde, S. Havlin, H.E. Roman, H.E. Stanley: Phys. Rev. Lett. **65**, 603 (1990)
4.80 B.B. Mandelbrot, C.J.G. Evertsz: Nature **348**, 143 (1990)

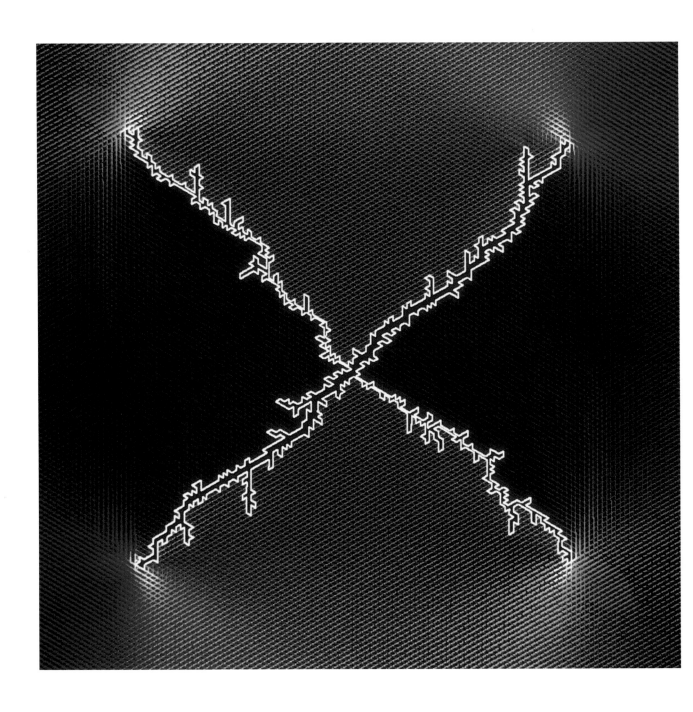

5 Fractures

Hans J. Herrmann

5.1 Introduction

The breaking of solids caused by an external load is evidently a problem of technological importance and has been intensively studied for the last hundred years [5.1]. On very small length scales ($\leq 10^{-6}$ m) fracture is a topic of materials science. From the electronic level [5.2] to the level of dislocations or grain boundaries [5.3] the mechanisms of fracture are highly material dependent. On very large length scales ($> 10^{-1}$ m) the prevention of fracture is studied by their engineers. There results are mainly based on experience and depend essentially on the application and the shape of the sample. On intermediate length scales the behavior of the solid can be described by the methods of applied mechanics, i.e., by continuous equations of motion. There exist on this level just a few types of different behaviors – for example elastic, plastic, viscoelastic – each given by its own set of differential equations containing some material-dependent parameters. The relatively general validity of the formalism makes the study of fracture in this intermediate (or mesoscopic) range of length scales particularly attractive to statistical physicists. If the reader wants to know more about recent developments in this direction I recommend [5.4].

In the present chapter we will restrict our attention to elastic media, i.e., brittle fracture, described by equations of motion valid on a mesoscopic level. We will show some novel concepts that statistical physics has contributed to the description of fracture. On one hand the introduction of disorder by considering material constants to be spatially random functions has been very useful. On the other hand, concepts in critical phenomena that have in the meantime be-

---

Fig. 5.0. Crack obtained under shear in a 200×200 triangular lattice using the central force model and having annealed disorder ($\eta = 1$). The crack consists of about 1000 bonds. The color coding is blue for compression and red for the stress field. Courtesy of A. Hansen, S. Roux and the Solid State Group of the University of Oslo

come standard, e.g., scaling, fractals and multifractals, have helped to formulate new, rather universal laws. For more on these subjects see Chaps. 1–4, 10. After a brief survey in the next section of the basic notions of elasticity, which should be skipped by experts, we consider the growth of one single crack. In Sect. 5.3, in analogy to fluid fingering, we formulate crack growth as a moving boundary problem. In Sect. 5.4 we present the various ways of modeling crack growth on a lattice. We show, in Sect. 5.5, how one obtains deterministic fractal cracks for stress corrosion. In Sect. 5.6 we introduce quenched disorder and show how the interaction of cracks results in the scaling of the breaking characteristics and multifractality in the local distribution of strain. Section 5.7 is devoted to some results on hydraulic fracture. We set down our conclusions in Sect. 5.8.

5.2 Some Basic Notions of Elasticity and Fracture

5.2.1 Phenomenological Description

In order to describe fracture one needs to know first how an unfractured solid responds to an externally applied force. For this purpose let us consider Young's experiment, i.e., a homogeneous block of size $\ell \times w^2$ subjected to a uniaxial force F. For small forces one expects a linear response (Hooke's law) of the type

$$\sigma = \frac{F}{w^2} = Y\frac{\Delta\ell}{\ell} = -\frac{Y}{\nu}\frac{\Delta w}{w}, \qquad (5.1)$$

where Y is called the Young modulus, ν the Poisson ratio, and σ the stress.

Usually, however, very small forces are not enough to fracture the solid. The behavior for an arbitrary force is given by the so-called constitutive law, shown schematically in Fig. 5.1. The linear regime, which is called *elastic*, is

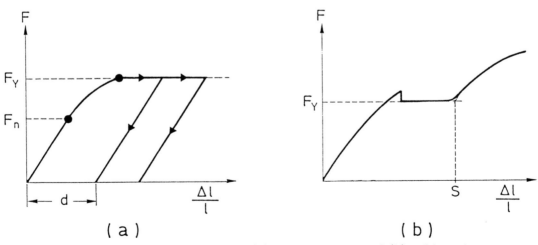

Fig. 5.1. Constitutive law for (a) ideal plasticity and (b) mild steel

valid up to a force F_n, beyond which one will find nonlinear deviations. As long as the force is less than the *yield point* F_Y the behavior is reversible, i.e., the original shape is recovered if the force is reset to zero. Beyond F_Y the behavior is *plastic*, i.e., the system deforms irreversibly so that when the force is reset to zero a finite elongation d remains. If, in the plastic regime, the material can flow without increasing the force, as shown in Fig. 5.1a, the behavior is called *ideal plasticity*, but most metals behave as shown in Fig. 5.1b, i.e., after an elongation S the material again has a finite toughness, called *strain hardening*. Plasticity, which in metals is due to dislocations and in polymers is due to chain reorientation, is described by a rather involved, nonlinear formalism. In this chapter we will treat only the linear, i.e., elastic, case and neglect plastic behavior.

In the current literature on fracture three families of materials are being studied fairly independently: metals, polymers, and, in a broad sense, "rock-alikes". The last family contains, among others, ceramics, concrete, tectonic plates, glass, and heavy clay. Typically, in this last family fracture occurs before the plastic regime sets in, which is called "brittle" fracture. In contrast, in the other two families, i.e., polymers and metals, fracture usually happens within the plastic regime, which is called "ductile" fracture. In reality materials are more complicated than stated above: many rocks or clays are in fact ductile and under some conditions metals are brittle. In particular one can find transitions from ductile to brittle behavior if one lowers the temperature or if one adds corrosive agents (corrosion induced cracking). We will only consider brittle fracture in this chapter.

Fracture strongly depends on the geometry of the sample and the direction of the externally applied load. In three dimensions one can actually distinguish three cracking modes, as shown in Fig. 5.2, depending on the direction of the applied stress. All real situations can be obtained as superpositions of these modes. These geometrical aspects are explained in great detail in the existing literature on fracture [5.5] so that in the following we will just consider some very simple and symmetrical geometries.

In many applications time dependence is important. First there can be a delay in the response of the system to a given force :

$$\sigma(t) = \frac{1}{\ell} \int_0^\infty C(\tau) \Delta \ell(t - \tau) d\tau, \tag{5.2}$$

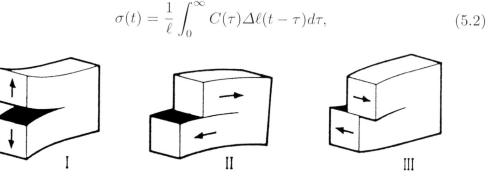

Fig. 5.2. The three modes of loading

which is called viscoelasticity, where $C(\tau)$ is a material-dependent delay kernel. This behavior, which is due to dissipation, introduces at least one characteristic time scale. Second, another time scale is given by the finite velocity of the elastic waves (which are usually waves on the crack surface, i.e., Rayleigh waves) that transport the elastic energy freed by the cracking process. These times must be compared to the speed of the crack growth. Only if crack propagation is very slow compared to the above processes can one neglect these time effects, and this is the limit we will assume in this chapter.

5.2.2 Elastic Equations of Motion

The formation and growth of a crack crucially depends on the local stress field. It is therefore important in the following to know the equations of motion of an elastic medium. A good variable to describe the medium is the displacement field $\boldsymbol{u}$, which gives the position of each volume element with respect to a reference position – usually the equilibrium position. Since a theory is required that is translationally invariant only the derivatives of $\boldsymbol{u}$, i.e., the tensor $(\partial_\alpha u_\beta)$, are going to appear. This tensor can be split into a symmetric part, called the strain $\varepsilon_{\alpha\beta}$, and an antisymmetric part. The antisymmetric part describes local rotations of a volume element. In a homogeneous medium a single volume element cannot be rotated with respect to the rest, so in this case only the symmetric part will be relevant. Porous or damaged media are not homogeneous and it is therefore useful to consider also the nonsymmetric contributions, known as Cosserat theory. We will return to this point later but first let us present the more common symmetric theory.

Since we want to describe linear elasticity the strain is given by

$$\varepsilon_{\alpha\beta} = \tfrac{1}{2}(\partial_\alpha u_\beta + \partial_\beta u_\alpha) \tag{5.3}$$

In order for the theory to be invariant under coordinate transformations it must depend only on the invariants $\sum_\alpha \varepsilon_{\alpha\alpha}$, $\sum_{\alpha\beta} \varepsilon_{\alpha\beta}\varepsilon_{\beta\alpha}$, and $\sum_{\alpha\beta\gamma} \varepsilon_{\alpha\beta}\varepsilon_{\beta\gamma}\varepsilon_{\gamma\alpha}$, if one assumes an isotropic medium. The energy of the theory must be a quadratic form of the strain if the theory is linear, and so it can be written as

$$E = \frac{1}{2}\left(2\mu \sum_{\alpha\beta} \varepsilon_{\alpha\beta}\varepsilon_{\beta\alpha} + \lambda \left(\sum_\alpha \varepsilon_{\alpha\alpha}\right)^2\right), \tag{5.4}$$

where μ and λ are positive dimensionless material constants called Lamé coefficients. They are in fact related to the Young modulus Y and the Poisson ratio ν of (5.1) through

$$\lambda = \frac{Y\nu}{(1+\nu)(1-2\nu)} \quad \text{and} \quad \mu = \frac{Y}{2(1+\nu)}. \tag{5.5}$$

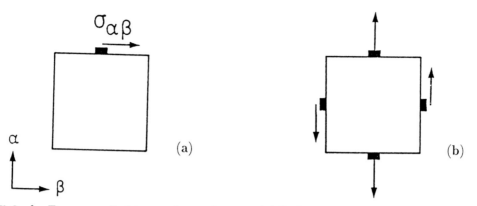

Fig. 5.3a,b. Forces applied to a volume element. (a) Definition of $\sigma_{\alpha\beta}$. (b) Balance of forces in the α direction

Since λ and μ are positive one finds $-1 \leq \nu \leq \frac{1}{2}$. The variable conjugate to $\varepsilon_{\alpha\beta}$ is the stress tensor $\sigma_{\alpha\beta}$, for which one obtains

$$\sigma_{\alpha\beta} = \frac{\partial E}{\partial \varepsilon_{\alpha\beta}} = 2\mu\,\varepsilon_{\alpha\beta} + \lambda\delta_{\alpha\beta}\sum_{\gamma}\varepsilon_{\gamma\gamma}. \tag{5.6}$$

This linear relation is the general form of Hooke's law. Physically $\sigma_{\alpha\beta}$ is the β component of the force that is exerted on the surface perpendicular to the α direction of a volume element, see Fig. 5.3a.

In order to obtain the equation of motion one must equate the inertial term with all the forces exerted on a volume element for each component β:

$$\rho\ddot{u}_{\beta} = \sum_{\alpha}\partial_{\alpha}\sigma_{\alpha\beta} + f_{\beta}, \tag{5.7}$$

where the first term on the r.h.s. is illustrated in Fig. 5.3b and where f_{β} are externally applied local forces. Since we do not consider time dependence the inertial term can be set to zero, so that in the absence of body forces f_{β} one has the vector equation

$$\nabla \cdot \bar{\bar{\sigma}} = \sum_{\alpha}\sigma_{\alpha\beta} = 0. \tag{5.8}$$

Inserting (5.3), (5.5), and (5.6) into (5.8) one obtains the equation of motion for the displacement field

$$\nabla(\nabla \cdot \boldsymbol{u}) + (1 - 2\nu)\nabla^2\boldsymbol{u} = 0, \tag{5.9}$$

which is called the Lamé or Navier equation. We see that this equation is just a consequence of the invariance under translations and coordinate transformations, imposing linearity and the homogeneity and isotropy of the medium.

In order to allow for local rotations in the medium, one considers in asymmetric (or Cosserat) elasticity [5.6] that each volume element has an additional angular variable φ. The strain is then defined as $\varepsilon_{\alpha\beta} = \partial_\alpha u_\beta - \sum_\gamma \in_{\alpha\beta\gamma} \varphi_\gamma$ where $\in_{\alpha\beta\gamma}$ is the totally antisymmetric tensor and one has in addition a torsion tensor $\kappa_{\alpha\beta} = \partial_\alpha \varphi_\beta$. Analogously to the symmetric elasticity, an energy can be constructed from invariants, and stress $\sigma_{\alpha\beta}$ and moment $\mu_{\alpha\beta} = \partial E / \partial \kappa_{\alpha\beta}$ can be defined as conjugated variables to strain and torsion. They fulfill the linear relations

$$
\sigma_{\alpha\beta} = (\mu + \alpha)\varepsilon_{\alpha\beta} + (\mu - \alpha)\varepsilon_{\beta\alpha} + \lambda\delta_{\alpha\beta} \sum_\gamma \varepsilon_{\gamma\gamma},
$$

$$
\mu_{\alpha\beta} = (\gamma + \delta)\kappa_{\alpha\beta} + (\gamma - \delta)\kappa_{\beta\alpha} + \eta\delta_{\alpha\beta} \sum_\gamma \kappa_{\gamma\gamma},
$$

(5.10)

with six material constants, of which α and δ describe the asymmetry. The equations of motion in equilibrium without internal body forces, called Cosserat equations, are then given by

$$
(\mu + \alpha)\nabla^2 \boldsymbol{u} + (\lambda + \mu - \alpha)\nabla(\nabla \boldsymbol{u}) + 2\alpha\nabla \times \boldsymbol{\varphi} = 0,
$$
$$
(\gamma + \delta)\nabla^2 \boldsymbol{\varphi} + (\beta + \gamma - \delta)\nabla(\nabla \boldsymbol{\varphi}) + 2\alpha\nabla \times \boldsymbol{u} - 4\alpha\boldsymbol{\varphi} = 0 \quad .
$$

(5.11)

We see that for $\alpha = 0$ one recovers the Lamé equation. In two dimensions $\boldsymbol{u} = (u_1, u_2, 0)$ and $\boldsymbol{\varphi} = (0, 0, \varphi)$ so that there are three independent variables per site determined by three equations of motion.

5.3 Fracture as a Growth Model

5.3.1 Formulation as a Moving Boundary Condition Problem

In order to study how a void in the material becomes a crack and how the crack grows it is useful to formulate fracture as a moving boundary problem in analogy to Laplacian growth as described in the previous chapter of this book. For simplicity we will do this here only for the case of symmetric elasticity.

Let us consider an elastic medium B subjected at the outer boundary A to some externally imposed displacement $\boldsymbol{u}_0$ (Fig. 5.4). Inside the medium one has a void (or crack) D so that on its surface C the elastic medium feels no restoring forces in the direction of the void. Consequently one has on the surface of the crack the boundary condition that the stress $\sigma_\perp$ perpendicular to the crack surface vanishes: $\sigma_\perp = 0$. As a result of (5.3) and (5.6), this is a condition on the derivatives of the displacement field. Since the Lamé equation is elliptic the two boundary conditions uniquely define one equilibrium solution for the displacement field.

The dynamics of crack growth can be described by the velocity v_n of the surface of the crack in the direction perpendicular to that surface. This velocity v_n depends on the microscopic rupture mechanisms of the material for which in general no explicit expression is known. It is probably not possible to derive such

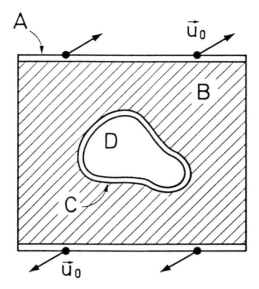

Fig. 5.4. Medium B with a void D of surface C. On the external boundary A a displacement u_0 is imposed

an expression from first principles or to find a material-independent form for it. Within the framework of symmetric elasticity the only nonzero first derivative of u on the boundary of the crack is $\sigma_{\|}$, the stress parallel to the surface of the crack. Therefore, to lowest order, v_n is a function $\mathcal{F}$ of $\sigma_{\|} - \sigma_c$ where σ_c is the cohesion force, which essentially means that one considers rupture through cleavage. In order to take into account the very common rupture due to bending, which occurs for instance when a cantilever is activated inside the material, one would have to take into account either higher derivatives of u or, better, asymmetric elasticity. For a concrete calculation one has to make assumptions about the explicit form of the function $\mathcal{F}$. In the following we shall use $v_n = c\left(\sigma_{\|} - \sigma_c\right)^{\eta}$, where η and c are material constants, and we will often set $\eta = 1$.

Let us consider the two-dimensional case. Then the moving boundary problem can be formulated as follows. The medium obeys the equation

$$\nabla\left(\nabla \cdot u\right) + (1 - 2\nu)\nabla^2 u = 0 \tag{5.12a}$$

with the boundary conditions

$$u_A = u_0 \tag{5.12b}$$

on the external boundary and

$$\begin{aligned} \partial_{\perp} u_{\|} + \partial_{\|} u_{\perp} &= 0, \\ (1 - \nu)\partial_{\perp} u_{\perp} + \nu \partial_{\|} u_{\|} &= 0 \end{aligned} \tag{5.12c}$$

on the internal boundary. The latter boundary, which is the crack's surface, moves with a velocity

$$v_n = c\left(\nu \partial_{\perp} u_{\perp} + (1 - \nu)\partial_{\|} u_{\|} - \sigma_c\right) \tag{5.12d}$$

normal to the surface when v_n is positive; otherwise it does not move. In (5.12c), which is identical to $\sigma_\perp = 0$, and (5.12d) we replaced stresses by displacements using (5.3) and (5.6). The symbols $\perp$ and $\parallel$ always denote the components perpendicular and parallel to the surface of the crack. Moving boundary problems similar to that of (5.12) appear in various contexts in physics, e.g., dendritic growth and viscous fingering, and are very difficult to treat mathematically, see also Chaps. 1, 4, and 10. This comes from the fact that after the boundary has moved, the internal boundary condition changes, changing the solution and therefore also the speed of the boundary. In other words a given solution is only valid during an infinitesimally short time and one must iteratively solve the equations and then move the boundary in principle an infinite number of times to see the precise development of the crack.

5.3.2 Linear Stability Analysis

Essentially two techniques have been applied to moving boundary problems: stability analysis and the numerical solution in discrete time steps. The stability analysis, which we discuss briefly in the following, considers only one iteration, i.e., what will happen after an infinitesimally short time. The numerical solution can be performed in various ways, as will be discussed in the next section.

Suppose one knows one particular, simple solution of the moving boundary problem in which the shape of the boundary does not change in time (stationary solution). The linear stability analysis consists in investigating in lowest order the effect of a small perturbation on this stationary solution. If the perturbation increases in time one has an instability which will give rise to more complicated but also more interesting solutions. Although these new solutions cannot be obtained explicitly one knows at least the motor driving its evolution, namely the instability. Some of these instabilities are very well known, such as the Saffman-Taylor instability in viscous fingering, the Mullins-Sekerka instability in dendritic growth, or, more recently discovered, the tip-splitting instability [5.7], which drives the growth of diffusion limited aggregates.

One type of stability analysis for fracture can be performed when the cohesive force is just compensated by an externally applied pressure inside the crack [5.8]. In this case all modes k of a perturbation of the form $\exp(ik\varphi + \omega t)$ on the radius of a circular hole will be unstable, i.e., their ω is positive and the perturbation increases with time.

Instead of considering a pressure inside the crack one can impose a radial traction at the outer boundary, so that the crack is like a hole in the stretched membrane of a drum. In that case all perturbations are stable [5.8].

The reason for this apparent contradiction to the linearity of elasticity theory is the strong nonlinearity at a finite cohesion strength σ_c given by (5.12d). A similar situation also appears in the Laplacian case when a threshold is introduced [5.9].

Another type of stability analysis for fracture can be applied to a sharp propagating crack tip [5.10]. The crack moves with a velocity v along the axis of an infinite strip that is pulled apart by imposing a displacement along its two borders. If the stress at the tip just overcomes the cohesion force the velocity v linearly increases with the excess strain. No velocity selection principle analogous to that found in dendritic growth is observed in this case.

5.4 Modelisation of Fracture on a Lattice

5.4.1 Lattice Models

A numerically tractable formulation of fracture is given by discretizing the continuum equations. In this case the medium is reduced to a set of points embedded in a grid. Only local laws, like the balance of forces and momenta, are considered and their implementation involves only a few neighbors for each point. Mathematically the calculation of collective properties such as the equilibrium displacement field is then reduced to numerically solving a set of coupled linear equations.

Evidently these methods do not pretend to describe nature on an atomic level, as for instance molecular dynamics does, but have their validity at much larger length scales where the medium can be described by continuous vector fields. Therefore one does not have to bother about realistic interatomic potentials because the elastic equations of motion are good enough. The behavior of the solid will be reproduced even quantitatively by these models if the (phenomenological) material constants, such as elastic moduli, yield thresholds, and relaxation kernels, are properly inserted in the equations of motion. On the other hand, the breaking of the lattice is not a natural consequence of the simulation but has to be put into the model by hand as an additional rule of the model. This rule can be based on experimental data or on phenomenological laws. Several examples will be discussed. One big advantage of lattice models is that they allow disorder to be introduced very naturally.

In engineering practice the most commonly used lattice models are finite element methods (FEMs). We do not want to discuss FEMs here in more detail because many textbooks [5.11] have been written on the subject. The implementation of a FEM is usually done using elaborate, commercially available computer codes. These codes are used like black boxes rather than being considered to have a physical interest of their own. The models that we want to discuss in the following are less sophisticated than FEMs, although some of them could be considered to be special cases of models situated within the framework of FEMs.

All the models we will consider here have a certain common framework which can be summarized in the following way. The variables that characterize the medium (the electric field for dielectric breakdown, the displacement vector and possibly a local rotation angle for elasticity) are placed on the sites of a lattice of N sites. The equations that describe the medium (e.g., Lamé or

Cosserat for elastic models) are discretized such that for each site one has one equation per variable which only involves m variables on the z neighbor sites. For example, discretizing the Laplace equation for an electric potential in this way, one would obtain the Kirchhoff equations. The continuous equations of the medium are therefore transformed into a set of $m \times N$ coupled linear equations. Since only nearest neighbors are involved in each of the equations a solution of the set of equations only involves the inversion of a sparse matrix. The boundary condition of the outer boundary on which the externally imposed constraint is applied is explicitly implemented. The boundary conditions on the internal boundary, namely the crack surface, are automatically fulfilled by the response of the bonds that constitute the crack, which will be discussed later.

After the boundary conditions have been implemented, the set of equations has a unique solution. Numerically there are many methods to find this solution and the method that is chosen for a particular case depends on the requirements one has on the solution. If very high precision is needed because, for instance, the local distribution of strain is of interest, a conjugate gradient method, perhaps speeded up by Fourier acceleration [5.12], seems ideal. If one wants to study dynamic effects such as viscoelasticity one can use Jacobi's method and give a physical sense to a relaxation step. If one does not need precision but wants to save computer time simple over-relaxation gives a fair compromise.

As already mentioned, the simulation of a rupture process must be done in an iterative way: the equations must be solved in order to determine which bond should be broken, but once the bond has been broken the (internal) boundary condition and therefore also the solution of the equations are changed. Consequently the equations must be solved again if one wants to know which bond to break next and so on. We see that the process is rather cumbersome because the set of linear equations must be solved as many times as bonds are broken. With the present computational means one can break one or two thousand bonds in a lattice of roughly 10^4 sites with reasonable computer effort (1–2 hours Cray, Y-MP).

The algorithm performed for each iteration can be decomposed into five steps:

1. The set of equations is solved.
2. The set of all the bonds that are eligible to be broken is determined.
3. For each bond of this set a certain quantity p is calculated using the solution of the equation on the sites adjacent to the bond.
4. According to a rule that depends on p one chooses, out of the set, one bond to be "broken".
5. The bond is broken, i.e., its elastic modulus is changed in the equations.

Each of these steps allows a large variety of options that can take into account many possible physical situations: step 1 describes the nature of the medium and the externally applied constraints, step 2 the describes connectivity of the crack, step 3 and step 4 describe the breaking rule and the disorder, and step

5 allows a residual strength to be incorporated. In the following we will discuss each of the steps in more detail.

5.4.2 Equations and Their Boundary Conditions

The problem of the breakdown of a network of electrical fuses [5.13] is very similar to fracture but with the vectorial displacement field $\boldsymbol{u}$ replaced by the scalar electrical potential ϕ. Since the scalar nature of the problem simplifies numerical and analytical work this analogous system is often studied too. In fact, in two dimensions, the "dual" [5.4] of the fuse problem is dielectric breakdown [5.14].

In scalar models one studies the Laplace equation $\nabla^2\phi = 0$, or in a more general form the equation $\nabla(\varepsilon\nabla)\phi = 0$. In the electric case ε is the material-dependent dielectric constant (or tensor). In principle, ε can vary from site to site, modeling spatial inhomogeneities of the material. Choosing a random function [5.15] for ε therefore constitutes a way of introducing disorder, but we will discuss this point in more detail later. In most cases ε is considered to be a constant so that the equation simplifies to the Laplace equation.

Until now vectorial models have been considered only for elasticity, although one could also imagine applying them to electrodynamics, fluids, magnetohydrodynamics, etc. So, only equations for the displacement field have been investigated. Unlike the scalar case, there are at least two possible sets of equations for elasticity, namely the Lamé equations (5.9) and the Cosserat equations (5.11). In addition, various rather different discretizations for the equations have been used: the central force model [5.16], the bond-bending model [5.17], and the beam model [5.18].

These models will be briefly discussed next. The beam model is a rather straightforward discretization of the Cosserat equations. In two dimensions there are, according to the remark after (5.11), three continuous degrees of freedom on each site i: the two coordinates x_i and y_i, and $z_i = l\vartheta_i$, where $\vartheta \in [-\pi, \pi]$ is the rotation angle and the lattice spacing l is usually set to unity (see Fig. 5.5b). Nearest-neighbor sites are connected through a "beam" in such a way that it joins site i forming an angle ϑ_i with respect to the underlying undistorted lattice. In Fig. 5.5a we see the form of the beams when just the site in the middle is rotated. Each beam is assumed to have a finite cross section A so that besides a traction force f along its axis one also has a shear s and moments m_i and m_j at its two ends i and j. These can be calculated [5.19] in the linear (elastic) approximation. For a horizontal beam one obtains in two dimensions

$$f = a(x_i - x_j), \tag{5.13a}$$

$$s = b\left(y_i - y_j + \tfrac{1}{2}(z_i + z_j)\right), \tag{5.13b}$$

$$m_i = cb(z_i - z_j) + \tfrac{1}{2}b\left(y_i - y_j + \tfrac{2}{3}z_i + \tfrac{1}{3}z_j\right) \tag{5.13c}$$

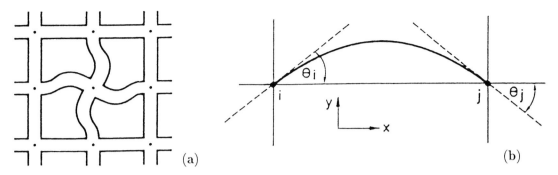

Fig. 5.5a,b. Schematic representation of the beam model. (a) Rotation of one site in the center of the lattice. (b) A beam is flexed due to rotations at its ends

where $a = YA$, $b = ((GA)^{-1} + (l^2/12)(YI)^{-1})^{-1}$, and $c = YI/l^2GA$ with a shear modulus G and a moment of inertia I. Analogous expressions are found for a vertical beam. To obtain three discretized equations for each site one imposes that the sum of horizontal and vertical force components and the sum of the moments must be zero at this site.

In the central force model each bond behaves like an elastic spring that can freely rotate around the site to which it is attached. The bond-breaking model is obtained from there by in addition reinforcing the angles between two bonds with another spring. Both models are believed also to describe Cosserat elasticity on very large scales. The central force model is simpler but it shows pathological behavior on certain lattices and has very soft modes, particularly on weakly connected graphs. Sometimes also hybrid models, such as the Born model [5.20], have been considered, which mix scalar and vectorial terms, but their physical significance is not very clear.

Elastic models allow a large variety of externally imposed constraints. At least five different types of external boundary conditions have been considered for fracture: shear [5.21] (Fig. 5.6a), uniaxial tension (Fig. 5.6b), uniaxial com-

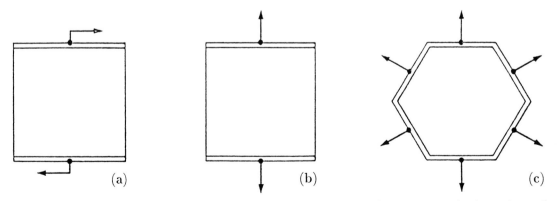

Fig. 5.6a–c. Three common ways of imposing an external displacement on the boundary of an elastic medium; (a) shear, (b) uniaxial tension, and (c) radial dilation. (a) and (b) are square lattices with periodic boundary conditions in the horizontal direction and (c) is a triangular lattice

pression, radial dilation [5.21] (Fig. 5.6c), and uniform dilatation [5.22]. Radial dilation can be considered to mimic a membrane spanned on a drum, while uniform dilation was conceived to describe the drying of a layer of paint on a surface. In this last case each site of the lattice is connected by a spring to another underlying lattice, representing the surface, which has a somewhat larger lattice constant than the original lattice. In this way all the bonds are equally overstretched without any force being applied on the boundary. In fact, no external boundary is needed in this case so that periodic boundary conditions can be considered in all directions. We have only discussed two-dimensional geometries since no serious three-dimensional work has been performed at present because of the numerical difficulties involved.

5.4.3 Connectivity

Basically, one can pursue two different points of view when studying crack propagation: either one wants to understand how *one* crack grows or one wants to see how cracks appear in a stressed medium, coalesce and break the system.

The first approach fits into the framework of cluster growth models and has similarities to dielectric breakdown [5.14] or viscous fingering. It can be cast into the formalism of a moving boundary condition problem as discussed in the previous section. In order to implement this approach in the general algorithm of lattice models described above one must impose the connectivity of the growing crack by allowing only bonds on the surface of the crack to be broken. This constraint, is particularly restrictive at the early stages of crack growth. The last stages of fracture are determined by the propagation of just one single crack and in this case a one-crack model can be justified; last-stage cracks are, however, usually rather fast and straight so that, on one hand, they are numerically difficult to handle by lattice models and, on the other, they do not generate very interesting, i.e., ramified or fractal, structures.

If the object is to study the growth of just one crack with a lattice model, one must determine which bonds are at the surface of the crack, i.e., the set of bonds eligible for being broken. The surface of a crack is, however, not uniquely defined on a lattice and so an arbitrary convention of *connectivity* must be made which states which bonds in the vicinity of a given broken bond i would, if broken, constitute together with i a connected crack. For instance, on the triangular lattice three different types of connectivities have been investigated [5.23], in which one broken bond has 4, 10, and 18 neighboring connected bonds, respectively, as seen in Fig. 5.7.

In the many-crack approach no restriction on connectivity is made so that at each time step all not-yet-broken bonds can in principle be broken. This approach corresponds to the majority of experimental situations.

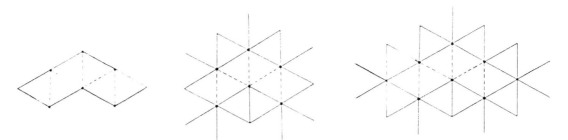

Fig. 5.7. Three possible definitions of connectivity on the triangular lattice. Broken bonds are marked by dashed lines. From [5.23]

5.4.4 The Breaking Rule

The subtleties of the microscopic physics of rupture are contained in the breaking rule. As observed before, the breaking rule has to be put into the model by hand, reflecting the phenomenological nature of lattice models. The freedom one has in the choice of the breaking rule and its parameters, however, allows careful exploration of the impact of a given microscopic mechanism on the macroscopic behavior of the system. In order to gain insight into the collective phenomena of rupture it is useful to consider particularly simple breaking rules, but one can also take an engineering point of view and try to model the material as realistically as possible. We focus on the first approach.

The effect of disorder on the breaking process is very important and for this reason its implementation within the breaking rule should be discussed in some detail. Real materials can exhibit the most diverse types of disorder, ranging from small deviations of the crystalline structure due to vacancies to the large-scale heterogeneities of composites. This disorder can also be in motion: dislocations can migrate, microcracks can form and heal, interstitials can diffuse, and so on. For our purposes the fundamental distinction is whether or not this motion is much slower than the investigated process (i.e., fracture). In the first case the disorder is called "quenched" and can be considered time independent. In the second case, called "annealed", the interplay of disorder fluctuations and fracture has to be taken into account.

On the mesoscopic length scales that we want to describe one can essentially reduce the description of disorder to a spatial dependence of the local density, elastic modulus or strength. In the case of quenched disorder this spatial dependence is fixed once and for all before the rupture process starts. Since we ignore the microscopic details and since we adhere to simplicity the spatial dependence is chosen to be random according to some probability distribution. This distribution contains (among other things) information about the degree of disorder, so that by considering various distributions the effects of disorder on rupture can be quantified.

In the context of lattice models the individual bond is characterized by a constitutive law, which in the simplest case should be linear up to a breaking threshold value as depicted in Fig. 5.8. The strength of the bond is given by the threshold value and its conductivity or elastic modulus by the slope in Fig. 5.8. The presence of disorder can now easily be implemented by allowing the bonds of the same lattice to have different constitutive laws. Strength fluctuations can be represented by choosing the threshold value for each bond randomly [5.24]. Spatial variations in the elastic moduli can be described by randomly choosing different slopes in Fig. 5.8 for each bond [5.15].

The most prominent example of spatial density fluctuations is that of porous media, where a given point in space is either solid or empty. This situation can be modeled by randomly removing a bond of the lattice with a probability $q = 1 - p$ before the breaking process starts. This approach has been extensively studied for electrical [5.13,25] and mechanical [5.26] networks. Particularly interesting is the behavior of these dilute systems close to the percolation threshold p_c (see Chaps. 2 and 3).

Although in the case of dilution the probability distribution is binary, i.e., the random numbers are either one or zero, the strength and conductivity fluctuations are generally chosen to be continuously distributed. Most often a uniform distribution,

$$P(x) = 1 \quad \text{with} \quad 0 \leq x \leq 1, \tag{5.14}$$

is chosen in this case, but power-law distributions,

$$P(x) = (1 - \alpha)x^\alpha \quad \text{with} \quad 0 \leq x \leq 1 \quad \text{and} \quad \alpha > -1, \tag{5.15}$$

and Weibull distributions,

$$P(x) = mx^{m-1}e^{-x^m} \quad \text{with} \quad 0 \leq x \leq 1, \tag{5.16}$$

have also been considered [5.27]. The constants α and m are parameters of the

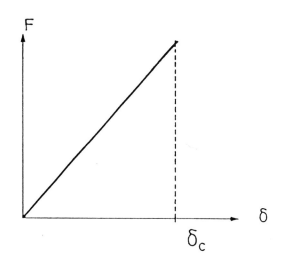

Fig. 5.8. The constitutive law (force F against displacement δ) chosen for each bond of a lattice model. The breaking threshold is at δ_c

distribution which determine how strong the disorder is. Distributions with a lower cut-off x_l, i.e., for which $0 < x_l < x$, are also of interest [5.24]. It is not easy to distinguish experimentally between the different types of distributions. For a Weibull distribution, values of $2 \leq m \leq 10$ have been fit to data measured on rocks.

In quenched systems the random variables are fixed at the beginning and the following rupture process is completely *deterministic*. To break a bond the equations are solved and the bond for which the ratio strain over threshold δ_c is largest is chosen to be broken. In contrast to this, annealed disorder requires the production of random numbers during the entire evolution of fracture, and the process is therefore *probabilistic*. It is not easy to know how to cast a given microscopic dynamical effect into an explicit expression for such a probabilistic breaking rule. If the dominant time dependence of the disorder stems from thermal fluctuations it seems reasonable to use a Boltzmann factor. So, one approach [5.22,28] is to calculate, before each breaking, the elastic energy ϵ_i of each bond i and from these $p_i = e^{-\beta\epsilon_i}$, where $\beta = 1/k_BT$. The bond to be broken, i_0, is then chosen with a probability proportional to its value p_{i_0}. This means that a random number z is chosen (uniformly distributed) between 0 and $\sum_{i=1}^N p_i$ and then the i_0 is determined for which $\sum_{i=1}^{i_0-1} p_i \leq z \leq \sum_{i=1}^{i_0} p_i$ is fulfilled. Instead of the energy ϵ_i other quantities have also been considered [5.29].

Since thermal fluctuations are not the only source of annealed disorder and since a first-principles derivation of the probabilistic breaking rule is not possible at present, another approach, inspired by dielectric breakdown models [5.14], has been taken. The bond to be broken is chosen with a probability p_i proportional to its elongation δ [5.23,30]. This can be generalized to $p_i \propto \delta^\eta$. The exponent η is phenomenological and can be adjusted to fit experimental results. In general η is a "relevant" parameter, i.e., the fractal dimensions depend continuously on η.

Unlike simple springs the elastic beams can break in various modes. A rubber band will tear apart when stretched but a glass rod will crack when bent. Try, instead, to break a rubber band by bending it or a glass rod by stretching it! In three dimensions a third mode of rupture appears, namely torsion.

Mechanical breaking rules should reflect these various rupture modes. This can be achieved by defining the quantity p that determines the breaking as a sum of terms, each corresponding to one mode. In the quenched case, for instance, one can consider that the beam that breaks is the one for which

$$p = \left(\frac{f}{t_f}\right)^2 + \frac{\max(|m_1|, |m_2|)}{t_m} \tag{5.17}$$

is largest [5.31]. In this expression, which is inspired by the *von Mises yielding criterion* [5.5], f is the force stretching the bond, m_1 and m_2 are the moments acting at the two ends of the bond, and t_f and t_m are two material-dependent threshold values. The first term in (5.17) describes breaking due to stretching and the second term takes into account the bending mode. If one has quenched disorder in the strength of the bonds the threshold values t_f and t_m are chosen randomly and one can in fact have for t_f a distribution between 0 and 1 while t_m may be distributed between 0 and r. In the case of (5.15) one would then have

$$P(t_f) = (1 - x)t_f^x \quad \text{with} \quad 0 < t_f \leq 1, \tag{5.18a}$$

and

$$P(t_m) = (1 - x)r^{x-1}t_m^x \quad \text{with} \quad 0 < t_m \leq r. \tag{5.18b}$$

Here r measures the susceptibility of the material to breaking through bending as compared to stretching. In the case of annealed disorder one can use

$$p = \left[f^2 + r \cdot \max(|m_1|, |m_2|) \right]^\eta \tag{5.19}$$

as the quantity to which the breaking probability is proportional.

Damage can be modeled within the breaking rule by lowering the breaking threshold of bonds in the vicinity of the crack by an amount proportional to the strain, but without breaking them. Bonds damaged in this way are more likely to break at one of the next iteration steps. Short-lived damage can be described by considering at iteration step t the expression [5.32]

$$p' = p(t) + f_0 p(t - 1), \tag{5.20}$$

where $p(t)$ is the usual quantity one would have used at step t without damage and f_0 is a parameter. For $p(t)$ one could for instance take expression (5.17) or (5.19). The second term in (5.20) simulates the memory effect due to damage that occurred at the previous iteration step and f_0 controls the strength of this memory. The quantity p' can be used as before either for quenched disorder by breaking the bond for which p' is largest or for annealed disorder by breaking a bond with probability proportional to p'.

Remanent damage occurring, for instance, in stress corrosion is an accumulation of the damage that has occurred during the entire breaking process. It can be modeled by putting counters $c(t)$ on the bonds susceptible to damage, for example the bonds on the surface of the crack. At the beginning, all counters are set to zero. At each iteration step t, after having calculated $p(t)$, one defines $\alpha(t) = [1 - c(t - 1)]/p(t)$ and chooses the bond which has the smallest $\alpha(t)$, namely $\alpha_{\min}$. In the quenched case this is the bond one will break. Then all the counters are updated using [5.32]

$$c(t) = \alpha_{\min} p(t) + f c(t - 1). \tag{5.21}$$

We see that $c(t)$ sums up the damage from all previous iteration steps, where the memory factor f controls how strongly previous damage remains. If $f = 1$ the damage is irreversible. The case $f \to 0$ corresponds to the criterion of (5.20) with $f_0 = 1$. These models will be discussed further in the next section.

5.4.5 The Breaking of a Bond

Once the bond to be broken has been chosen, its actual removal or replacement has to be implemented. In the case of fracture (or fuse networks) one can simply take the bond out, i.e., remove the terms corresponding to this bond from the set of equations describing the medium. This is equivalent to setting the elastic modulus of the bond to zero. If one studies dielectric breakdown [5.14] (or DLA [5.33]), which in two dimensions is the dual of a fuse network, one will instead set the conductivity of the broken bond to infinity (superconductivity). For more details on DLA, see Chaps. 1, 4, and 10.

It is, however, also possible to consider effects of residual strength by reducing the elastic modulus of the bond by a certain amount instead of setting it to zero. In this case the interesting phenomenon of crack arrest has been observed [5.34].

5.4.6 Summary

We have seen that a large variety of lattice models can be defined, many of them describing realistic situations to a certain degree. Is there a standard model which contains the relevant features? The answer is yes and no. If one wants to study just the typical mechanisms of crack dynamics of brittle fracture, random fuse models [5.13,24,25], which will be discussed in Sect. 5.6 are probably the simplest and best-understood candidates. But if one wants to explain qualitatively a particular experiment this is not enough, because of various effects that considerably influence real fracture. To understand, for instance, the experiments performed on clay in the previous section, the connectivity of the crack is essential. Or, if one wants to reproduce the crack patterns of drying paint, the vectorial nature of elasticity is necessary. It is consequently the richness of the phenomenology of fracture mechanics that forces us to consider different lattice models.

5.5 Deterministic Growth of a Fractal Crack

Let us next describe a specific example of the growth of one connected crack, but without including any disorder yet. We consider a finite square lattice of linear size L with periodic boundary conditions in the horizontal direction. On the top and on the bottom an external shear is imposed. We remove one beam in the center of the lattice, which represents the initial microcrack. Next we consider the six nearest-neighbor beams of this broken beam. These include the two beams that are parallel to the broken beam and the four perpendicular beams that have an end site in common with the broken beam. This choice of nearest-neighbors comes from the fact that the actual crack consists of the bonds that are dual to the set of broken beams. The beam model is solved by a conjugate gradient method [5.12] to very high precision (10^{-20}) and the ps of (5.19) are calculated for each of the nearest-neighbor beams. We set $p = 0$ for a beam that is not a nearest-neighbor to the crack. Now various criteria for breaking are possible. (I) One breaks the beam with the largest value of p. (II) One breaks the beam for which $p' = p(t) + f_0 p(t-1)$ is largest, see (5.20). (III) On each beam of the lattice we put a counter $c(t)$ which is set to zero at the very beginning and updated according to (5.21). Then one breaks the beam for which $\alpha(t) = [1 - c(t-1)]/p(t)$ is smallest.

All three breaking criteria described above are deterministic. Criterion III corresponds for $f = 1$ to the limit of infinite noise reduction [5.35], which was invented to reduce statistical noise in DLA simulations. Noise reduction has also been applied recently to central-force breaking [5.36]. Physically the three breaking criteria defined above correspond to three different situations. Criterion I describes ideally brittle and fast rupture. Criterion II contains the short-term memory one would expect in cracks that produce strong local deformations at the tip of the crack, as happens in most realistic situations, and where this local damage heals after some time. Criterion III could be applied to situations of stress corrosion or static fatigue. The memory factors f_0 and f measure the strength of these time correlations. In criterion III the limit $f \to 0$ gives criterion II with $f_0 = 1$, and in criterion II the limit $f_0 \to 0$ gives criterion I.

Let us next discuss the results [5.32] for the above models. If the beams break according to criterion I, as shown in Fig. 5.11 for $L = 50$ and $r = 0.28$, cleavage tends to make the crack grow in the diagonal direction while the bending mode favors a horizontal rupture. The competition between these two effects can lead to complex branched structures. The exact shape of these cracks strongly depends on r and the system size. For any finite r the horizontal

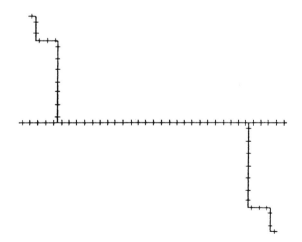

Fig. 5.9. Crack grown in a 50×50 system with an external shear applied. Beams break either under traction or with an affinity of $r = 0.28$ in the bending mode. Only the bond with the largest value of p breaks (criterion I). The first broken beam was vertical. (From [5.32])

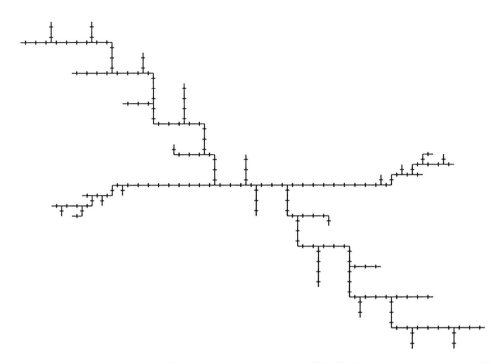

Fig. 5.10. Crack grown under the same conditions as in Fig. 5.9 except that criterion II with $f_0 = 1$ was used instead of criterion I

rupture will eventually win if the system is large enough, while for $r = 0$ one obtains diagonal cracks, possibly with kinks. For this reason the cracks will not be fractal. In the analogous scalar model (i.e., dielectric breakdown), however, only straight lines will be formed in criterion I; the different behavior here is due to the fact that competing directions are possible in a vectorial model.

Let us now consider cracks grown using criterion II. We see in Fig. 5.10 a crack grown under the same conditions as in Fig. 5.9, except that criterion II with $f_0 = 1$ was used instead of criterion I. A branch of a larger crack is seen in the lower part of Fig. 5.11 and one clearly recognizes a self-similar structure.

If we count the number of broken beams inside a box of length l around the first broken beam and plot it as a function of l in a log-log plot ("sand box method") we find straight lines with slopes larger than unity, which means that the cracks are fractals. In system sizes of $L = 118$ we find for the fractal dimensions d_f values that depend on η: $d_f = 1.3$ for $\eta = 1.0$, $d_f = 1.25$ for $\eta = 0.7$, $d_f = 1.15$ for $\eta = 0.5$, and $d_f = 1.1$ for $\eta = 0.2$. The structures are self-similar around the origin and are likely to be directed fractals [5.37]. Changing the elastic constants (i.e., the Lamé coefficients) just ·changes the opening angle of the crack but not d_f. If one uses an exponential instead of a power law in (5.19), the structures seem to be dense.

The case of criterion II giving fractal structures is very different from what is seen in the scalar case of dielectric breakdown. It shows that neither noise nor long-range time correlations are necessary in order to obtain fractals. The

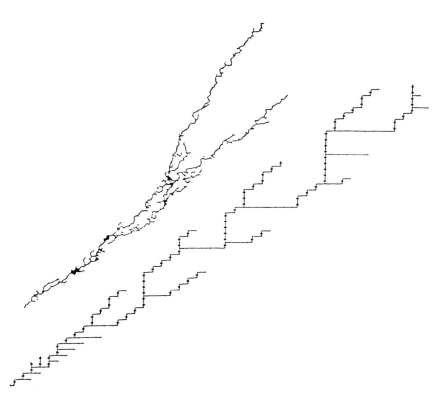

Fig. 5.11. Numerical and experimental cracks. The lower shape is the upper half of a crack grown in a 118×118 system under external shear with $r = 0$ and vertical first-broken beam using criterion II with $f_0 = 1$ and $\eta = 0.2$. The lower picture shows the morphology of cracking in Ti–11.5 Mo–6 Zr–4.5 Sn aged 100 hrs at 750K and tested in 0.6M LiCl in methanol at 500mV under increasing stress intensity. (From [5.40])

origin of fractality is the competition between a global stress perpendicular to the diagonal and a local stress that tends to continue along a given straight crack due to tip instability. Again we see the important role of the interplay of different directions, which is only possible in a truly vectorial model. The significance of a short-term memory in criterion II indicates that there might be a relation between this case and the models that have been put forward for snowflakes [5.38].

In Fig. 5.12 we show a crack grown using criterion III for $r = 0$, $f = 1$, $\eta = 1$, and $L = 118$. The physical situation is similar to that seen for criterion II, except that the fractal dimension is a little higher. This case can be directly compared to results obtained for DLA in the limit of infinite noise reduction [5.33,39], where needles not fractals are predicted (see Chap. 1).

In Fig. 5.11 we compare a deterministic crack with an experimental example of stress corrosion cracking in an alloy [5.40]. The fractal dimension of the crack obtained for different alloys has been measured using a "sand-box" method for digitized images, and a rather universal value of $d_f \approx 1.4$ was obtained [5.41] which would correspond to $\eta = 1.3$ in our model. It also seems clear that the inhomogeneities of the medium are important during the growth of the

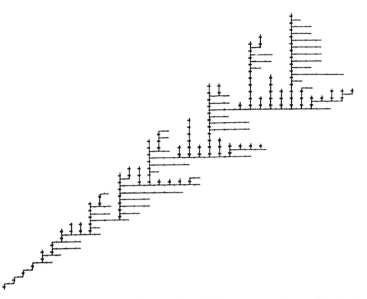

Fig. 5.12. Upper half of a crack grown in a 118×118 system using criterion III with $f = 1$, $\eta = 1$, $r = 0$, and vertical first-broken beam when the beams break only under traction

experimental crack so that disorder should be included to model real cracks. For one-crack models a probabilistic approach similar to the one used for DLA [5.33] has been applied in the central force model [5.21,23,30]. The cracks obtained in this way seem to be fractal with a fractal dimension d_f that depends on the external boundary condition. Roughly speaking, in two dimensions $d_f \approx 1.6$ has been found under dilation (i.e., if the system is pulled radially), $d_f \approx 1.2$ was obtained under uniaxial tension, and $d_f \approx 1.4$ under shear. The latter case, however, may just be a superposition of two clusters of dimension 1.2 [5.30]. Unfortunately, the clusters from which these data were measured are nowhere nearly as large as state-of-the-art DLA clusters. This is mainly due to the fact that no algorithm using random walkers has yet been implemented for elastic problems, although in principle the Lamé equations can be described by a stochastic process [5.42]. Progress in this direction has been reported recently in [5.43].

5.6 Scaling Laws of the Fracture of Heterogeneous Media

The models discussed in the last section produced one connected cluster, which in most cases is rather unrealistic. In fact, the possibility of having many cracks attracting or screening each other is crucial in fracture if one is interested in a quantitative comparison with real forces or displacements. In the following we will consider a finite two-dimensional lattice $L \times L$ with periodic boundary conditions in the horizontal direction and fixed bus bars on top and bottom to which the external strain (elongation or shear) will be applied. Each bond is supposed to be ideally fragile, i.e., to have a linear elastic dependence between the force f and displacement δ with unit elastic constant up to a certain threshold force f_c where it breaks (see Fig. 5.8). We will discuss a model [5.24] in which the thresholds are randomly distributed according to some probability distribution $P(f_c)$. Once a force beyond f_c is applied to a bond, this is irreversibly removed from the system. As the external strain is increased one can watch bonds breaking one by one until the system falls apart altogether.

For the beam model we introduce two random thresholds t_f and t_m and break the beam for which the p of (5.17) is largest. We distribute the thresholds according to (5.18). If the bonds are electrical fuses [5.27] or springs which can only be subjected to a central force [5.44], the breaking criterion is to choose the bond with the maximum value of $|i|/i_c$ where i is the current or the central force, respectively. The threshold is again randomly distributed and for the electrical case we will also consider a Weibull distribution as given in (5.16).

Each time a bond is broken we monitor the external force F and the external displacement λ, both averaged for a fixed number n of bonds cut. The relation between the two gives the breaking characteristics of the entire system as shown in Fig. 5.13 for the beam model. We see that, unlike the single bond, which was ideally elastic, the macroscopic characteristic is nonlinear and after a maximum force F_b has been applied, the system can still be elongated considerably before becoming disconnected, a regime that is experimentally accessible only if a displacement rather than a force is imposed. After reaching the maximum, the

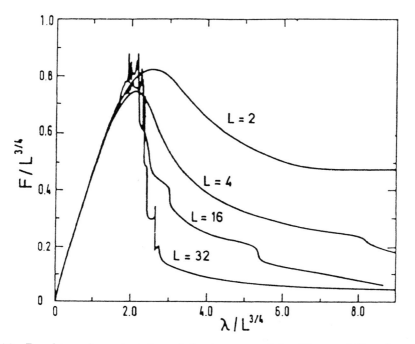

Fig. 5.13. Breaking characteristics of the beam model with $x = 0.5$ and $r = 1$ with both axis scaled by $L^{-3/4}$ for different sizes L. The data have been smoothed to reduce statistical fluctuations. (From [5.31])

breaking characteristic is subject to strong statistical fluctuations. For weak disorder ($m > 2$) the characteristic even bends back, i.e., both F and λ decrease. This behavior is called "class II" in rock mechanics and can only be measured with very stiff and quick, servo-controlled testing machines.

Before the maximum is reached, there is an initial regime with fewer statistical fluctuations and dominated by the disorder. This regime shrinks for decreasing disorder. In this regime Fig. 5.13 verifies the scaling law

$$F = L^\alpha \phi(\lambda L^{-\beta}) \tag{5.22}$$

with $\alpha \approx \beta \approx 0.75$. This law can be checked for all three models [5.27,31,44], for all distributions ($0.8 \leq x \leq -1, \quad 2 \leq m \leq 5$), and for both external extension and shear (in the elastic case) with exponents that agree with 3/4 to within 5%–10%. For the same range of forces one finds the following scaling law for the number n of bonds cut:

$$n = L^\gamma \psi(\lambda L^{-\beta}) \tag{5.23}$$

with $\gamma \approx 1.7$ and the same universal range of validity as for (5.22).

When the force reaches the maximum, the number n_b of bonds that have been cut, scales again for most cases as $n_b \sim L^{1.7}$, as seen in Fig. 5.14a. Only

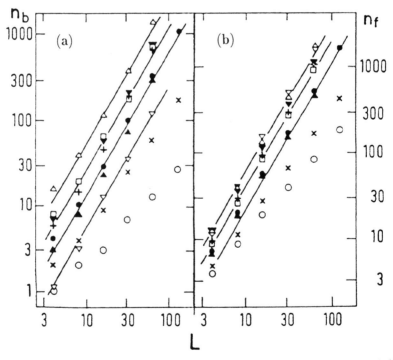

Fig. 5.14. Number of bonds cut (a) when the maximum force is reached and (b) when the system falls apart in a log-log plot against L for the scalar model with $x = 0(\square)$, $x = 0.5(\triangle)$, $m = 2(\circ)$, $m = 5(\times)$, and $m = 10(\circ)$, and for the beam model ($r = 1$) for an external elongation with $x = 0(+)$, $x = 0.5(\triangledown)$, $x = -1(\triangle)$, and for an external shear with $x = 0(\triangledown)$. The full lines are guides to the eye of slope 1.7. (From [5.31])

when the disorder becomes very small, i.e., for $m = 5$ and 10, does there seem to be a crossover to $n_b \sim L^0$ as expected. Force and displacement at the maximum do not seem to obey a power-law relation, at least, not for the small sizes considered.

Finally, after n_f bonds are cut the system breaks apart altogether. Again a behavior $n_f \sim L^{1.7}$ is reasonably well followed by the data except for small disorder, where for $m = 5$ and $m = 10$ a crossover to the expected $n_f \sim L$ is observed (Fig. 5.14b). Moreover, this finding seems universal with respect to the three models [5.27,31,44], the distribution of randomness, and the external boundary conditions within our error bars (5%–10% depending on the model). For the scalar model it can also be verified that the length of the largest crack, which causes the failure of the system, scales proportional to L for all distributions considered.

The scaling relation for the number of bonds cut, valid throughout the whole breaking process up to its end with an exponent of about 1.7, is not has unexpected result since [5.24] predicts $n \sim L/\ln L$ for the case $x = 0$ using the "dilute crack" approximation. This approximation is based on extreme value statistics [5.45]: it is the longest crack that will determine the ultimate failure and its length is proportional to $\ln L$. Rather powerful results can be obtained using these arguments, we refer the reader to [5.25] and [5.46].

Let us analyze next the distribution $n(i)$ of local forces (shears and moments can also be considered) or, in the case of electrical fuses, of local currents. The moments of this distribution are defined as $M_q = \sum_{\text{bonds}} i^q n(i)$. In Fig. 5.15b M_0 and the quantities $m_q = (M_q/M_0)^{1/q}$ are plotted as functions of L for the scalar model at the point when the last bond is cut before the system breaks apart. We see that with varying q the m_q scale like $m_q \sim L^{y_q}$ with different exponents y_q. This is in sharp contrast to what happens if the same analysis is made for the $n(i)$ at the maximum of the breaking characteristics (Fig. 5.15a). Here the m_q fall on parallel straight lines for different q and so all y_q are the same (constant gap scaling).

The phenomenon of y_q varying with q shown in Fig. 5.15b is a manifestation of *multifractality* that has recently been observed in various contexts [5.47] (for more details see Chaps. 1, 3, 4, and 10). Another analysis, namely the investigation of the $f(\alpha)$ spectrum [5.48], also leads to the conclusion that $n(i)$ is multifractal just before the last bond is cut for the central force model [5.44] and the beam model [5.31] as well.

Physically the multifractality means that the regions with the highest variation in local strains, i.e., the regions that are finally responsible for rupture, lie on a fractal subset of the system. The fractal dimension of this subset depends on the strength of the local variations. In practical terms this means that the larger the system the more pronounced the contrast between highly strained and practically unstrained regions. This effect only occurs just before the system breaks and not during the whole process as seen in Fig. 5.15a. Our data permit us to quantify this statement. The appearance of multifractality is more

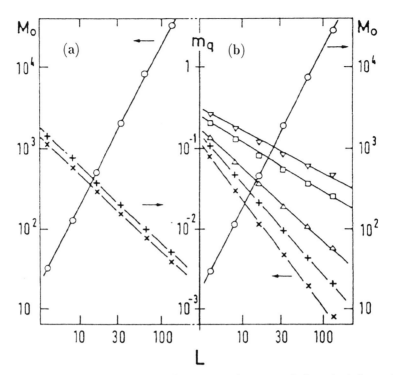

Fig. 5.15. Moments of the current distribution in the case of electrical fuses for $m = 2$ (a) at the maximum of the characteristics and (b) just before the system falls apart in a log-log plot against L. The symbols are $\circ$ for M_0, $\times$ for m_1, $+$ for m_2, $\triangle$ for m_3, $\square$ for m_6 and ∇ for m_9, where $m_q = (M_q/M_0)^{1/q}$. (From [5.27])

astonishing if one considers that only a negligible number of bonds $(n \sim L^{1.7})$ has been cut, in contrast to the case of percolation where multifractality [5.47] only appears at the percolation threshold p_c (i.e., $n = p_c L^2$), see Sect. 3.7. Local strains can be studied by photoelasticity and this might also be the best means of verifying the multifractal properties.

5.7 Hydraulic Fracture

Hydraulic fracture is a technology used to fracture soils for improving geothermal recovery or oil production, which consists in injecting water under high pressure into a borehole. In Orlèans [5.49] beautiful two-dimensional experiments were performed by pushing water or air into a hole in the center of a Hele-Shaw cell filled with clay. In Fig. 5.16 we see the penetration pattern for water into clays of different concentrations. On the left we have viscous fingers with a finite radius of curvature at the tips and on the right we see fracturing. The crack tips are sharp because, as discussed in Sect. 5.3.2, even perturbations of very small wavevector $\boldsymbol{k}$ are unstable. The fractal dimension for the left-hand pattern is about 1.6, whereas for the pattern on the right it is about 1.4. If the

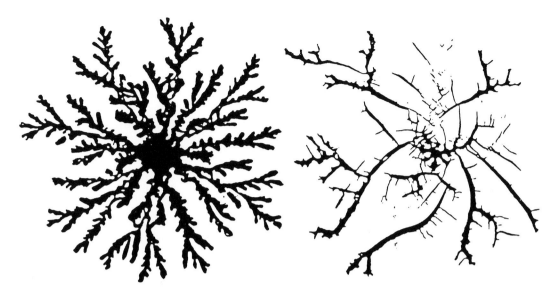

Fig. 5.16. Patterns produced by water pushed into a bentonite/water paste at concentrations of 0.08 (*left*) and 0.20 (*right*). From [5.49]

concentration is increased, a morphological transition between the two patterns can be witnessed which also depends on the injection velocity since clay is a viscoelastic fluid [5.50].

The interesting new ingredient in hydraulic fracturing is the fact that the externally imposed load, namely the pressure acting on the inner surface of the hole, increases as the crack grows and its direction, which is perpendicular to the surface, will follow the orientation of an eventually fractal crack shape. A simple beam model [5.51] shows the consequences. In the center of a square lattice with free boundary conditions a crack is initiated by removing one vertical beam. At each endpoint of the removed beam a force p is applied pointing into the bulk. The equilibrium forces acting along each beam along the surface of the crack are calculated through a conjugate gradient and among the beams under traction one is chosen randomly and removed next. This model resembles the Eden model but has the additional restriction that surface elements under compression cannot break. Fig 5.17 shows a typical crack generated with this model. One sees that the crack opening is larger for main branches than for side branching. In fact, the crack volume increases like the square of the radius R [5.51] showing that fluid really enters into the soil. The grey scales in Fig. 5.17 are the difference of the eigenvalues of the stress tensor modulated sinusoidally and thus represent the interference pattern observed experimentally through photoelasticity, the density of lines being a measure for the elastic stress. For the surface S of the crack one finds [5.51]

$$S \propto R^{d_f}, \tag{5.24}$$

with a fractal dimension of $d_f = 1.56 \pm 0.05$.

Fig. 5.17. Typical crack obtained for constant presure on a square lattice of 150×150 with free boundary conditions

In a more realistic model [5.52] fluid is injected at a constant rate into the crack such that during each time step Δt the crack volume is increased by ΔV. Also the disorder is quenched by randomly giving to each beam a breaking threshold according (5.15). As before the stress distribution at a fixed (fictitious) pressure p_0 can be calculated using a conjugate gradient. Exploiting the linearity of the elastic equations, we obtain the real pressure p, since the real crack volume is known. All beams on the crack surface for which the longitudinal force exceeds the threshold are removed. If ΔV is to be chosen sufficiently small, beams break in well-separated clusters or bursts as seen in Fig. 5.18. The lifetime of the bursts and the time between breakings have power-law distributions, the later corresponding to Omori's law [5.53] for earthquakes.

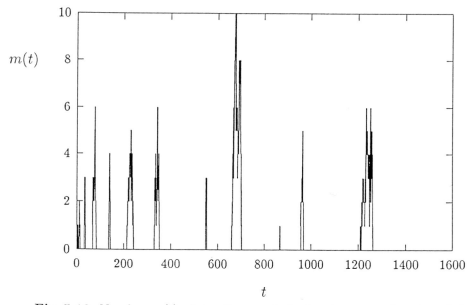

Fig. 5.18. Number $m(t)$ of simultaneously broken beams as function of time

In hydraulic fracturing these events can be recorded through acoustic emission [5.54]. As measured at boreholes in the field the pressure p increases during the periods of quiescence and abruptly chutes during the bursts. The cracks have a fractal dimension of about 1.4 [5.52] in very good agreement with the experiments [5.49].

5.8 Conclusion

We have seen in this chapter how fractals appear in fracture if it is formulated as a growth problem similar to viscous fingering. The vectorial nature of elasticity, however, generates very complex patterns even for deterministic models.

The influence of disorder on fracture is very strong because fracture is dominated by the extreme moments of the distribution of local strain. In the case of quenched disorder we gave evidence for the existence of scaling laws in the breaking characteristics and multifractality just before the system breaks apart. A formulation of fracture as a critical phenomenon should therefore be possible.

The experimental reality of fracture is very rich and our description is at present restricted to some sorts of brittle cracks. In order to compare the results from simulations to experimentally observed cracks [5.55] one must, however, still overcome numerical difficulties and obtain more reliable data.

Acknowledgements. I thank my collaborators Lucilla de Arcangelis, Alex Hansen, Janos Kertész, Stephane Roux, and F. Tzschichholz.

References

5.1 H. Liebowitz, ed.: *Fracture*, Vols. I-VII, (Academic Press, New York 1986)

5.2 C.L. Briant, R.P. Messmer: Phil. Mag. B **42**, 569 (1980);
     M.E. Eberhart, D.D. Vvdensky: Phys. Rev. Lett. **58**, 61 (1987)

5.3 E.C. Aifantis: J. Eng. Mater. Technol. **106**, 326 (1986)

5.4 H.J. Herrmann, S. Roux, eds.: *Statistical Models for the Fracture of Disordered Media*
     (Elsevier, Amsterdam 1990)

5.5 D. Broek: *Elementary Engineering Fracture Mechanics* (Martinus Nijhoff, Dordrecht
     1986)

5.6 W. Nowacki: *Theory of Micropolar Elasticity*, CISM Int. Center for Mechanical Sci.,
     Courses and Lectures, No. 25 (Springer, Heidelberg 1970)

5.7 D.A. Kessler, J. Koplik, H. Levine: Adv. Phys. **37**, 255 (1988)

5.8 H.J. Herrmann and J. Kertész: Physica A **178**, 409 (1991)

5.9 H.J. Herrmann, in *Growth Patterns in Physical Sciences and Biology*, eds. J.M. Garcia,
     E. Louis, P. Meakin and L.M. Sander (Plenum Press, New York, 1993) p.299

5.10 M. Barber, J. Donley, J.S. Langer: Phys. Rev. A **40**, 366 (1989)

5.11 S.S. Rao: *The Finite Element Method in Engineering* (Pergamon, Oxford 1982);
     P. Tong, J.N. Rossettos: *Finite Element Method: Basic Technique and Implementation*
     (MIT Press, Cambridge 1977)

5.12 G.G. Batrouni, A. Hansen: J. Stat. Phys. **52**, 747 (1988)

5.13 L. de Arcangelis, S. Redner, H.J. Herrmann: J. de Phys. Lett. **46**, L585 (1985)

5.14 L. Niemeyer, L. Pietronero, H.J. Wiesmann: Phys. Rev. Lett. **52**, 1033 (1984)

5.15 H. Takayasu, in: *Fractals in Physics*, ed. by L. Pietronero, E. Tossatti (Elsevier, Amsterdam 1986), p. 181;
     H. Takayasu: Phys. Rev. Lett. **54**, 1099 (1985)

5.16 S. Feng, P.N. Sen: Phys. Rev. Lett. **52**, 216 (1984)

5.17 Y. Kantor, I. Webman: Phys. Rev. Lett. **52**, 1891 (1984)

5.18 S. Roux, E. Guyon: J. de Phys. Lett. **46**, L999 (1985)

5.19 R.J. Roark, W.C. Young: *Formulas for Stress and Strain* (McGraw-Hill, Tokyo 1975),
     p. 89

5.20 H. Yan, G. Li, L.M. Sander: Europhys. Lett. **10**, 7 (1989)

5.21 E. Louis, F. Guinea, F. Flores, in: *Fractals in Physics*, ed. by L. Pietronero, E. Tossatti
     (Elsevier, Amsterdam 1986);
     E. Louis, F. Guinea: Europhys. Lett. **3**, 871 (1987)

5.22 P. Meakin: Thin Solid Films **151**, 165 (1987)

5.23 P. Meakin, G. Li, L.M. Sander, E. Louis, F. Guinea: J. Phys. A **22**, 1393 (1989)

5.24 B. Kahng, G.G. Batrouni, S. Redner, L. de Arcangelis, H.J. Herrmann: Phys. Rev. B
     37, 7625 (1988)

5.25 P.M. Duxbury, P.D. Beale, P.L. Leath: Phys. Rev. Lett. **57**, 1052 (1986);
     P.M. Duxbury, P.L. Leath, P.D. Beale: Phys. Rev. B **36**, 367 (1987)

5.26 P.D. Beale, D.J. Srolovitz: Phys. Rev. B **37**, 5500 (1988)

5.27 L. de Arcangelis, H.J. Herrmann: Phys. Rev. B **39**, 2678 (1989)

5.28 Y. Termonia, P. Meakin, P. Smith: Macromolecules **18**, 2246 (1985)

5.29 P. Meakin, in: Ref. [5.4]

5.30 E.L. Hinrichsen, A. Hansen, S. Roux: Europhys. Lett. **8**,1 (1989)

5.31 H.J. Herrmann, A. Hansen, S. Roux: Phys. Rev. B **39**, 637 (1989);
     L. de Arcangelis, A. Hansen, H.J. Herrmann, S. Roux: Phys. Rev. B **40**, 877 (1989)

5.32 H.J. Herrmann, J. Kertész, L. de Arcangelis: Europhys. Lett. **10**, 147 (1989)

5.33 T.A. Witten, L.M. Sander: Phys. Rev. Lett. **47**, 1400 (1981)

5.34 Y.S. Li, P.M. Duxbury: Phys. Rev. B **38**, 9257 (1988)

5.35 C. Tang: Phys. Rev. A **31**, 1977 (1985);
     J. Szép, J. Cserti, J. Kertész: J. Phys. A **18**, L413 (1985);
     J. Nittmann, H.E. Stanley: Nature **321**, 661 (1986);
     J. Kertész, T. Vicsek: J. Phys. A **19**, L257 (1986)

5.36 J. Fernandez, F. Guinea, E. Louis: J. Phys. A **21**, L301 (1988)

5.37 B.B. Mandelbrot, T. Vicsek: J. Phys. A **22**, L377 (1989)

5.38   F. Family, D.E. Platt, T. Vicsek: J. Phys. A **20**, L1177 (1987)

5.39   J.P. Eckmann, P. Meakin, I. Procaccia, R. Zeitak: Phys. Rev. A **39**, 3185 (1989)

5.40   M.J. Blackburn, W.H. Smyrl, J.A. Feeney, in: *Stress Corrosion in High Strength Steels and in Titanium and Aluminium Alloys*, ed. by B.F. Brown (Naval Research Lab., Washington 1972), p. 344

5.41   V.K. Horvath and H.J. Herrmann: Chaos, Solitons and Fractals **1**, 395 (1992)

5.42   S. Roux: J. Stat. Phys. **48**, 201 (1987)

5.43   P. Ossadnik, Phys. Rev. E, to appear

5.44   A. Hansen, S. Roux, H.J. Herrmann: J. Physique **50**, 733 (1989)

5.45   E.J. Gumbel: *Statistics of Extremes* (Columbia University, New York 1958)
        J. Galambos: *The Asymptotic Theory of Extreme Order Statistics* (Wiley, New York 1978)

5.46   P.M. Duxbury, Y.S. Li, in: Proc. of the SIAM Conf. on *Random Media and Composites*, ed. by R.V. Cohn and G.W. Milton (SIAM, Philadelphia 1989)

5.47   G. Paladin, A. Vulpiani: Phys. Rep. **156**, 147 (1987);
        L. de Arcangelis, S. Redner, A. Coniglio: Phys. Rev. B **31**, 4725 (1985)

5.48   T.C. Halsey, P. Meakin, I. Procaccia: Phys. Rev. Lett. **56**, 854 (1986)

5.49   H. Van Damme, E. Alsac, C. Laroche: C.R. Acad. Sci., Ser.II **309**, 11 (1988):
        H. Van Damme, in: *The Fractal Approach to Heterogeneous Chemistry*, ed. by D. Avnir (Wiley, New York 1989), p. 199

5.50   L.D. Landau, E.M. Lifshitz: *Elasticity* (Pergamon, Oxford 1960), p. 130

5.51   F. Tzschichholz, H.J. Herrmann, H.E. Roman and M. Pfuff: Phys. Rev. B **49**, 7056 (1994)

5.52   F. Tzschichholz and H.J. Herrmann: preprint

5.53   F. Omori: J. Coll. Sci. Imper. Univ. Tokio **7**, 111 (1894)

5.54   I.G. Main, P.R. Sammonds and P.G. Meredith: Geophys. J. Int. **115**, 367 (1993)

5.55   B.B. Mandelbrot, D.E Passoja, A.J. Paulley: Nature **308**, 721 (1984);
        C.S. Pande, L.E. Richards, N. Louat, B.D. Dempsey, A.J. Schwoble: Acta Metall. **35**, 1633 (1987)
        S.R. Brown, C.H. Scholz: J. Geophys. Res. **90**, 12575 (1985)

6 Transport Across Irregular Interfaces: Fractal Electrodes, Membranes and Catalysts

Bernard Sapoval

6.1 Introduction

How do irregular surfaces operate? This chapter is devoted to this general question, which has been revived by the concept of fractal geometry.

Many natural or industrial processes take place through surfaces or across the interfaces between two media. In this manner the roots of a tree exchange water and inorganic salts with the earth through their surface. Oxygen in the air is exchanged with blood hemoglobin through the surface of the pulmonary alveoli. These are natural processes.

In order for a car to start, the battery must deliver enough power to turn the engine, and a large current is needed for this purpose. In fact, the electrochemical process which supplies the current takes place at the interface between the electrode and the electrolyte. At this interface, the current is determined by elementary physical and chemical processes which limit its density. To obtain a large current it is necessary to increase the area of the exchange surface, and it is for this reason that *porous* electrodes are used. A porous medium is one which has a maximum surface area for a given volume. We know that fractal geometry qualitatively meets this criterion (the Von Koch curve has infinite length even though it is entirely contained within a finite surface [6.1–3]).

In this last example, the transport of the materials reacting in the electrolyte is due to electric current but it is possible to envisage similar processes involving no electric charge. Let us take the example of the lung: in the upper bronchial tube, air circulates hydrodynamically, but beyond this, in the alveoli, no air flows. The transport of oxygen to the cells which coat the pulmonary alveoli, where the gas exchange takes place, is a phenomenon of diffusion (see Sect. 3.2). Close to the membrane, the oxygen concentration in the air falls as the oxygen

◀ **Fig. 6.0.** Photograph of a soft coral, Solomon Islands, by Carl Roessler. Courtesy of C. Roessler

is trapped, and it is the diffusion of oxygen molecules in air which finally brings further molecules into contact with the membrane. Pulmonar alveoli have an irregular space-filling distribution [6.4–7] and fractal geometry can be used as a guideline.

We will show that these two problems, the current in a battery and breathing, can be solved using the same equations. The same applies to the absorption of a fertilizer through the roots of a plant, and other examples suggest that nature itself has perfected these systems by giving them a fractal geometry: "The problem of energy interchange in trees can be simplified by considering the tree as a system in which as large an area as possible must be irrigated with the minimum production of volume while at the same time guaranteeing the evacuation of absorbed energy" [6.8].

Again the same type of problem arises in heterogeneous catalysis. Solid catalysts are employed widely in the chemical and petroleum industries to promote many important chemical reactions. Porous catalysts are preferred since they can provide an enormous surface in a very small volume (up to several hundred square meters per gram) [6.9]. Reactants diffuse into the porous structure, reaction takes place, and the products formed diffuse back out to the ambient fluid.

The problem of transport to and across an irregular interface is thus a problem of general interest. It is of concern in studies in very different fields. We do not claim here that all irregular interfaces are indeed fractals. We first consider the question: if the interface is fractal what do we know about the transfer across such a surface? Once we have the answer to that question, we will see that it is possible to understand irregular interfaces in general. In the past fifteen years this field of research has been moving rapidly [6.10–60] but only recently can it be considered as settled and open to applications. It could appear in the future that the main applications of these results will exist in fields which are far from electrochemistry itself. Today, we can consider the following five conclusions as confirmed, and this is what I am going to describe and to discuss in this chapter.

1. The frequency response of a fractal electrode depends on the electrochemical regime.
2. In the so-called diffusion limited regime, the impedance of a fractal electrode depends directly on its Minkowski-Bouligand dimension. This will be shown in the following.
3. In the so-called blocking regime the response of a fractal electrode generally exhibits a nontrivial Constant Phase Angle (CPA) behavior. This behavior is directly related to the fractal hierarchy but the response of self-affine and self-similar electrodes are very different. In particular they do not depend on the fractal dimension in the same manner.
4. The macroscopic response of fractal interfaces is not proportional to the microscopic transport coefficients but is a power-law function of these pa-

rameters. This general property may be important because it may alter experimental measurements of the microscopic transport parameters.

5. We will show that there is an exact correspondence between the ac and dc electrical response of an electrode, the diffusive response of a membrane, and the steady-state yield of a heterogeneous catalyst of the same geometry. This is perhaps the most important result of this chapter. It unifies and widens the scope of applications of the study of transport to and across irregular interfaces. In particular we will see that the results can be applied to a semi-quantitative understanding of the absorption of oxygen into the blood in the terminal part of the respiratory apparatus of mammalians. In that sense the blocking electrode problem can be considered as a "model problem" offering the possibility of impedance spectroscopy measurements on model electrodes. It allows for a test of the various approximations which are necessary to understand simply the behavior of irregular interfaces. It is a means by which to validate ideas which may have their main future applications outside of the field of electrochemistry.

6.2 The Electrode Problem and the Constant Phase Angle Conjecture

The problem of transfer across a fractal surface was first addressed in the study of the impedance of batteries. It has been observed for a long time that rough or porous electrodes do not have a simple frequency response even in the small-voltage linear regime [6.61–66]. The equivalent circuit of a cell with planar electrodes of area S in principle consists of a surface capacitance C in parallel with the Faradaic resistance R_f, both being in series with the resistance of the electrolyte R_{el}. Figure 6.1 represents the electrical equivalent circuit of an electrochemical cell with planar electrodes.

The Faradaic resistance R_f is equal to (rS^{-1}) where r is the inverse rate of electrochemical transfer at the interface, for instance in a redox reaction such as

$$Fe^{3+} + e^- \longleftrightarrow Fe^{2+}. \tag{6.1}$$

In the absence of such a process the Faradaic resistance R_f is infinite and the electrode is said to be *blocking* or *ideally polarizable*. The resistance of the electrolyte R_{el} is proportional to the electrolyte resistivity ρ and depends upon the geometry of the cell. The surface capacitance $C = \gamma S$ is proportional to the surface area and to the specific capacitance per unit area γ. It corresponds to the charge accumulation across the interface. In the presence of such a Faradaic reaction it may happen that diffusion of the species in the liquid plays a role: such a case is termed a *diffusive regime* and appears at very low frequencies.

In fact, the impedance of rough or porous electrodes is generally found to be of the form

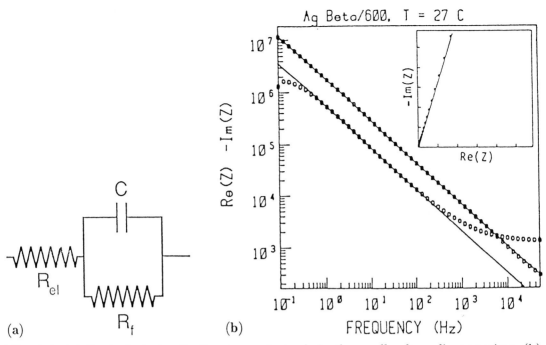

Fig. 6.1. (a) Equivalent circuit of a planar electrode in the small-voltage linear regime. (b) Experimental evidence for constant phase angle behavior obtained with metallic contacts on a solid electrolyte: sodium β-alumina [6.12]. Real and imaginary parts are measured as a function of frequency. The insert is an impedance diagram in the complex plane; it exhibits CPA behavior

$$Z \sim (i\omega)^{-\eta}, \quad 0 < \eta < 1, \tag{6.2}$$

in series with a pure resistance R_{el} which represents the resistance of the electrolyte. An example of such an experimental behavior is shown in Fig. 6.1b. This behavior is known as the constant phase angle or CPA behavior, because a plot of $-\mathrm{Im}Z(\omega)$ versus $\mathrm{Re}Z(\omega)$ in the complex plane gives a straight line with a "constant phase angle" with the x axis (as shown in the insert). There is a constant phase angle between the real and the imaginary parts. A smooth surface exhibits $\eta = 1$ (and a $\pi/2$ angle) whereas η decreases when the roughness of the surface increases.

Le Méhauté first proposed considering a rough or porous electrode as fractal, and many studies, the majority of them theoretical, have been devoted to this subject. These publications have mainly considered the properties of a blocking or ideally polarizable electrode. One should, however, keep in mind that there exist other interpretations of the CPA behavior which are not based on geometry. A suitable distribution of relaxation parameters on a flat surface can possibly account for CPA behavior [6.66 67]. Of course, both hierarchical geometry and these distributions can occur in real systems. We are interested here only in the role of the geometry because it is applicable to other phenomena: transport across membranes and heterogeneous catalysis.

At the beginning of these studies the emphasis was on searching only for a relation between the CPA exponent and the fractal dimension through scaling arguments. We will see below that we now have analytical results from theory not only about exponents but also describing the response as a function of the geometrical and physical parameters which determine the system behavior. Although these results cannot be considered as "exact" from their derivation, which bears on a few approximations, they are in very good agreement with numerical simulations and explain quantitatively from first principles experiments on model electrodes.

6.3 The Diffusion Impedance and the Measurement of the Minkowski-Bouligand Exterior Dimension

We start with a consideration of a blocking electrode with a very small Faradaic transfer resistance. We recall that an electrochemical reaction (*Faradaic process*) may become diffusion limited whenever *indifferent* charged species (i.e., ions which are not participating in the electrochemical reaction) are present in the electrolyte simultaneously with the *electroactive* participating species. If a large concentration of indifferent charge species exists and a small potential step is applied, the interface capacitance will be charged after a very short time, because the solution is highly conductive, and the externally applied potential will then appear directly across the interface. A large initial Faradaic current will then flow, but this will change the concentrations of the electroactive species in the vicinity of the interface. Because the solution is highly conductive no electric field will be left and the only driving force bringing new electroactive species to the interface will be the concentration gradient, through a diffusion process. Consequently, a corresponding *diffusion impedance* appears in series with the (small) Faradaic resistance of Fig. 6.1a.

The concentration varies locally from c_0 to $c_0 + \delta c(x, t)$. The concentration profile $\delta c(x, t)$ will be governed by the diffusion equation in the sinusoidal regime $(\delta c(x, t) = \delta c(x) e^{i\omega t})$, which for a planar electrode is

$$i\omega(\delta c(x)) = D \frac{d^2(\delta c(x))}{dx^2}, \tag{6.3}$$

where D is the diffusion coefficient. The electrochemical current will be

$$J = -zeD \frac{d(\delta c(x))}{dx} \bigg|_{x=0} = ze(i\omega D)^{1/2} \delta c(x = 0), \tag{6.4}$$

where $\delta c(x = 0)$ is the variation of concentration at the surface, and z is the number of elementary charges per ion. Physically, this means that the charge passing through the interface during a cycle corresponds to the number of species at a concentration $\delta c(x = 0)$ contained in a "diffusion layer" of thickness $\Lambda_D \sim (D/\omega)^{1/2}$. The associated admittance Y_D can be calculated by relating the concentration $\delta c(x = 0)$ to the local potential by the Nernst law (see for example [6.55] and references therein). For a planar electrode of area S one finds

$$Y_D = S\Gamma(D/i\omega)^{1/2}i\omega. \tag{6.5}$$

The quantity Γ is a capacitance per unit volume; we call it the specific diffusive capacitance [6.55]. This relation tells us that the diffusive admittance is simply that of the capacitance of the diffusive volume $S\Lambda_D \sim S(D/\omega)^{1/2}$. This can be readily extended to fractal surfaces.

The general form of the diffusion admittance at a fractal electrode can be found if we know the charge $\Gamma V(\Lambda_D)$ which is contained in the vicinity of the fractal electrode surface up to a distance of the order of $\Lambda_D \sim (D/\omega)^{1/2}$. We then need to know the volume $V(\Lambda_D)$ located within a distance Λ_D from the fractal surface. This is precisely what is measured by the *exterior Minkowski-Bouligand* dimension (here "exterior" means exterior to the electrode).

Let us now define this dimension. Let ϵ be a (small) positive length. Let V_ϵ denote the volume of electrolyte which lies within the distance ϵ of the electrode. Then

$$d_f = \lim_{\epsilon \to 0}\left(3 - \frac{\ln V_\epsilon}{\ln \epsilon}\right). \tag{6.6}$$

This definition is illustrated in Fig. 6.2.

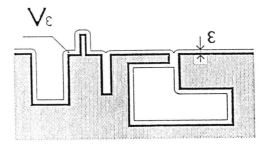

Fig. 6.2. Definition of the exterior Minkowski-Bouligand dimension. The contour of the electrode is the solid line. The electrode is shown in gray. The contour is "fattened" in the electrolyte by taking all points a distance smaller than ϵ away from the electrode (*thin line*). The volume V_ϵ is proportional to ϵ^{3-d_f} for an ordinary fractal surface

It is known that, for specific geometries, d_f may differ from the Hausdorff dimension of the common boundary of the electrode and electrolyte [6.68].

From (6.6) the volume $V(\Lambda_D)$ located within a distance Λ_D of the fractal varies as $\Lambda_D^{3-d_f}$, hence like $\omega^{(d_f-3)/2}$. The diffusion admittance of the fractal electrode is then $|Y_D| \sim \Gamma V(D/\omega)^{1/2}\omega \sim \omega^{(d_f-1)/2}$.

More precisely, if a fractal electrode has a macroscopic surface S, the content of the neighboring volume $V(\Lambda_D)$ is equal to $S^{d_f/2}\Lambda_D^{3-d_f}$, and one obtains

$$Y_D \sim S^{d_f/2}D^{(3-d_f)/2}c\,\omega^{(d_f-1)/2}. \tag{6.7}$$

Note that the response is given by a noninteger power law related to the fractal geometry. The above discussion fails at very low frequencies where the diffusion length may be larger than the size of the fractal electrode and reaches the size of the electrochemical cell. (For a solution of the problem in dc conditions see below.)

The very general reason allowing one to directly relate the ac diffusive regime to the fractal dimension is that, due to the presence of the support electrolyte, the volume of the electrolyte is electrically equipotential. The electrochemical potential is nonuniform on a scale of the order of the diffusion length. The reason for this nonuniformity is the electrical potential drop at the exchange surface. Since the electrical potential is uniform, this excitation is constant over the surface. As a consequence, the perturbation occurs only very close to the surface and the flux at some point on the surface is a local response. The ionic diffusion mechanism itself always takes place in the Euclidean space occupied by the electrolyte and is not perturbed by the presence of the fractal surface, which acts only as a boundary. It is therefore not surprising that this flux can be related to some dimension through a *fattening* of the surface as done in the Minkowski-Bouligand approach. More generally, the content of a Minkowski-Bouligand layer of thickness Λ_D permits one to describe the impedance of any irregular, rough or porous, fractal or nonfractal interface in the diffusive regime. These effects have been thoroughly studied together with the time response (Cotrell law) of these electrodes and verified by numerical simulations and experiments [6.46,47,49–51,55].

On the other hand, if the resistance of the electrolyte plays a role, as in the case discussed previously of blocking electrodes, the electrolyte is no longer equipotential and the response is no longer local. In that case, the admittance is not related to the local properties of the interface as characterized by the fractal dimension. This will be shown first in a specific case where the response of a model electrode can be calculated exactly: the *generalized modified Sierpinski electrode*. Our calculation would also be applicable to a catalyst of the same geometry as will be shown later.

6.4 The Generalized Modified Sierpinski Electrode

This electrode is the metallic electrode shown in Fig. 6.3. It is made in a decimation process. In the first step, a square pore of side a_0 is made in the electrode of side a. Then, at each step, N smaller pores of side a_0/α are added around a pore of a given size and so on. Here $N = 4$ and $\alpha = 3$. A cross section of the electrode is a modified Sierpinski carpet with a fractal dimension in the plane $d_{f,p} = \ln N/\ln \alpha$. We consider the case where:

1. The other electrode of the electrochemical cell is planar. It is very near the fractal electrode so that one can neglect the resistance of the thin layer of electrolyte between the two electrodes.

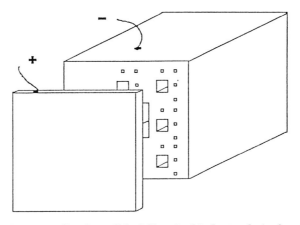

Fig. 6.3. Picture of the generalized modified Sierpinski electrode in front of a planar counter electrode. The fractal object that we consider is made by a decimation process. In the first step a square pore of side a_0 and depth L is made in the electrode of side a. Then, at each step, N smaller pores of side a_0/α and depth L/α_z are added around a pore of a given size, and so on. Here $N = 4$ and $\alpha = 3$. For this model electrode, exact results can be obtained for various electrochemical regimes [6.55]

2. The metallic volumes are assumed to have zero resistance and the external surface is coated with an insulating material so that we neglect conduction through this surface.

3. The bottom of each pore is insulating. All the pores are linked in parallel and the admittance is simply the sum of the admittances of all the pores.

4. The lengths of the pores also scale (either as the side of the pores or differently). Hence, the largest pore has a length L, the next pores have a length L/α_z and so on. For example, thinner pores can be shorter than thicker pores.

This electrode really has a self-affine geometry (see Chap. 7). It has an infinite surface area included in a finite volume. Its exterior Minkowski-Bouligand dimension can be easily calculated as we now show. We have N^n pores of length $L\alpha_z^{-n}$ and of side $a_0\alpha^{-n}$. The fact that the electrode occupies a finite volume implies $N \leq \alpha^2$. To calculate the dimension, we choose a small positive length ϵ and separate the pores with sides smaller and greater than ϵ. For this we find the order of decimation ν of the pore side ϵ. The quantities ϵ and ν are related by

$$\alpha^{-\nu-1}a_0 < \epsilon/2 < \alpha^{-\nu}a_0. \tag{6.8}$$

We then have

$$V_\epsilon \sim \sum_{n=\nu}^{\infty} N^n a_0^2 L(\alpha^2\alpha_z)^{-n} + 4\epsilon \sum_{n=0}^{\nu-1} N^n a_0 L(\alpha\alpha_z)^{-n}. \tag{6.9}$$

There are two possibilities. First $N > \alpha\alpha_z$. Then the second term is of the order of magnitude of $\epsilon N^\nu a_0 L(\alpha\alpha_z)^{-\nu}$, and the first term is equivalent to $N^\nu a_0^2 L(\alpha^2\alpha_z)^{-\nu}$. Remembering (6.8), ν is of order of $\ln(a_0/\epsilon)/\ln\alpha$ and we get $V_\epsilon \sim \epsilon^2\epsilon^{-\ln(N/\alpha_z)/\ln\alpha}$, and from (6.6)

$$d_f = 1 + \frac{\ln N - \ln \alpha_z}{\ln \alpha}. \tag{6.10}$$

Note that if $\alpha_z = 1$, all the pores have the same depth and $d_f = 1 + \ln N / \ln \alpha$. In the case where $N \leq \alpha \alpha_z$, the dimension is 2. As a result of the simple geometry, several electrochemical regimes have been calculated for the generalized modified Sierpinski electrode [6.55].

We show now that the generalized modified Sierpinski electrode exhibits CPA response in the blocking regime, but the CPA exponent is not a function of the fractal dimension. The pores are branched in parallel and the total admittance is simply the sum of the admittance of the individual pores,

$$Y = \sum_{n \geq 0} N^n Y_n. \tag{6.11}$$

In a first approximation, a single pore can be considered as a resistance R_n in series with a capacitance C_n, if one neglects propagation effects. The quantities R_n and C_n are the resistance and the surface capacitance of a pore of order n. These pores have a length $L/(\alpha_z)^n$ and a side a_0/α^n. Hence:

$$R_n = \rho(L/\alpha_z^n)(a_0^{-2}\alpha^{2n}) \tag{6.12}$$

and

$$C_n = 4\gamma(a_0/\alpha^n)(L/\alpha_z^n). \tag{6.13}$$

There are N^n such circuits in parallel for each stage of decimation. It can be easily verified that at a given frequency ω the admittance is dominated by the pores with $R_n C_n \omega = 1$, that is, those for which the order of decimation is $n(\omega)$ such that [6.55]

$$4\rho\gamma(L^2/a_0)(\alpha/\alpha_z^2)^{n(\omega)} = \omega^{-1} \tag{6.14}$$

or

$$n(\omega) = \frac{\ln(a_0/4\rho\gamma\omega L^2)}{\ln(\alpha/\alpha_z^2)}. \tag{6.15}$$

For these pores the resistive and capacitive admittances are equal and the admittance is then of the order of

$$|Y| \sim N^{n(\omega)} R_{n(\omega)} \sim (a_0^2/\rho L)(a_0/4\gamma\rho L^2)^{-\eta}\omega^\eta, \tag{6.16}$$

with

$$\eta = \frac{\ln N + \ln \alpha_z - 2\ln \alpha}{2\ln \alpha_z - \ln \alpha}. \tag{6.17}$$

Hence there exists a CPA response but the exponent η is not a function of the fractal dimension (6.10) of the interface. This constitutes an *exact* result. In that very special geometry there is no direct relation between the dimension and the phase angle. Note that the function (6.16) is a nontrivial function of a_0, ρ, γ, L, and ω. The admittance is dominated by the $n(\omega)$ pores of characteristic frequency ω and the energy is dissipated in these particular pores. On such fractal interfaces the power dissipation is nonuniform.

Keddam and Takenouti [6.31] have studied the frequency response of a two-dimensional Koch electrode made of anodized aluminum. That is a case where

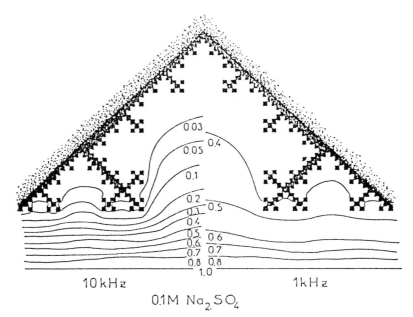

Fig. 6.4. Experimental map of the ac electric potential for two different frequencies in a fractal electrochemical cell, after Keddam and Takenouti [6.31,74]. The potential is 1 on the bottom horizontal line. Equipotential lines are shown for 10kHz in the left part and for 1Hz in the right part. The electrode is made of an oxidized aluminum profile and is blocking in the Na_2SO_4 solution where the measurement is made. One observes a stronger penetration of the ac potential in the structure at lower frequencies. This is related to the higher value of the surface capacitive impedance at low frequencies. These maps indicate qualitatively that the different parts of a fractal object do not play the same role in the blocking regime [6.60,83,84,86]. This must be contrasted with the diffusive regime response that we have studied above and where the active zone is evenly distributed on the surface as shown in Fig. 6.2

the electrode is really a "blocking" electrode. They have studied the potential distribution in that case and have demonstrated that the equipotential lines in the electrolyte penetrate the fractal object in a nonuniform manner which depends on the frequency. This is shown in Fig. 6.4.

The idea of a nonuniform role of the fractal surface was independently discussed by Wang [6.60] and will be discussed below.

6.5 A General Formulation of Laplacian Transfer Across Irregular Surfaces

We describe now a simple way to consider and to compute the impedance of irregular interfaces using a simple and general argument [6.69]. It applies to any irregular electrode and in particular to self-similar electrodes. In addition it permits us to compute the response from the geometry of the interface only.

In the electrochemical cell the potential obeys the Laplace equation [6.70,71]. The method is to substitute the problem of Laplacian transfer across the real

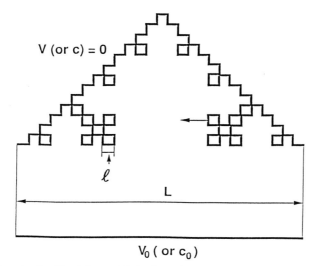

Fig. 6.5. Electrochemical cell with a self-similar electrode. The irregular electrode of interest (the working electrode in electrochemistry) has an inner cut-off ℓ and size or diameter L. The arrow indicates the orientation of the normal. This chapter deals with the electrochemical problem in which the applied voltage is V_0 on the planar counter electrode and 0 on the working electrode. An equivalent diffusion problem exists in which a planar source of diffusion is maintained at a constant concentration c_0 and particles diffuse towards an irregular membrane with finite permeability W

electrode (which presents a finite transfer rate) by a problem of a Laplacian field obeying the Dirichlet boundary condition ($V = 0$) but with a different geometry, obtained by coarse graining. The coarse-graining scale is directly related to the transport coefficients of both the electrolyte and the electrode. Using general properties of Dirichlet-Laplace fields one finds an effective screening factor which gives the size of the zone being really active on the initial geometry, and hence its admittance.

To calculate the response of an electrochemical cell with an irregular electrode, as in Fig. 6.5, one has to solve the Laplace equation ($\Delta V = 0$) which governs the electric potential distribution in the electrolyte with a boundary condition that reflects the electrochemical process at the working surface. This surface possesses a finite admittance and what is known about the properties of Laplacian fields on surfaces without impedance (with $V = 0$) cannot be applied directly. This is unfortunate since the Dirichlet-Laplace problem on an irregular electrode has been thoroughly studied, at least in $d = 2$. More specifically an important theorem, Makarov's theorem, describing the properties of the charge distribution on an irregular (possibly fractal) electrode capacitor can be applied [6.72]. This theorem states that the information dimension of the harmonic measure (for instance the electrostatic charge for the capacitor case) on a singly connected object in $d = 2$ is exactly equal to 1. This very special property of the Laplacian field can be illustrated in the following manner: whatever the shape of an capacitor electrode, the size of the region where the charge accumulates is proportional to the overall size (or diameter) L of the electrode under a dilation transformation.

This result generalizes, to an arbitrary geometry, a fact which has been known for a long time for simple geometries. It has a profound meaning in terms of the screening efficiency of the irregularity and this is what we use. For this, we consider the simplest description of an irregular surface: the length of the perimeter L_p divided by its size or diameter L [6.73].

$$S = L_p/L. \tag{6.18}$$

This number S has a direct significance: it measures the screening efficiency of the irregularity for Dirichlet-Laplacian fields: If whatever the geometry the active zone has a size L, then as

$$L = L_p/S \tag{6.19}$$

the factor $1/S$ can be considered to be the "screening efficiency" due to the geometrical irregularity. This is the physical significance of Makarov's theorem. (Note that when we discuss fractal lines we consider only physical objects with a finite inner cut-off ℓ so that S is always finite.)

The above result cannot be applied directly to the screening of the current in an electrochemical cell because the boundary condition on the electrode is not $V = 0$. In the simplest linear regime, a "flat" element of an electrode surface with unit area behaves as a resistor r across a capacitance γ. The Faradaic resistor r describes the finite rate of the electrochemical reaction if the interface is not blocking. Due to charge conservation, the current $j_\perp = -V(r^{-1} + j\gamma\omega)$ crossing the electrode surface must be equal to the Ohmic current $j_\perp = -\nabla_\perp V/\rho$ reaching it from the bulk (ρ being the electrolyte resistivity). As a consequence the dc boundary condition can be written as

$$V/\nabla_\perp V = \Lambda \quad \text{with} \quad \Lambda = r/\rho. \tag{6.20}$$

This boundary condition then introduces a physical length scale Λ in the problem. The procedure that we describe now is to switch from the real geometry obeying the real boundary condition to a coarse-grained geometry obeying the Dirichlet boundary condition, with the coarse-graining depending on Λ. In that new geometry we will apply equation (6.19) to obtain the effective screening, hence the size of the working zone of the real electrode.

To be more specific we describe this analysis for the situation of an electrode in a planar $d = 2$ cell as represented in Fig. 6.5. Consider a part i of the surface with a perimeter length $L_{p,i}$. If the thickness of the cell is b, this surface possesses an admittance $Y_i = bL_{p,i}/r$. The admittance to access the surface is of order $Y_{\text{acc}} = b/\rho$ because in $d = 2$ the admittance of a square of electrolyte with thickness b is equal to b/ρ whatever its size. Depending on the size of the region i there exists two situations: $Y_i < Y_{\text{acc}}$ or $Y_i > Y_{\text{acc}}$. If $L_{p,i}$ is small, $Y_i < Y_{\text{acc}}$ and the current is limited by the surface admittance. On the contrary if $L_{p,i}$ is large enough we have $Y_i > Y_{\text{acc}}$ and the current is limited by the resistance to access the surface. But in the latter situation we are, in a first approximation, back to the case of a pure Laplacian field with the boundary condition $V = 0$. The idea then is to coarse-grain the real geometry to a scale $L_i = L_{cg}$ such that the perimeter $L_{p,cg}$ in a region of size (diameter) L_{cg} is given by the critical condition $Y_i = Y_{acc}$ or

$$L_{p,cg} = \Lambda. \tag{6.21}$$

A coarse-grained site is then a region with a perimeter equal to $\Lambda = r/\rho$. Because of its definition, such a region can be considered as acting uniformity. At the same time, in the new coarse-grained geometry we are dealing with a pure Dirichlet Laplacian field and we can then use the screening factor $1/S_{cg}$ of this object to find its effective active surface. Note that if we did the coarse-graining to a scale larger than L_{cg} it would no longer be correct to consider a uniform distribution of the current within a macrosite and that consequently we would not be able to find the size of the active zone. If N_p is the number of yardsticks of length L_{cg} needed to measure the perimeter of the electrode, the number S of the coarse-grained object is simply

$$S_{cg} = N_p/N, \tag{6.22}$$

where $N = L/L_{cg}$ is the number of yardsticks needed to measure the size (or diameter) of the electrode. The quantity $1/S_{cg}$ is the effective fraction of the surface which is active and the admittance of the electrode will simply be given by

$$Y(r) = Y_p(r)/S_{cg}, \tag{6.23}$$

where $Y_p(r)$ would be the surface admittance of a "stretched" electrode with a length L_p. In this frame the number S_{cg} of the coarse-grained object determines directly how the admittance of the total surface is reduced by the screening effects. If we consider a self-similar electrode with an inner cut-off ℓ and a fractal dimension d_f there exists a simple relation between the size L_{cg} of the coarse-graining and the length of the perimeter:

$$\Lambda = \ell(L_{cg}/\ell)^{d_f}. \tag{6.24}$$

Using (6.21–24) one obtains for the admittance of a self-similar electrode of macroscopic size L and thickness b the value

$$Y = Lb(\ell\rho)^{(1-d_f)/d_f} r^{-1/d_f}. \tag{6.25}$$

The general (nonblocking) ac response is obtained by substituting r by $(r^{-1} + j\gamma\omega)^{-1}$

$$Y(\omega) = Lb(\ell\rho)^{(1-d_f)/d_f} (r^{-1} + j\gamma\omega)^{1/d_f}. \tag{6.25a}$$

For blocking electrodes with $r^{-1} = 0$ we have

$$Y(\omega) = Lb(\ell\rho)^{(1-d_f)/d_f} (j\gamma\omega)^{1/d_f}. \tag{6.26}$$

The dc form of this result has been verified by numerical simulation and the ac form has been verified by experiments on model electrodes as described in detail in reference [6.74].

At very low frequency, where the size of the coarse-grained site is larger than the diameter of the system, the admittance is limited by the capacitance of the total surface and

$$Y(\omega) = Y_p(\omega) = \ell b(L/\ell)^{d_f}(j\gamma\omega). \qquad (6.27)$$

The high-frequency fractal regime and the low-frequency capacitive regime meet at a cross-over frequency ω_c where the size of the coarse-graining is the size or diameter L of the electrode itself. For a blocking electrode $\Lambda = (\rho\gamma\omega)^{-1}$ and (6.24) gives for this crossover

$$\omega_c = (\rho\gamma\ell)^{-1}(L/\ell)^{-d_f}. \qquad (6.28)$$

The above simple arguments are general and permit, as we show below, the computation of the response of irregular electrodes even for nonscaling geometries. First we show how this method can be used from the image of the electrode. All we need is to have a "flexible" measuring rod of length $\Lambda = r/\rho$ (or $|\Lambda(\omega)| = 1/\rho\gamma\omega$ for blocking electrodes) that we use to measure the length of the perimeter from one end to the other. To perform this task we need a number N_p of flexible rods. Note that here, we do not measure the irregular object with a rigid yardstick as usually considered in measuring fractals. On the contrary, starting at one end of the irregular object, we map the object with the flexible rod of length Λ and the distance in real space between the ends of this rod determines a distance L_1. This length is a "local yardstick" associated with Λ and the local geometry. Then we place a second flexible rod of length Λ from the end of the first rod and find a new yardstick length L_2 and so on. The total number of rods (of length Λ) needed to map the object is N_p and the number S of the coarse-grained electrode is N_p/N where N is the number of yardsticks which measure the size. The real electrode has a perimeter $N_p\Lambda$ and a total admittance $Y_p(r) = N_p\Lambda b/r = N_p b/\rho$. Dividing by S as in (6.23) we find

$$Y = Nb/\rho. \qquad (6.29)$$

We then obtain that the modulus of the admittance of an irregular electrode in $d = 2$ is simply the square admittance b/ρ multiplied by the number of yardsticks needed to measure the size (or diameter) L of the electrode. This shows that deterministic and random fractals with the same inner and outer cut-offs and the same fractal dimension have the same response. The reason is now trivial because what really matters is the total number of macrosites, whatever their individual size, which may be distributed over some range of sizes.

We now apply this method to nonscaling geometries as shown in Fig. 6.6. The dc admittance of the perimeter surface is now $Y_p = (N_p' + N_p'')b\Lambda/r$ with

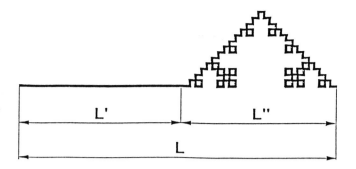

Fig. 6.6. Example of an electrode with a nonscaling geometry. This electrode is build by the association of a planar electrode of length L' with a self-similar electrode of length L''

$N'_p = L'/L'_{cg}$ and $N''_p = (L''/L''_{cg})^{d_f}$. To obtain the admittance of this electrode we have to divide this value by S of the entire electrode. The macrosites do not have the same size if they correspond to the planar part, for which the yardstick is simply $L'_{cg} = \Lambda$, or if they corresponds to the fractal part for which the yardstick is given by $L''_{cg} = \ell(\Lambda/\ell)^{1/d_f}$ from (6.24). The total number of yardsticks on the object is now $N_p = N'_p + N''_p$, whereas the number of yardsticks needed to measure the size is $N = N' + N''$ with $N' = L'/L'_{cg}$ and $N'' = L''/L''_{cg}$. Applying (6.23) with $S = (N'_p + N''_p)/(N' + N'')$ one finds for the admittance $Y = (N' + N'')b/\rho$ or $Y = L'b/r + L''b(\ell\rho)^{(1-d_f)/d_f}r^{-1/d_f}$, which is the sum of the admittances of the two electrodes in parallel.

This general argument then restores the essential property of Laplacian transfer: the admittance of two electrodes in parallel is the sum of the individual admittances. This indicates that this method is general. The ac response of blocking electrodes is obtained by replacing r by $(j\gamma\omega)^{-1}$. Note that we have used the number N of yardsticks needed to measure the total size as equal to the sum $N' + N''$ of the number of yardsticks needed to measure the sizes of the two different parts. If we wish to use equation (6.29) we have to use an average yardstick $\langle L_{cg} \rangle$ defined by $L/\langle L_{cg} \rangle = L'/L'_{cg} + L''/L''_{cg}$. The average yardstick $\langle L_{cg} \rangle$ must be obtained by this harmonic mean.

In the known cases of self-affine electrodes, such as the Cantor bar electrode [6.38–40] or the Sierpinski electrode studied above, the pores have different aspect ratios. For this case a coarse-grained pore can be still defined by $Y_i = Y_{acc}$ but not by $L_{p,cg} = \Lambda$. As explained above in a somewhat different language the impedance can be found from the total surface of these coarse-grained pores, which are all accessible at the same time due to the particular geometry.

We then have shown how to find, from first principles, the response of a nonscaling irregular object. Apart from the image, all that is needed are the values of the microscopic transport coefficients (here, for instance, the electrolyte resistivity and the Faradaic resistance). The same method will apply to the equivalent problem of the steady-state diffusion rate towards an irregular membrane with finite permeability W as indicated below.

The simplicity of this method probably makes it a good candidate for the study of the response of irregular electrodes in the nonlinear regime where the local current across the electrode is related to the local voltage by a non nonlinear relation $j = f(V)$ [6.47,69,75].

The case of self-similar electrodes with $2 < d_f < 3$ (in $d = 3$) has been treated by an equivalent circuit method in [6.74] and gives, respectively, for the high-frequency fractal and low-frequency capacitive regime

$$Y(\omega) = L^2 (\ell\rho)^{(d_f - 2)/(1 - d_f)} (j\gamma\omega)^{1/(d_f - 1)}, \qquad (6.30)$$

and

$$Y(\omega) = \ell^2 (L/\ell)^{d_f} (j\gamma\omega). \qquad (6.31)$$

The crossover frequency ω_c is now

$$\omega_c = (\rho\gamma\ell)^{-1} (L/\ell)^{1 - d_f}. \qquad (6.32)$$

Note that in this case the crossover frequency occurs when the perimeter of a cut of the fractal surface by a plane is equal to the length Λ. These theoretical predictions allow us to explain quantitatively the high-frequency and low-frequency response of the electrode shown in Fig. 6.7 [6.76,77].

We have presented here a simple approach based on first principles. These results have been verified by the existing numerical simulations in $d = 2$ and permit us to understand quantitatively the experimental results in $d = 2$ and $d = 3$. A different and more complex approach has been proposed in [6.78–82]. These more detailed studies present slightly different values for the exponent η but do provide analytic expressions for the admittance. In this situation, one should consider that the above results (6.25-28) and (6.30-32) allow us to grasp the essential features of that question.

Another significant result from the equivalent circuit approach of [6.74] and from [6.83,84,86] is the notion of information fractal and of active zones. The active-zone studies give a direct insight on the localization of the regions which are effectively working and those which are passive for given physical conditions. This is a further step in the detailed understanding of how irregular interfaces operate in the linear regime. This notion could be of great help in the understanding of how irregular interfaces operate in a nonlinear regime [6.69].

The above discussion about screening indicates that if one considers self-affine electrodes in the case where the aspect ratio of the groves is smaller than 1 there is no effect and no CPA. The experiments of Bates, Chu, and Stribling [6.13] are, from this point of view, experiments on self-affine electrodes with small aspect ratios. From the discussion about S numbers in that case, the response should be capacitive. This is why they do not obtain any relation between fractal dimensions and the CPA exponents. In their case the CPA may be due to microscopic effects.

In this chapter we have assumed that no dc polarization current is flowing in the system, which is far from being the case in very many practical elec-

Fig. 6.7. Wood's metal ramified electrode. This object is approximately 6 cm high and has its smallest branching in the range of 20μm. It is a metallic moulding of the penetration space of water injected into plaster and was obtained by G. Daccord and R. Lenormand [6.77]. In a blocking situation the admittance of this electrode approximately obeys relations (6.30–32)

trochemical situations. The presence of a dc polarization current would make the (potential-dependent) Faradaic admittance vary from one point of the interface to another. This would complicate the problem considerably, and most probably the results would be affected.

6.6 Electrodes, Roots, Lungs, ...

The dc admittance given by (6.25) is proportional to $r^{-\eta}$. This result is not trivial. It means, for instance, that dividing the surface resistance at the surface of a porous electrode by a factor of two will not double the current. Also the current is not proportional to the electrolyte conductivity but is a power law of the conductivity. As a result, the macroscopic response coefficient across a fractal interface is not proportional to the microscopic transport coefficients. A power law depending on the geometrical hierarchy relates these factors. This conclusion could have applications in several systems found in nature or those built to have large-surface porous structures.

The same kinds of properties should be observed in the study of bulk and membrane diffusion that exists in biological or physiological fractal systems. We now illustrate the correspondence between the above study and the description of the flux of a neutral species across a membrane when the overall process is limited by the diffusion from a source at a constant concentration. Consider the mathematical problem of finding the dc response in the electrochemical cell of Fig. 6.8a.

The potentials are V_0 on the counter electrode and $V = 0$ on the electrode of interest. As we have seen, the equations to be solved are the Laplace equation and the current equation in the bulk of the electrolyte with the boundary

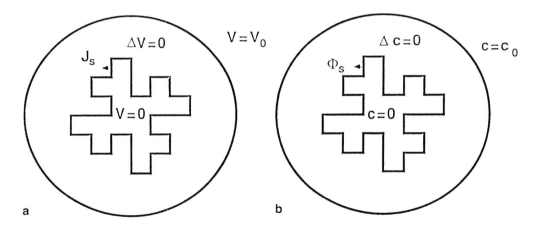

Fig. 6.8a,b. Equivalence between the electrode problem and the membrane problem in the same geometry. If one replaces Φ by J, c by V, D by ρ^{-1}, and W by r^{-1} one obtains the same mathematical problem for a given geometry

condition for the electrical current normal to the surface. Instead of an electrochemical problem we imagine the problem of the transfer of neutral species diffusing across a membrane of the same geometry. Instead of a counterelectrode as in Fig. 6.8a we imagine that some process maintains a constant concentration c_0 of the species of interest. This is the situation of Fig. 6.8b. We call Φ the flux vector at point x. There are two flow processes in our system. First, diffusion, which in a steady state obeys Fick's law,

$$\Phi = -D\nabla c(x), \tag{6.33}$$

where c is the concentration of the particles of interest and D is the diffusion coefficient, as before. Together with the conservation law

$$\frac{\partial c}{\partial t} = -\nabla \Phi \tag{6.34}$$

this leads to the diffusion equation, which in a steady state is

$$\Delta c = 0. \tag{6.35}$$

Second, there is a transfer equation across the membrane,

$$\Phi_n = -Wc(x), \tag{6.36}$$

where W is the probability per unit time, surface, and concentration of a particle crossing the membrane. In the last equation we have neglected a back transfer, assuming that the concentration on the other side of the membrane is maintained equal to zero by some forced flow mechanism (blood circulation for example in the case of lungs). Provided that we replace Φ by J, c by V, D by ρ^{-1}, and W by r^{-1}, relations (6.33,35–36) are exactly equivalent to the dc current and potential equation in the electrochemical cell. The same results do then apply. If we define a diffusion admittance Y_D by a linear relation between the total flux Φ_T and the concentration c_0 at the entrance of the system we can write

$$\Phi_T = Y_D c_0 \tag{6.37}$$

Note that one should not confuse the steady-state diffusion regime that we study here and the so-called diffusion electrochemical regime that we have discussed in Sect. 6.3. This last regime was the time-dependent response of the electrode in the special case where the electric field is screened by the presence of a large concentration of inert (indifferent) electrolyte. The equivalence that we use here is between a steady-state diffusion regime and the Laplacian response of an electrode with the same geometry.

By transposition of (6.30–32) to the steady-state diffusion case one obtains a low-permeability regime with a diffusion admittance, which trivially is

$$Y_D = \ell^2 (L/\ell)^{d_f} W, \tag{6.38}$$

and a high-mobility "fractal" regime with a diffusion admittance equal to

$$Y_D = L^2 \ell^{(d_f - 2)/(1 - d_f)} D^{(d_f - 2)/(d_f - 1)} W^{1/(d_f - 1)}. \tag{6.39}$$

Here also the macroscopic response is a power law of the transport coefficients D and W. The value of the length Λ is now $\Lambda = D/W$ and the crossover between the two regimes occurs when

$$\Lambda = \ell (L/\ell)^{(d_f - 1)}, \tag{6.40}$$

which is when the perimeter of a cut of the fractal surface by a plane is equal to the length Λ.

Note that at crossover, the value of the membrane admittance needed is equal to the value of the admittance to reach the surface, which, in $d = 3$ is typically the (bulk) admittance of a cube of size L: $Y_B = LD$. The statement that, at the crossover point, the value of the classical admittance of the fractal membrane is equal to the access resistance is very simple. It may then be of more general value and may apply to many irregular geometries. That is probably why the above considerations seem to apply to the "acinus" of several animals, as discussed now, although their real geometry is not a simple fractal [6.6,7].

The airways of mammalians are made of two successive systems, the bronchial tree in which oxygen is transported with air and the terminal alveolar system, called the acinus, in which the air does not move. In the acinus the transport of oxygen towards the alveolar membrane is purely diffusive. In this system the transport may be limited both by the (bulk) diffusion admittance to reach the membrane and the admittance needed for the air of the membrane itself as given by (6.38,39). In $d = 3$ the bulk admittance is equal to LD and the total admittance is

$$Y_T = ([LD]^{-1} + Y_D^{-1})^{-1}. \tag{6.41}$$

This admittance increases with the diameter L and large exchangers should trivially be better exchangers. But the "best" exchange system should be that for which it is not the admittance itself but the admittance per unit volume, as the thorax volume of animals is limited. The above discussion indicates that the admittance per unit volume or specific admittance, defined as $Y_T/L^3 = ([LD]^{-1} + Y_D^{-1})^{-1}/L^3$, varies as $L^{d_f - 3}$ for small L and as L^{-2} for large L. If the structure of the alveolar surface is dense, $d_f = 3$ and the specific admittance does not depend on the size of the fractal up to the critical size discussed above but decreases with the power L^{-2}. From this discussion a geometry with a dense (space filling with $d_f = 3$) arrangement of alveolar surface of smaller size ℓ, up to a diameter L satisfying condition (6.40) should be optimum because lower losses in the higher airways ask for the larger size compatible with a large specific admittance. In that sense the "best possible acinus" size should be that for which the perimeter of a cross section through the acinus should have a length of the order of Λ. The physiological and anatomical data obtained by E. Weibel [6.5–7] allows us to compare the above considerations with what is observed in real lungs. The transport coefficients D and W governing the diffusion of oxygen in air and its capture by the alveolar membrane are known and it seems that the value of Λ (a few cm) is indeed close to the perimeter length of a cross section through the acinus for several animals such as the mouse, the rat, the rabbit, and the human. Although discrepancies exist, the general agreement can be considered as satisfactory if one takes into account the oversimplified model that is used [6.85].

It seems that these considerations can be extended to the gills of fish, in which the geometry is regular. Because of the smaller diffusion coefficient of oxygen in water, the value of Λ is of the order of a few tens of micrometers. The size of the gills is comparable to that value. The gills generally have a regular structure which corresponds to $d_f = 2$ and not to $d_f = 3$ [6.87]. In that case the optimum value for L is equal to Λ, now of the order of tens of microns, as observed.

6.7 Fractal Catalysts

The problem of heterogeneous reactions is of major importance in the chemical industry. The old Wenzel's law [6.88] states that, for heterogeneous reactions, the larger the interface, the faster the reaction. "This is why in industry and in the kitchen one grinds the material in order to speed up the reaction" [6.89].

If a chemical reaction is accelerated on the surface of a catalyst one should use porous catalysts to increase the overall efficiency of the reaction. A number of questions are raised by the dynamics of these complex processes. Very frequently catalyst reactions exhibit noninteger order. Very often the activity of a catalyst follows a power law as a function of the size of the catalyst grain [6.90–94].

Of course catalysts are not necessarily fractal, but the fractal geometry gives a possible hint to the understanding of some part of the catalytic process in specific cases. On the other hand, some very basic growth mechanisms, such as random aggregation or diffusion limited aggregation, build a fractal geometry so that the existence of a fractal aspect in the geometry of irregular materials cannot be considered as exceptional [6.3]. Several chapters of this book discuss the growth of fractal objects.

Several papers have been published on the relation between heterogeneous reactions and fractality (see [6.89–94] and references therein). We present here a few simple ideas on that matter. In many cases the catalytic activity A of a catalyst particle of size L obeys a power law

$$A \sim L^{d_R}, \tag{6.42}$$

where d_R is called the reaction dimension. Values of d_R ranging from 0.2 to 5.8 have been quoted, but in practice they range from 1 to 3 [6.91,92]. A very simple way to interpret d_R is the notion of active sites. This idea stresses that on a given surface only specific sites are active for catalysis. Suppose that the grain shape is cubic but only the corners of the cube are active for catalysis. Then two grains of different sizes have the same number of active sites and $d_R = 0$.

If the active sites are located on the edges of the cube the activity is proportional to L and $d_R = 1$. If the faces of the cube are active, $d_R = 2$. Experimental values of $d_R < 2$ may indicate that only a fractal subset of the surface is active. Experimental values of $d_R > 2$ may indicate that the catalyst itself has a fractal geometry.

If all sites are active then the activity is proportional to the mass, $A \sim L^{d_f}$ and $d_f = d_R$. These ideas apply if the reactants are present evenly on the catalytic sites, i.e., if the diffusion of reacting species from the source of reactant to the active surface is infinitely fast.

This may not be true and there are many situations where the diffusion of reactants to the active surface (or its active sites) is too slow and so controls the reaction kinetics. Such a process is called an Eley-Rideal mechanism. A third situation exists where the two reactants are on the surface and have to meet after diffusion onto the surface to react. This last mechanism is called the Langmuir-Hinshelwood mechanism. It has specific properties on fractal catalysts. For example, segregation of reactants may appear. This is discussed in [6.89].

Our purpose here is only to show that the results we have described for diffusion to membranes can be transposed to the case of the Eley-Rideal mechanism for heterogeneous catalysis [6.95]. Consider for instance a gas–solid catalysis in the simple case of the reaction

$$A + B \rightarrow AB. \tag{6.43}$$

Normally in a homogeneous phase the reaction rate is

$$\frac{d[AB]}{dt} = K[A][B]. \tag{6.44}$$

In such a case the reaction rate is directly proportional to the first power of the concentration (or pressure) of A and B. The sum of these exponents is 2 and the reaction is said to be second order. It is observed that a large number of heterogeneous reactions follow fractional-order kinetics under different experimental conditions. We show here that the above diffusion scheme gives rise to that result. Let us consider a case where the efficient reaction process takes place on the surface after adsorption of B atoms. The first step for the reaction is then

$$B_{\text{gas}} \rightarrow B_{\text{ads}}, \tag{6.45}$$

from which the reaction

$$A_{\text{gas}} + B_{\text{ads}} \rightarrow AB_{\text{gas}} \tag{6.46}$$

follows. If the chemical reaction is slow, as in the case of a poor catalyst, the diffusion from the exterior to the walls of the catalyst is accomplished by a small concentration gradient, and the concentration throughout is nearly the same as the external concentration. Fast reactions, however, may take place in the pores very near the exterior, and the internal pore surfaces contribute little. If the activity of the entire surface of the catalyst is compared (if diffusion were infinitely rapid), it is usual for the effectiveness of a poor catalyst to be high and that of an excellent catalyst to be low. Diffusion into the pores involves a decrease in the concentration of the diffusing reactants, and the concentration effective in promoting the chemical reaction at the active sites is everywhere less than if diffusion were not involved. The catalyst, therefore, is less effective than if all the surface were in contact with the reactants at the concentrations maintained in the external or ambient fluid. This loss of catalyst effectiveness due to diffusion is the subject of very many publications.

In general, diffusion within the pores may be ordinary molecular diffusion, Knudsen diffusion, surface diffusion, or a combination of all three. We restrict ourselves here to the consideration of molecular diffusion and we discuss in some detail the equations which describe the population of adsorbed B atoms $[B_{\text{ads}}]$. Without chemical reaction the surface concentration varies by adsorption of molecules from the gas and desorption from the surface, and the rate of absorption is equal to the net flux

$$\frac{d[B_{\text{ads}}]}{dt} = \Phi_n(B) = W_A[B_{\text{gas}}] - W_D[B_{\text{ads}}], \tag{6.47}$$

where $[B_{\text{gas}}]$ is the local partial concentration or partial pressure near the surface and W_A and W_D are the probability per unit time for adsorption and desorption. In the case where reaction (6.46) takes place there is an additional decrease due to the reaction rate proportional to $[A]$ and to $[B_{\text{ads}}]$. Consequently

$$\frac{d[B_{\text{ads}}]}{dt} = W_A[B_{\text{gas}}] - W_D[B_{\text{ads}}] - K_S[A][B_{\text{ads}}], \tag{6.48}$$

where K_S is the reaction coefficient. The steady-state local concentration on the surface is

$$[B_{\text{ads}}] = \frac{W_A[B_{\text{gas}}]}{W_D + K_S[A]}. \tag{6.49}$$

The net flux (6.47) to the surface is simply

$$\Phi_n(B) = \frac{W_A K_S[A][B_{\text{gas}}]}{W_D + K_S[A]}. \tag{6.50}$$

This equation is analogous to (6.36) if we replace W by $W_A K_S[A]/(W_D + K_S[A])$. The catalytic problem is then exactly identical to the membrane problem with the same geometry. A self-affine Sierpinski geometry could be a model for a porous catalyst with hierarchical pore structure on which the reactants are kept at a constant concentration at the entry of the pores. One can then use relation (6.16) or its equivalent for diffusion, which is $Y_p \sim a_0^{(2-\eta)} L^{(2\eta-1)} D^{(1-\eta)} W^\eta$. If a given partial pressure $[B]_{\text{ext}}$ of B exists outside the porous system, the net production of AB is the total flux

$$\frac{d[AB]}{dt} \sim a_0^{(2-\eta)} L^{(2\eta-1)} D^{1-\eta} \left(\frac{W_A K_S[A]}{W_D + K_S[A]} \right)^\eta [B]_{\text{ext}}. \tag{6.51a}$$

Note that this *"fractal"* speed of reaction is not a linear function of the microscopic sticking probability W_A. Note also that D is the diffusion coefficient of B in A. If B is diluted in A, D is for normal pressures inversely proportional to the concentration or pressure of A. Finally, the rate of reaction will be proportional to

$$\frac{d[AB]}{dt} \sim a_0^{(2-\eta)} L^{(2\eta-1)} [A]^{1-\eta} \left(\frac{W_A K_S[A]}{W_D + K_S[A]} \right)^\eta [B]_{\text{ext}}. \tag{6.51b}$$

We have thus presented a model of a catalytic reaction on a fractal surface with a noninteger reaction rate. If $W_D \ll K_S[A]$ (high pressure) the reaction order is η. If $W_D \gg K_S[A]$ (low pressure) the reaction order is 2η. The speed of reaction is a noninteger power-law function of the parameters L and a_0 which describe the macroscopic sizes of the catalyst grain. Although the modified Sierpinski geometry is artificial, it brings to light the idea that reaction order and reaction dimension could be related through simple relations in specific cases.

For a self-similar catalyst one would also find a noninteger order of reaction but the reaction dimension will be equal to 2 from (6.39). For reactions and transport on fractals, see also Chap. 3.

6.8 Summary

This chapter has briefly presented what was known at the end of 1994 about two different but partially related problems: the relation between fractality and the response of irregular electrodes, and the general question of the response of irregular interfaces to Laplacian fields. As these problems have been the subject of several conflicting (or apparently conflicting) results it seems useful to discuss the actual situation.

Let us first consider this situation from the point of view of electrochemistry. The first problem itself can be considered in two ways: First, do fractal electrodes exhibit CPA behavior and how? As we have seen, fractality will generally lead to CPA and the answer to the first part is "yes" and this can be considered as sufficiently documented.

The second, different way is to ask: to what extent can experimental CPA behavior of rough or irregular electrodes be related to fractality? The answer to this question depends on the system. As we have mentioned, a suitable distribution of microscopic parameters could also explain (and contribute to) CPA. In a discussion of porous platinum electrodes, T. Pajkossy has shown that in this specific case microscopic effects are really the source of CPA behavior. Generally speaking, there exist two main objections to fractality as a general explanation for CPA:

1. Fractality leads to Cole-Davidson [6.96] behavior (of the form $Y(\omega) \sim (1+j\gamma\omega\tau)^\eta$ as indicated from relation (6.25a)), whereas experimental CPA are often of the Cole-Cole form $Y(\omega) \sim 1 + (j\gamma\omega\tau)^\eta$.
2. The frequency range of experimental CPA is often very large, exceeding the range $(L/\ell)^{(d_f-1)}$ that one can expect.

Also, fractal CPA can hold only for objects in which the length $|\Lambda(\omega)| = 1/\rho\gamma\omega$ is larger than the smaller cut-off and smaller than the perimeter. This condition will not be met for rough electrodes (with irregularities in the 10 micron range) in concentrated liquid electrolytes. In contrast, this condition is met for larger electrodes with irregularities on the mm or cm range or in solid electrolytes (where ρ is smaller).

The conclusion is that geometrical fractality although leading to CPA cannot be used exclusively as the "general" explanation for experimental CPA without checking for the above physical conditions. One should also recall that there exist several electrochemical regimes (diffusion + faradaic or diffusion + resistive as discussed in [6.55] for Sierpinski electrodes) for which no general prediction exists for self-similar electrodes. Also, in the real world both microscopic and geometrical macroscopic effects will probably exist simultaneously. (This was the case in the experiments described in [6.74,76].) For this very reason, exponents alone are not sufficient to ascertain a fractal model. One should check the compatibly of the measurements with relations (6.26–28).

Even more important for the future are the simple results that we have obtained on the general question of the response of irregular interfaces to Laplacian fields. There are three main results:

1. It is possible to find the (linear) response of an irregular interface from its image through an appropriate coarse-graining process.
2. It is a reasonable approximation to consider that there exists a homogeneously active zone and a totally passive zone [6.69,83,84]. In this simplifying approximation it is possible to deal simply with nonlinear responses [6.69].
3. The semiquantitative agreement between the anatomical and physiological data and the expressions that we predict for the "best possible acinus" seems to indicate that the design for the oxygen exchanger in the lung of several animals is optimized. The statement that for the optimized situation the value of the classical admittance of the fractal membrane is equal to the access admittance is very simple. It may thus apply to many situations which are not "simply" fractal.

This brings out the idea of a "perfect exchanger" or "smart filter". One should note that the lung is a simple gas exchanger in the sense that only two gases are exchanged, oxygen and carbon dioxide. (They have approximately the same Λ.) If one considers the transfer of several species with very different transport parameters, the simultaneous optimization of the transfer should imply that different Λs should correspond to the same morphology. This cannot be realized by an homogeneous membrane. An optimized fractal multi-species filter or exchanger membrane system could nevertheless be realized by a suitable distribution of specific pores (permeable to specific species) on the membrane. In that sense it is possible that the micro-geometry of the distribution of specific cells (or of cells with active transfer) in inhomogeneous membranes permits the optimization of the transfer of very different species in the same organ. These notions could possibly lead to a better understanding of the morphology and physiology of complex exchangers or filters such as the kidney.

References

6.1   B.B. Mandelbrot: *The Fractal Geometry of Nature* (Freeman, San Francisco 1982)
6.2   J. Feder: *Fractals* (Plenum, New York 1988)
6.3   B. Sapoval: *Fractals* (Aditech, Paris 1990)
6.4   M. Rodriguez, S. Bur, A. Favre, E.R. Weibel: Am. J. Anat. **180**, 143 (1987)
6.5   B. Haeffli-Bleuer, E.R. Weibel: Anat. Record **220**, 401 (1988)
6.6   E.R. Weibel, in: *Respiratory Physiology, an Analytical Approach*, ed. by H.K. Chang, M. Paiva (Dekker, Basel 1989), p. 1
6.7   E.R. Weibel: in *Fractals in Biology and Medicine,* ed. by T.F. Nonnenmacher, G.A. Losa and E.R. Hulin (Birkhäuser, Basel, 1994), p.68

6.8   F. Hallé, R.A.A. Oldeman, P.B. Tomlinson: *Tropical Trees and Forests* (Springer, New York 1978)

6.9   S.W. Benson: *The Foundation of Chemical Kinetics* (McGraw-Hill, New York 1960); A.W. Adamson: *Physical Chemistry of Surfaces* (Wiley, New York 1982)

6.10 R. Ball, M. Blunt: J. Phys. A: Math. Gen. **21**, 197 (1988)

6.11 J.B. Bates, Y.T. Chu: Solid State Ionics **28-30**, 1388 (1988)

6.12 J.B. Bates, J.C. Wang, Y.T. Chu: Solid State Ionics **18/19**, 1045 (1986)

6.13 J.B. Bates, Y.T. Chu, W.T. Stribling: Phys. Rev. Lett. **60**, 7, 627 (1988)

6.14 R. Blender, W. Dieterich, T. Kirchhoff, B. Sapoval: J. Phys. A: Math. Gen. **23**, 1225 (1990)

6.15 M. Blunt: J. Phys. A **22**, 1179 (1989)

6.16 E. Chassaing, B. Sapoval, G. Daccord, R. Lenormand: J. Electroanal. Chem. **279**, 67 (1990)

6.17 Y.T. Chu: Solid State Ionics **23**, 253 (1987)

6.18 Y.T. Chu: Solid State Ionics **26**, 299 (1988)

6.19 L. Fruchter, G. Crepy, A. Le Méhauté: J. Power Sources **18**, 51 (1986)

6.20 W. Geertsma, J.E. Gols, L. Pietronero: Physica A **158**, 691 (1989)

6.21 W. Geertsma, J.E. Gols: J. Phys. C: Condens. Matter **1**, 4469 (1989)

6.22 P.G. de Gennes: C.R. Acad. Sci. (Paris) **295**, 1061 (1982)

6.23 L.J. Gray, S.H. Liu, T. Kaplan, in: *Scaling Phenomena in Disordered Systems*, ed. by R. Pynn, A. Skjeltorp (Plenum, New York 1985)

6.24 T.C. Halsey: Phys. Rev. A **35**, 3512 (1987)

6.25 T.C. Halsey: Phys. Rev. A **36**, 5877 (1987)

6.26 R.M. Hill, L.A. Dissado: Solid State Ionics **26**, 295 (1988)

6.27 T. Kaplan, L.J. Gray: Phys. Rev. B **32**, 11, 7360 (1985)

6.28 T. Kaplan, S.H. Liu, L.J. Gray: Phys. Rev. B **34**, 4870 (1986)

6.29 T. Kaplan, L.J. Gray, S.H. Liu: Phys. Rev. B **35**, 5379 (1987)

6.30 M. Keddam, H. Takenouti: C.R. Acad. Sci. (Paris) **302** Sér. II, 281 (1986)

6.31 M. Keddam, H. Takenouti: Electrochim. Acta **33**, 445 (1988) and 40th I.S.E. Meeting, Kyoto, Japan (1989) Ext. Abst. d2, 1183

6.32 M. Keddam, H. Takenouti, in: *Fractal Aspects of Materials*, ed. by R.B. Leibowitz, B.B. Mandelbrot, D.E. Passoja (Material Research Society, Pittsburgh 1985), p. 89

6.33 A. Le Méhauté: Electrochim. Acta **34**, 4, 591 (1989)

6.34 A. Le Méhauté, G. Crépy: Solid State Ionics **9/10**, 17 (1983)

6.35 A. Le Méhauté, G. Crepy, A. Hurd, D. Schaefer, J. Wilcoxon, S. Spooner: C.R. Acad. Sc. (Paris) **304**, 491 (1987)

6.36 A. Le Méhauté, A. Dugast: J. Power Sources **9**, 35 (1983)

6.37 R. de Levie: J. Electroanal. Chem. **261**, 1 (1989)

6.38 S.H. Liu: Phys. Rev. Lett. **55**, 529 (1985)

6.39 S.H. Liu, T. Kaplan, L.J. Gray, in: *Fractals in Physics*, ed by L. Pietronero, E. Tossati (North-Holland, Amsterdam 1986), p. 383

6.40 S.H. Liu, T. Kaplan, L.J. Gray: Solid State Ionics **18/19**, 65 (1986)

6.41 E.T. McAdams: Surface Topography **2**, 107 (1989)

6.42 A. Maritan, A.L. Stella, F. Toigo: Phys. Rev. B **40**, 9267 (1989)

6.43 A.M. Marvin, F. Toigo, A. Maritan: Surf. Sci. **211/212**, 422 (1989)

6.44 W.H. Mulder, J. H. Sluyters: Electrochim. Acta **33**, 303 (1988)

6.45 L. Nyikos, T. Pajkossy: Electrochim. Acta **30**, 1533 (1985)

6.46 L. Nyikos, T. Pajkossy: Electrochim. Acta **31**, 1347 (1986)

6.47 L. Nyikos, T. Pajkossy: Electrochim. Acta **35**, 1567 (1990)

6.48 T. Pajkossy, L. Nyikos: J. Electrochem. Soc. **133**, 2061 (1986)

6.49 T. Pajkossy, L. Nyikos: Electrochim. Acta **34**, 171 (1989)

6.50 T. Pajkossy, L. Nyikos: Electrochim. Acta **34**, 181 (1989)

6.51 T. Pajkossy: J. Electroanal. Chem. **300**, 1 (1991)

6.52 B. Sapoval: Solid State Ionics **23**, 253 (1987)

6.53 B. Sapoval: *Fractal Aspects of Materials, Disordered Systems* (Materials Research Society, Pittsburgh 1987), p. 66

6.54 B. Sapoval, J.-N. Chazalviel, J. Peyrière: Solid State Ionics **28-30**, 1441 (1988)

6.55 B. Sapoval, J.-N. Chazalviel, J. Peyriere: Phys. Rev. A **38**, 5867 (1988) and references therein

6.56 B. Sapoval: Acta Stereologica **6/III**, 785 (1987)

6.57 B. Sapoval, E. Chassaing: Physica A **157**, 610 (1989)

6.58 J.C. Wang, J.B. Bates: Solid State Ionics **18/19**, 224 (1986)

6.59 J.C. Wang: J. Electrochem. Soc, **134**, 1915 (1987)

6.60 J.C. Wang: Solid State Ionics **28-30**, 143 (1988)

6.61 W. Scheider: J. Phys. Chem. **79**, 127 (1975) and references therein

6.62 R. de Levie, Electrochim. Acta **8**, 751 (1953); **10**, 113 (1965)

6.63 R.D. Armstrong, R.A. Burnham: J. Electroanal. Chem. **72**, 257 (1976)

6.64 P.H. Bottelberghs, in: *Solid Electrolytes*, ed. by P. Hagenmuller, W. Van Gool (Academic, New York 1978)

6.65 P.H. Bottelberghs, H. Erkelens, L. Louwerse, G.H.J. Broers: J. Appl. Electrochem. **5**, 165 (1975); P.H. Bottelberghs, G.H.J. Broers: J. Electroanal. Chem. **67**, 155 (1976)

6.66 J.R. Macdonald: J. Appl. Phys. **58**, 1955 (1985); J. Appl. Phys. **62**, R51 (1987) and references therein

6.67 T. Pajkossy: J. Electroanal. Chem. **364**, 111 (1991)

6.68 C. Tricot: Phys. Lett. A**114**, 430 (1986)

6.69 B. Sapoval: Solid State Ionics **75**, 269 (1995)

6.70 B. Sapoval: J. Electrochem. Soc. **137**, 144C (1990); Extended Abstract, Spring Meeting of the Electrochemical Society, Pennington 1990, p.772

6.71 P. Meakin, B. Sapoval: Phys. Rev. A **43**, 2893 (1991)

6.72 N.G. Makarov: Proc. London Math. Soc. **51**, 369 (1985)

6.73 R.P. Wool, J.M. Long: Macromolecules: **26**, 5227 (1993); R.P. Wool: *Structure and Strength of Polymer Interfaces* (Hanser publ. 1994). In these references the number S is used to describe the relative irregularity of fractal "diffusion front"

6.74 B. Sapoval, R. Gutfraind, P. Meakin, M. Keddam, H. Takenouti: Phys. Rev. E **48**, 3333 (1993)

6.75 W.H. Mulder, J.H. Sluyters, T. Pajkossy and L. Nyikos: J. Electroanal. Chem. **285**, 103 (1990)

6.76 E. Chassaing, B. Sapoval: J. Electrochem. Soc. **141**, 2711 (1994)

6.77 G. Daccord, R. Lenormand: Nature (London) **325**, 41 (1987)

6.78 T.C Halsey, M. Liebig: Europhys. Lett. **14**, 815 (1991) and Phys. Rev. A **43**, 7087 (1991)

6.79 R.C. Ball: in *Surface Disordering, Growth, Roughening and Phase Transitions*, ed. by R. Julien, P. Meakin, D. Wolf (Nova Science Publisher, 1993), p.277

6.80 T.C Halsey and M. Liebig: Annals Phys. **219**, 109 (1992)

6.81 M. Liebig, T.C Halsey: J. Electroanal. Chem. **358**, 77 (1993)

6.82 H. Ruiz-Estrada, R. Blender and W. Dieterich: to be published in J. Phys.: Condens. Matter (1994)

6.83 B. Sapoval: in *Surface Disordering, Growth, Roughening and Phase Transitions*, ed. by R. Julien, P. Meakin, D. Wolf (Nova Science Publisher, 1993), p.285

6.84 R. Gutfraind, B. Sapoval: J. Physique I **3** 1801 (1993)

6.85 B. Sapoval in *Fractals in Biology and Medicine,* ed. by T.F. Nonnenmacher, G.A. Losa and E.R. Hulin (Birkhäuser, Basel, 1994), p.241

6.86 R. Gutfraind, B. Sapoval: in *Fractals in Biology and Medicine,* ed. by T.F. Nonnenmacher, G.A. Losa and E.R. Hulin (Birkhäuser, Basel, 1994), p.251

6.87 A. Beaumont, P. Cassier: *Biologie Animale* (Dunod Université, Paris, 1978), p.443

6.88 C.F. Wenzel: *Lehre von der Verwandtschaft der Körper* (Dresden, 1777)

6.89 R. Kopelman, in: *The Fractal Approach to Heterogeneous Chemistry*, ed. by D. Avnir (Wiley, New York 1989), p. 295 and references therein

6.90 P. Pfeiffer, D. Avnir, D. Farin: J. Stat. Phys. **36**, 199 (1984)

6.91 D. Farin, D. Avnir: J. Am. Chem. Soc. **110**, 2039 (1988)

6.92 D. Farin, D. Avnir: J. Phys. Chem. **91**, 5517 (1987)

6.93 D. Farin, D. Avnir, in: *The Fractal Approach to Heterogeneous Chemistry*, ed. by D. Avnir (Wiley, New York 1989), p. 271 and references therein

6.94 D. Avnir, R. Gutfraind and D. Farin: in *Fractals in Science,* ed. by A. Bunde and S. Havlin (Springer-Verlag, 1994), chap. 8

6.95 B. Sapoval: C.R. Acad. Sci. (Paris) **312**, Sér. II, 599 (1991)

6.96 D.W. Davidson and R.H. Cole: J. Chem. Phys. **19**, 1484 (1951)

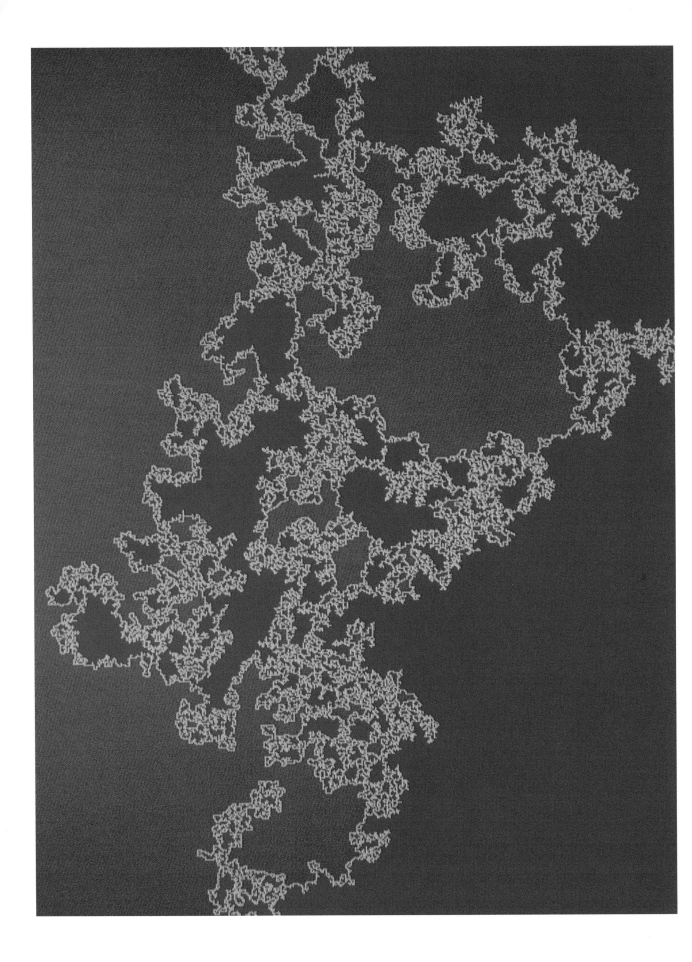

7 Fractal Surfaces and Interfaces

Jean-François Gouyet, Michel Rosso and Bernard Sapoval

7.1 Introduction

Irregular surfaces and interfaces are ubiquitous in nature. They exist on all scales from that of atoms to that of mountains. Their study requires three levels of consideration: (1) a description of their geometry, (2) an explanation of their existence, and (3) an understanding of their properties. Fractal geometry may play a role at all three of these levels.

An exhaustive presentation of fractal surfaces and interfaces and of their topological, physical, chemical properties, and soon would require a complete book and is beyond the scope of this chapter. We just present here a selected list of examples, which covers a rich variety of physical situations with special attention being paid to the question of the formation of fractal interfaces. Some of the questions concerning fractal surfaces are treated in detail in other chapters of this book, for instance, how to measure the fractal dimension, how aggregation builds fractal surfaces, and how electrode properties can be related to the fractal character of some surfaces.

Following Pfeifer and Obert [7.1] we shall distinguish three classes of fractal surfaces (Fig. 7.1). A dense object with a fractal surface is called a *surface fractal*. If the object itself is fractal, and hence also its surface, the object is called a *mass fractal*. Finally, a dense object in which there exists a distribution of holes or pores with a fractal structure is a *pore fractal*. Usually mass and pore fractals belong to the class of self-similar objects whereas surface fractals are self-affine fractals. Self-affinity will be defined in the next section.

Although a surface is in principle the boundary of an object embedded in the three-dimensional ($d = 3$) Euclidean space, we shall also consider as a surface the boundary of an object in $d = 2$ space, which should be called a

◄ **Fig. 7.0.** Diffusion front generated on a 600×400 square lattice. The front is shown in yellow. The particles are diffusing from left to right

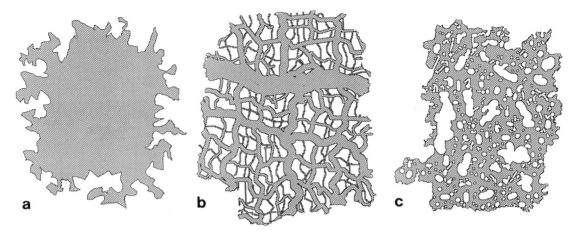

Fig. 7.1a,b. The Pfeifer classification: (a) a surface fractal, (b) a mass fractal, (c) a pore fractal. Mass regions are shown in black

"contour". (Theoretical studies are generally easier for contours than for real surfaces and many models have been considered for the former case.)

The examples that will be given are classified, somewhat arbitrarily, according to the following scheme:

- rough surfaces of solids,
- solid–solid or fluid–fluid interfaces,
- membranes.

7.2 Rough Surfaces of Solids

One may schematically distinguish two classes of irregular surfaces of solids: rough surfaces and surfaces of porous materials and powders. Here we shall mainly concentrate on rough surfaces, some aspects of porous materials will be discussed in Sect. 7.4 (see also Chap. 6).

Solids with rough surfaces are extremely common in nature, and they appear in many industrial processes. They may be created by various physical or chemical mechanisms such as deposition, erosion, corrosion and wear, dissolution, machining, and fracture. The study of these surfaces and of their physical and structural properties is consequently of great interest for many practical purposes.

A number of rough surfaces show fractal structures. Prominent examples are the surface of mountains [7.2–4], deposited thin films, surfaces formed by the fracture of a material, and sand-blasted metal surfaces. Two examples are shown in Fig. 7.2 [7.5,6].

In this section, we shall present the basic elements of the fractal description and main properties of rough surfaces: self-affinity, scaling hypothesis for grow-

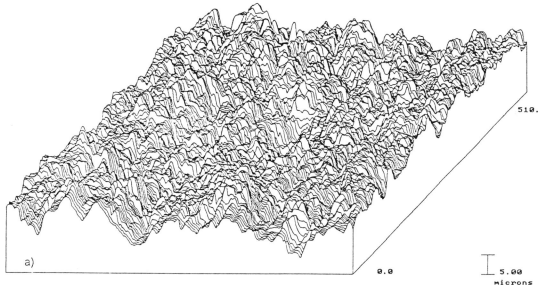

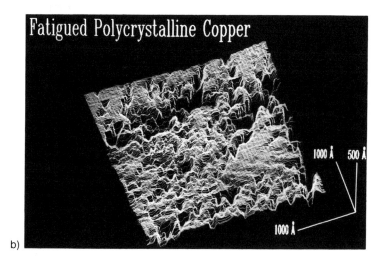

Fig. 7.2a,b. Two rough surfaces, observed on very different scales. (a) Sand-blasted brass surface [7.5], where the roughness is observed at a scale of about 50μm, (b) fatigued poly-crystalline copper [7.6], where the roughness is found at a scale of a few hundred ångstrøms

ing rough surfaces, simple deposition models, and the Kardar, Parisi, Zhang equation. For important new developments in the theoretical and experimental studies of self-affine interfaces, the interested reader is referred to the book by Barabási and Stanley [7.7] and the recent reviews of Kertész and Vicsek, Chap. 4 in [7.8], Meakin [7.9], and Halpin-Healey and Zhang [7.10].

7.2.1 Self-Affine Description of Rough Surfaces

We use the term *rough* surface for an irregular surface on which there are no overhanging regions or where at least these regions do not dominate the scaling properties. Fatigued polycrystalline copper (Fig. 7.2b) is an example of a rough surface without overhangs. The fact that the overhangs do not dominate the scaling behavior means that when the size of the surface increases, the relative weight of these regions becomes negligible. As a consequence, the surface can be correctly described by a function $h(\boldsymbol{r})$, which specifies the "height" of the surface at the coordinate $\boldsymbol{r}(x, y)$ in an appropriate reference plane. For the relief of a mountain, the reference plane is the sea level and $h(\boldsymbol{r})$ is the altitude at location $\boldsymbol{r}$ (latitude and longitude). Overhangs such as caves are neglected.

For many rough surfaces this function $h(\boldsymbol{r})$ is characteristic of a *self-affine fractal*. Self-affine fractals (see also Chaps. 1, 4, 8, and 10) have been defined by Mandelbrot [7.2] as geometrical objects that are (statistically) invariant under an *anisotropic* dilation (*self-similar fractals* are invariant under *isotropic* dilation, see, for example, Figs. 2.5 and 7.16). This means in the present case that there exists an exponent H smaller than 1 such that the transformation $\boldsymbol{r} \to \lambda \boldsymbol{r}$, $h \to \lambda^H h$ leaves the surface statistically invariant (Fig. 7.3). In contrast to the self-similar case, one must distinguish two different dimensions, a "global" and a "local" dimension [7.11–13]. The global dimension, which is observed above a certain crossover scale, is simply $d - 1$. This means that, at large distances, the self-affine fractal looks essentially smooth. The local fractal dimension d_f can be determined by classical methods, such as the box-counting method [7.2,12] (see also Chaps. 1 and 8). It is related to the exponent H by

$$d_f = d - H \qquad (d - 1 \le d_f \le d).\tag{7.1}$$

Note that in the case $H = 1$, the graph reduces trivially to a simple straight line.

In the example shown in Fig. 7.3, which is invariant under the transformation $\boldsymbol{r} \to 5\boldsymbol{r}$ and $h \to 3h$, $H = \log 3 / \log 5 \cong 0.68$ and $d_f \cong 1.32$. As in

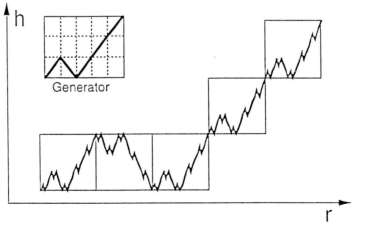

Fig. 7.3. A deterministic self-affine fractal generated by iterating the generator in the inset. Mathematically the iteration is continued ad infinitum on both the small and the large scales. There exists a local fractal dimension $d_f \cong 1.32$ and a global dimension $d - 1 = 1$. In a real surface the self-affine structure is restricted to a range $a < r < \xi$

general, cuts parallel to the substrate have a dimension $d_f - 1$ and can be used to measure d_f. These cuts are self-similar.

In addition to the local fractal dimension we need three lengths to characterize a rough surface of lateral size L: (1) the *average position* of the surface (mean height $\langle h \rangle$ of the surface points), (2) its *thickness* σ (root-mean-square fluctuation around the averaged position), which can be identified with a *correlation length* $\xi_\perp$ normal to the surface,

$$\sigma = \langle (h(\boldsymbol{r}_0) - \langle h \rangle)^2 \rangle^{1/2}, \tag{7.2}$$

and (3) a *correlation length* $\xi_\parallel$ parallel to the surface.

As usual, the correlation length can be defined from a correlation function. As the *irregularity* of the surface is characterized by the fluctuations of $h(\boldsymbol{r})$, we consider the *height correlation* function $\Gamma_h(\boldsymbol{r})$,

$$\Gamma_h(\boldsymbol{r}) = \langle h(\boldsymbol{r}_0)\, h(\boldsymbol{r}_0 + \boldsymbol{r}) \rangle - \langle h(\boldsymbol{r}_0) \rangle^2, \tag{7.3}$$

where the average is over all vectors $\boldsymbol{r}_0$ in the reference plane. $\Gamma_h(\boldsymbol{r})$ is a pair correlation function in the sense that it relates the heights at two positions $\boldsymbol{r}_0$ and $\boldsymbol{r}_0 + \boldsymbol{r}$. It goes to zero when the two heights become uncorrelated at a distance of the order of the correlation length $\xi_\parallel$. The $\xi_\parallel$ can be determined from the average

$$\xi_\parallel = \frac{\int r \Gamma_h dr}{\int \Gamma_h dr}. \tag{7.4}$$

Instead of $\Gamma_h(\boldsymbol{r})$, very often the height-difference correlation function

$$C_h(\boldsymbol{r}) = \langle [h(\boldsymbol{r}_0 + \boldsymbol{r}) - h(\boldsymbol{r}_0)]^2 \rangle, \tag{7.5}$$

is considered. The two functions are trivially related by $C_h = 2(\sigma^2 - \Gamma_h)$. If the surface is isotropic in the reference plane, $\Gamma_h(\boldsymbol{r})$ and $C_h(\boldsymbol{r})$ depend only on $r = |\boldsymbol{r}|$.

For a self-affine rough surface, $C_h(r)$ follows a power law for small-distance r ($r \ll \xi_\parallel$),

$$C_h(r) \sim r^{2H}. \tag{7.6}$$

This kind of power law characterizes a scaling behavior, which is due to the absence of a characteristic length scale below $\xi_\parallel$. Above $\xi_\parallel$, Γ_h tends to zero and C_h becomes a constant (see Fig. 7.4). (An introduction to scaling theory is given in Chaps. 2 and 3.)

Before we go further into the description of self-affine rough surfaces, it may be enlightning to make a detour to the fractional Brownian motion introduced by Mandelbrot [7.2] (see also [7.12]). We first consider a simple random walk (Brownian motion) along an axis h. At time $t = 0$, the walker is at position $h = 0$; the walker then steps randomly forward and backward, with constant step length (see Sect. 3.2). The trail of this walk is described by a *Brownian function* $h(t)$ that roughly looks like the ridge of a mountain. (Here t plays the

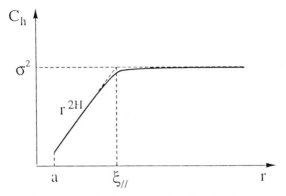

Fig. 7.4. Sketch of $C_h(r)$ in a double logarithmic plot: in the range $a < r < \xi_\|$, the scaling behavior (7.6) holds; above $\xi_\|$, the heights become uncorrelated ($\Gamma_h = 0$) and $C_h(r)$ saturates at a constant value $2\sigma^2$. The crossover then fixes both correlation lengths σ and $\xi_\|$

role of r.) The correlation function $C_h(t)$ for such a profile is identical to the mean square displacement along the h axis, and hence $C_h(t) \sim t$.

To generalize this simple random walk, we next consider a correlated random walk along h. The correlation between successive steps of the walk is chosen such that $C_h(t) \sim t^{2H}$, a relation comparable to (7.6) for the self-affine rough surfaces. This generalized random walk is called *fractional Brownian motion* [7.2,12]. The corresponding profile $h(t)$ is a self-affine fractal with fractal dimension $d_f = d - H$ ($H = 1/2$ and $d_f = 3/2$ for the simple random walk). The parameter H is called the Hurst exponent after the first person to introduce these considerations to characterize the water-level fluctuations of rivers [7.14]. The Hurst exponent varies between 0 and 1. When $H = 0$, the heights h are uncorrelated; when $H \to 1$, there is an increasing probability for the trail of the fractional Brownian motion to continue in the same direction. More generally speaking the successive increments δh between t and $t + \delta t$ are negatively correlated (anticorrelated) when $H < 1/2$, and positively correlated when $H > 1/2$.

The fractional Brownian motion can be generalized easily to $d > 2$ dimensions. The self-affine surface generated in this way is a function $h(\boldsymbol{r})$, where $\boldsymbol{r}$ now belongs to a $d - 1$ reference plane. Fractional Brownian motion in $d = 3$ represents one of the best ways to generate artificial reliefs of mountains and landscapes [7.3].

Equations 7.2–6 describe rough surfaces, which do not evolve with time or which are considered at a given fixed time. If we are concerned with a growing surface, the time enters into the above expressions, which then show interesting scaling behavior.

7.2.2 Growing Rough Surfaces: The Dynamic Scaling Hypothesis

In deposition, corrosion, or growth phenomena, the surface structure is time dependent, and this time variation can be studied by numerical simulations. Recently some progress has also been achieved by applying analytic methods. A review on theoretical models of surface growth can be found in [7.15] and in Chap. 4 of [7.8].

We now have [7.2,16] to consider a *time-dependent height correlation* function $C_h(\boldsymbol{r}, t)$ (a generalization of (7.5)), which is defined as

$$C_h(\boldsymbol{r}, t) = \left\langle \left[\tilde{h}(\boldsymbol{r}_0 + \boldsymbol{r}, t_0 + t) - \tilde{h}(\boldsymbol{r}_0, t_0) \right]^2 \right\rangle, \qquad (7.7)$$

where $\tilde{h}(\boldsymbol{r}, t) = h(\boldsymbol{r}, t) - \langle h(\boldsymbol{r}, t) \rangle$. The average is taken over $\boldsymbol{r}_0$ and t_0. In the cases considered here, the bulk of the growing object is compact (with dimension d) and the average height $\langle h(t) \rangle$ is simply controlled by the growth rate (which is the flux of particles arriving at the surface). At constant rate, $\langle h(t) \rangle$ varies linearly with time.

We can now define a time-dependent width $\sigma(t)$, which plays the role of a correlation length in the growth direction. Because of the basic anisotropy of the surface, $\sigma(t)$ will vary differently in time from the correlation length $\xi_\parallel(t)$ parallel to the surface. Consider a growth process starting from a flat surface at $t = 0$. At small times, due to the absence of a characteristic time scale, we can assume that both correlation lengths $\xi_\parallel(t)$ and $\sigma(t)$ can be written as powers of t (see Chap. 3),

$$\xi_\parallel(t) \sim t^{1/z}, \qquad (7.8)$$

and

$$\sigma(t) \sim t^\beta, \qquad (7.9)$$

where z is a conventional exponent [7.13] for finite size effects (see below); in general $1/z \neq \beta$. According to (7.8) and (7.9), the time dependence of the correlation function $C_h(r, t)$ (we consider only isotropic surfaces) is known around $r = 0$: expanding (7.7) for $r = 0$, we obtain $C_h(0, t) = \sigma^2(t) + \sigma^2(0)$ so that, at least as long as $\xi_\parallel(t)$ has not reached L,

$$C_h(0, t) \sim t^{2\beta}. \qquad (7.10)$$

On the other hand, the spatial dependence of the correlation function $C_h(r, t)$ is also known at fixed t, via (7.6), so that a complete scaling behavior of $C_h(r, t)$ can in principle be obtained. From (7.8,9) we see that the two correlation lengths $\xi_\parallel(t)$ and $\sigma(t)$ are related by

$$\sigma(t) \sim \xi_\parallel(t)^{\beta z}. \qquad (7.11)$$

As mentioned above, relations (7.8–10) are only valid at *short* times below a crossover time τ, at which the correlation length $\xi_\parallel(t)$ reaches the size L of the sample. Notice that, experimentally, rough surfaces are mostly found in a situation corresponding to this short time regime, where $\xi_\parallel(t)$ in much smaller than the lateral size L of the sample surface.

At $t = \tau$, we have $\xi_\parallel(t) = L$, from which $\tau \sim L^z$ follows. For times larger than τ, the growth will exhibit finite size effects as $\xi_\parallel(t)$ obviously cannot become larger than L. Replacing $\xi_\parallel(t)$ by L in (7.11), we find, setting $\alpha = \beta z$

$$\sigma \sim L^\alpha. \tag{7.12}$$

The growth evolves into a steady state for which there is no characteristic length scale below L: the whole sample surface has become a self-affine fractal.

To bridge between (7.10) and (7.12), the following scaling form has been proposed [7.17,18]:

$$\sigma(L,t) = L^\alpha f\left(\frac{t}{L^z}\right) \tag{7.13a}$$

with

$$f(x) \sim \begin{cases} x^\beta & \text{for } x \ll 1, \\ \text{const} & \text{for } x \gg 1, \end{cases} \tag{7.13b}$$

where α, β and z are related by $z = \alpha/\beta$ [from (7.9) and (7.12)]. Comparison of $C_h(r,0)$ with the static or fixed time correlation function $C_h(r)$ shows that the roughness exponent α is equal to the Hurst exponent H. Hence $d_f = d - \alpha$, and it is possible to determine d_f from (7.12).

7.2.3 Deposition and Deposition Models

Vapor deposition and sedimentation are natural examples of *deposition*. A very common industrial process is thin film deposition, which plays an important role in integrated circuit technology.

Irregular surfaces are frequently obtained. For example, beautiful cauliflower-like surfaces (see also Fig. 8.0) have been observed by Messier and Yehoda on vapor deposited graphite [7.18]. Frequently, the deposit shows a columnar structure (Fig. 7.5) [7.20–23]; for more references to experimental work we refer to [7.24].

Several deposition models have been explored. In these models, each particle falls from a random point at the top of a box, vertically or with a fixed oblique direction, until it reaches the growing deposit. Different rules may be chosen to determine how and where the particle comes to rest and sticks to the deposit. We present a few simple deposition models below. The simplest are *random deposition, random deposition with surface diffusion*, and *ballistic deposition*. Other models have also been investigated: *single-step ballistic deposition, restricted solid-on-solid model*, and *spatially correlated deposition*. A goal of these studies is to classify these models and the underlying physical phenomena into universality classes associated with the same set of critical ex-

Fig. 7.5. Columnar structure of iron films evaporated at oblique incidence on a glass plate at $300°$C [7.20]

ponents. Very different phenomena may belong to the same universality class. For instance, ballistic deposition, single-step ballistic deposition, and the restricted solid-on-solid model are believed to belong to the same universality class as the Eden model (see Sect. 4.3).

a) Random deposition. Random deposition (Fig. 7.6) is the simplest (but most unrealistic) deposition model. The particles fall vertically at a constant rate independently of one another and stick when they reach the top of a column of deposited particles [7.25].

As there is no horizontal correlation between neighboring columns, the surface is extremely rough. The heights of the columns follow a Poisson distribution and in any dimension d we have

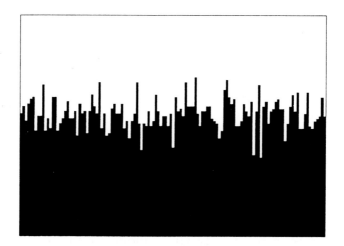

Fig. 7.6. Structure of the interface in the random deposition model, from a numerical calculation in $d = 2$ [7.15]

Fig. 7.7a,b. Random deposition model with surface diffusion: (a) schematic representation of the model in the triangular lattice, (b) structure of the interface from a numerical calculation in the square lattice [7.15]

$$\langle h \rangle \sim t \,,$$
$$\sigma \sim t^{\beta} \sim t^{1/2} \,. \tag{7.14}$$

This behavior does not depend on the sample size ($\alpha = 0$). The surface structure is compact, i.e., $d_f = d$.

b) **Random deposition with surface "diffusion".** In vapor deposition or in sedimentation a local diffusion is in fact always present. Family [7.25] has proposed a model (Fig. 7.7) in which a particle reaches the surface as in the random deposition model, but then is allowed to "diffuse" on the surface. The diffusion continues until the particle finds the column of minimum height inside a domain of finite size around the initial contact.

The surface diffusion generates a nontrivial correlation between the heights of the columns. As a consequence, the thickness of the surface grows with an exponent β different from the random deposition model. Using a Langevin equation approach, Edwards and Wilkinson [7.26] proposed the relations

$$\alpha = (3 - d)/2, \quad \beta = (3 - d)/4, \quad \text{and} \quad z = 2 \,. \tag{7.15}$$

Numerically, this result is well verified, with $\beta \cong 1/4$ and $\alpha \cong 1/2$ in $d = 2$. The computer simulations show that the surface is smoothed by the diffusion process, $d_f \cong 1.5$. It seems that the scaling behavior of the surface fluctuations does not depend on the distance allowed for diffusion on the surface.

c) **Ballistic deposition.** In the ballistic deposition model, first introduced by Vold [7.27], the particles fall onto the substrate following straight lines normal to the substrate. When a particle reaches the surface, it sticks on the first site encountered that is a nearest neighbor of a deposit site (Figs. 7.8, 7.9). This constraint allows for growth parallel to the substrate. As is seen in Fig. 7.9, there are many vacancies in the bulk, but as in the preceding models, the bulk is not fractal. Overhangs exist but do not affect the scaling behavior. Numerical calculations in $d = 2$ give [7.17,28]

$$\alpha \cong 1/2, \quad \beta \cong 1/3 \quad (d = 2) \,. \tag{7.16}$$

In practice there is no reason for the particles to fall vertically and a varia-

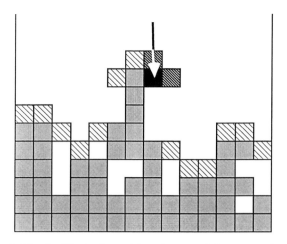

Fig. 7.8. Simplest model for ballistic deposition (*vertical deposition*). Gray squares represent deposited particles on a substrate (*the first row*). A new (*black square*) particle is added at one of the growth sites (*hatched squares*). After the addition of this new particle, two new growth sites are available (*dark hatched squares*)

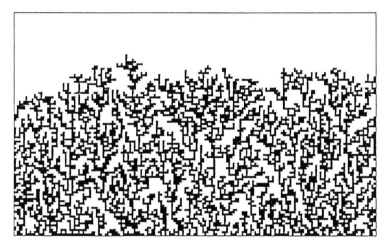

Fig. 7.9. Deposit obtained using the model of ballistic deposition described in Fig. 7.8 (numerical calculation in $d = 2$ [7.15]). Overhanging is present but does not dominate the scaling behavior

tion of ballistic deposition [7.29] is a model with an oblique angle of incidence (Fig. 7.10). In this case, which is very close to many experiments, columnar structures are observed [7.19–24,30]. The exponents β and α seem to vary continuously with the angle of incidence. As the angle increases from zero (normal incidence) to 90°, α approaches a value close to 1 and β a value close to $1/2$. For details of the scaling behavior we refer to [7.29,31].

d) Other models. *Single-step ballistic deposition* [7.28] (see Fig. 7.11) and the *restricted solid-on-solid* model (RSOS) [7.32] both belong to the universality

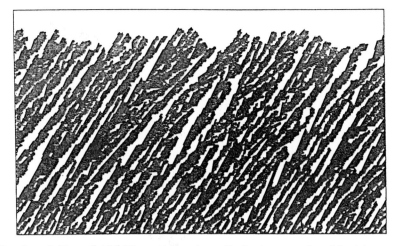

Fig. 7.10. Simulated film of 40 000 particles deposited at an angle of incidence of 50° with respect to the substrate normal [7.24]. This numerical 2*d* calculation can be compared with the *d* = 3 experimental structure in Fig. 7.5

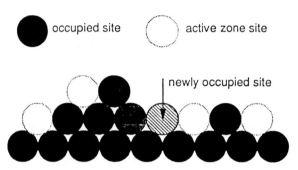

Fig. 7.11. Single-step ballistic deposition [7.28]. In this model a particle is added at a randomly selected site *i* such that the added particle is in a stable situation with respect to a vertical gravity field

class of ballistic deposition, but have the advantage of being simpler to analyze theoretically.

The single-step model can be mapped on a spin 1/2 system with non-Hermitian evolution operators. It has been solved exactly in $d = 1 + 1$ dimensions by Dhar [7.33] (the notation $d = 1 + 1$ indicates that the h coordinate plays a particular role). For a simpler physical approach see also [7.28]. In the RSOS model, neighboring columns have a height difference of at most one step and therefore the model can be mapped on a spin-1 system $(\Delta h = h(x) - h(x + \delta) = -1, 0, 1)$.

Horizontal correlations between falling particles can be superimposed on these processes, in the same way they have been superimposed in cluster growth processes [7.34] (see also Chap. 1). Such spatially correlated deposition has been studied by Medina et al. [7.35], Meakin and Jullien [7.36], and Margolina and Warriner [7.37]. How its behavior and universality class depends on the correlation range is not fully understood at present. Recently, a new class of growth models with surface diffusion has been proposed [7.38]. In these models the physical mechanism behind the surface smoothening is surface diffusion controlled by the nearest bonding energy. This means that the contribution of the surface diffusion to the height variation $\partial h/\partial t$ has the form $(\partial h/\partial t)_{\mathrm{dif}} = -K\nabla^2(\nabla^2 h)$. Actually, $\partial h/\partial t$ must obey a continuity equation $\partial h/\partial t = -\nabla \cdot \boldsymbol{j}$; the diffusion current $\boldsymbol{j} = -M\nabla\mu$, is the gradient of the chemical potential μ times a mobility M, and due to the nearest bonding energy, μ is proportional to the local curvature $\nabla^2 h$.

e) Theoretical and numerical calculations. To describe surface growth processes, Kardar, Parisi, and Zhang suggested the nonlinear Langevin equation (KPZ equation)

$$\frac{\partial \tilde{h}(\boldsymbol{x},t)}{\partial t} = \gamma\nabla^2\tilde{h}(\boldsymbol{x},t) + \lambda\left(\nabla\tilde{h}(\boldsymbol{x},t)\right)^2 + \eta(\boldsymbol{x},t), \quad \tilde{h} = h - \langle h\rangle, \qquad (7.17)$$

which contains a relaxation term with a coefficient γ associated with a surface tension, and a lateral growth contribution with a coefficient λ; η is a Gaussian noise [7.13,38].

For ballistic deposition and the Eden model (which is described in detail in Sect. 4.3) the nonlinear term is important ($\lambda \neq 0$) [7.39], while for random deposition one has simply $\gamma = \lambda = 0$. Random deposition with diffusion is less particular ($\lambda = 0$, $\gamma \neq 0$). It has been shown by Kardar et al. [7.39] that for nonzero λ, the exponents α and $z = \alpha/\beta$ are simply related by $\alpha + z = 2$.

Despite of the fact that the KPZ equation leads to interesting theoretical (and unsolved) problems, its experimental importance is very limited: there is presently no really convincing experimental confirmation of this equation. Many factors play a role in experimental growth: nonlinear effects, gravity effects, finite size effects, etc. Other nonlinear continuous equations have been proposed to describe the growth of various self-affine surfaces. For example, the Wolf-Villain-Lai-Das Sarma-Tamborenea model [7.38] and the "Montreal model" [7.40] (for further details see Chap. 4 in [7.8]).

It has also been realized recently that in many phenomena where rough interfaces are generated, such as fluid invasion in a porous medium [7.42,43] or growth of magnetic domains in thin films [7.44], the noise is "quenched". This means that the disorder (generating the fluctuations) is attached to the sample itself, and does not vary with time. To study the effects of quenched noise on interface growth, a DPD (directed percolation depinning) model based on directed percolation and directed surfaces has been proposed [7.42]. Note that the presence of quenched noise in the KPZ equation changes its universality class: for instance, $\alpha \cong (5 - d)/4$ [7.45]. This gives $\alpha \cong 0.75$ in $d = 1 + 1$ dimensions, a result which better agrees with experiments. It was suggested recently, that KPZ with quenched noise is in the same universality class as the DPD model [7.46].

7.2.4 Fractures

Everyone has noticed the close similarity that exists between the surface of a broken stone and that of a rock face, in spite of the considerable difference in scales. Scale invariance is clearly present. We show in Fig. 7.12a a fractured surface with a size of a few micrometers (TiB_2-AlN composite [7.47]) and in Fig. 7.12b a fracture surface of a few hundred meters (mountain rock face [7.48]).

Structural analysis of fractured surfaces on the basis of fractal geometry was initiated by Mandelbrot et al. [7.49], who found that the fractures of steel samples exhibited fractal geometry. Similar results have been reported since for various materials, including titanium [7.50], aluminum-particulate SiC composite [7.51], and aluminum alloys [7.52].

Fig. 7.12a,b. Fractures observed at two very different length scales: (a) air fracture of TiB_2–AlN composite [7.47]: the length of the white bar is about 50 μm; (b) rock face in the "Dames Anglaises" (Mont-Blanc): the overall size here is a few hundred meters [7.48]

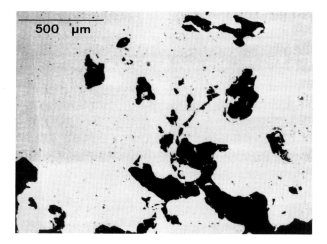

Fig. 7.13. Fracture of an aluminum alloy sample. The cuts of the surface were observed with a scanning electron microscope, using backscattered electron contrast. On the figure, the cut of the fractured aluminum alloy is in black, the white regions correspond to a nickel coating. (Courtesy of E. Bouchaud [7.52])

Figure 7.13 shows a cut of a fractured surface of an aluminum alloy [7.52]. A basic question is whether any correlation exists between the fractal dimension of the fracture and the physical parameters involved in the fracture process. Mandelbrot et al. [7.49] have found such a correlation between impact energy and fractal dimension. Other authors find no clear relation between the fractal dimension and mechanical properties. In particular, Davidson [7.51] failed to find any correlation between the fracture surface roughness and fracture toughness.

Interesting experimental studies have been performed by Bouchaud et al. [7.52]. They found the same fractal dimension, $d_f \cong 2.2$ over three decades of length scales, in three samples with different toughness values. This implies a universal value $\alpha \cong 0.8$ for the roughness exponent, i.e., independent of the type of fracture and the material (the roughness exponent α is sometimes called the roughness index and is denoted by ζ).

However, other experiments lead to significantly smaller values of α, $\alpha \cong 0.5$–0.6 either at very small length scales, or in the case of small crack propagation [7.53]. It has been suggested [7.54] that in this case the path chosen by a crack in a random environment should be such that the overall fracture energy is minimized. In a recent letter [7.55], Bouchaud and Navéos suggested the existence of a crossover length ξ such that at distances shorter than ξ, where the material has time to minimize its energy, the roughness exponent is $\alpha \cong 0.45$; at larger distances, the fracture is in a rapid crack regime and $\alpha \cong 0.8$.

Most numerical results in fractures have been obtained only in $d = 2$ [7.56]. Louis et al. [7.57] found, for instance, that the broken bonds of an elastic $d = 2$ network form a self-affine fractal with a fractal dimension $d_f \cong 1.25$. Very little is known in three dimensions. A major difficulty comes from the fact that even in an ideal homogeneous material a complex nonlocal stress-strain field develops as the material begins to break. This makes numerical calculations in $d = 3$ extremely time consuming, and no analytical approach is available at present.

Chapter 5 of this book is devoted to fracture models, and a review on fractals in fractography can be found in [7.58].

7.3 Diffusion Fronts: Natural Fractal Interfaces in Solids

Diffusion is a common phenomenon, which is the basis of many natural and industrial processes. Diffusion is generally expected to build smooth interfaces, and only one decade ago it was realized that diffusion could generate fractal interfaces [7.59]. The average progression of diffusing particles may follow a classical diffusion law. At atomic length scales, diffusion is the result of the Brownian motion of particles, and because of the random character of that motion, diffusion may create complex structures.

7.3.1 Diffusion Fronts of Noninteracting Particles

In Fig. 7.14, we show a schematic picture of diffusion into a two-dimensional triangular lattice. Particles shown as white disks invade the dark medium from a source kept at concentration one at the top of the picture. Each particle can hop at random to one of its six nearest-neighbor positions. There are six possible jump directions on condition that the final site is empty. After a given time, one obtains a random object with a complex geometry.

The related physical properties may be quite different from those of a smooth interface. On defining a *connection* between the particles within the lattice, different sets are obtained. Consider for instance particles which are

a)   b)

Fig. 7.14a,b. Schematic representation of diffusion into a triangular lattice. (a) At $t = 0$ all the particles are in the top row, which is kept filled ($p = 1$) at $t > 0$. (b) The particles diffuse by hopping at random onto their nearest-neighbor sites, the jump being forbidden if the arrival site is already occupied. A connection between particles is also introduced: here two particles are said to be connected (or in electrical contact) if they are first neighbors. The gray particle in (b) is connected to the two particles below it

Fig. 7.15. The diffusion front in two dimensions. The diffusion has been performed on a triangular lattice (which preserves the symmetry between occupied and empty sites). A steady state is considered and the concentration of diffusing particles varies from 0.8 at the top of the figure to 0.2 at the bottom. The particles connected to the source (at the source potential if the particles are conducting) are in green, isolated clusters of particles are in orange, and isolated clusters of empty sites in light blue. The diffusion front is in yellow

metallic atoms. One can say they are electrically linked if they are in contact, for example, with their nearest neighbors. This is the case for the light gray particle in Fig. 7.14b, which is *connected* to the two particles below it. Other types of connections can be involved, such as magnetic coupling or overlap of Rydberg orbitals.

This system can be conveniently pictured in a geographical description in which the set of particles *connected* to the source is called the "land". The land is shown in green in Fig. 7.15. In this geographical language the set of connected empty sites not surrounded by land is called the "sea", shown in blue. There naturally exist groups of particles not connected with the land; they are islands, shown in orange. There are also connected empty sites surrounded by land. They are lakes, which are shown in light blue in the figure. In this geographical description, the part of the land in contact with the sea is the seashore: it is colored yellow. We call this line the *diffusion front*. It is a very intricate line. Its geometry is that of a *self-similar fractal*.

The self-similar property of the front is evident in Fig. 7.16, where a front with a very large size is shown at different magnifications. In all these pictures

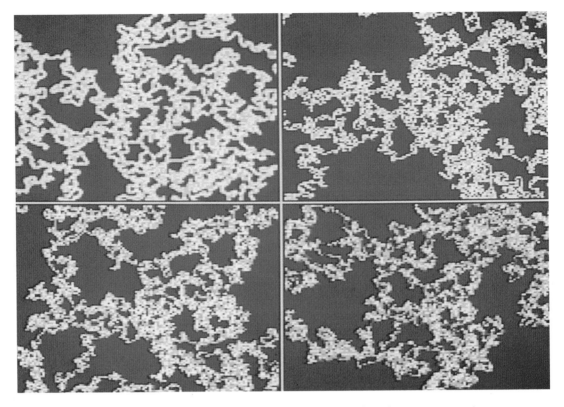

Fig. 7.16. Self-similarity of the diffusion front. The front has been generated on a square lattice with a very low concentration gradient. Only the front is shown in this picture. The difficulty of ordering these four different magnifications ($\times 1$, $\times 2$, $\times 4$, and $\times 8$) of the front is an indication of its self-similarity (from [7.60])

the visual impression is the same, which is qualitative evidence of the self-similarity of the front, or its fractal geometry. From numerical simulations [7.59], the fractal dimension of the front is found to be $d_{\mathrm{fr}} = 1.74 \pm 0.02$.

The diffusion problem is closely related to the site-percolation problem on the same lattice (see Chap. 2), with the same connection rules. As the concentration p of occupied sites varies with x, we called it *gradient percolation*. For large systems ($L \to \infty$), the structure of the diffusion front is identical to the structure of the hull of a large percolation cluster [7.61], see also Chaps. 1 and 2. The concentration of particles at the mean front position x_f is identical to the percolation threshold p_c,

$$p(x_f) = p_c. \tag{7.18}$$

This relation can be used to determine p_c very accurately: actually it represents the most powerful method of computing p_c in $d = 2$ [7.62], see also Sect. 2.7.3.

Numerical and theoretical studies [7.59,63] in $d = 2$ have indicated that the fractal dimension d_{fr} of the front is related to the critical exponent ν in percolation by $d_{\mathrm{fr}} = (\nu + 1)/\nu = 7/4$. This relation was proved by Saleur and Duplantier for the percolation hull [7.64]. As has been discussed in Sect. 2.2,

ν characterizes the divergence of the correlation length ξ at the percolation threshold [7.65], $\xi \sim |p - p_c|^{-\nu}$ (see (2.2)).

The diffusion front is conveniently described by its average width σ_f and the total number N_f of atoms which constitute it. Both quantities can be related easily to the concentration gradient dp/dx at the position of the front as we will show now.

First we consider σ_f. We see in Fig. 7.15 that, far from the front, islands or lakes are very small, whereas, near the front, their size becomes comparable to the width of the front. The islands correspond to the finite clusters in a percolation system, and the lakes correspond to the finite holes. The typical linear size of both quantities scales as ξ (see Sects. 2.2, 2.3). Relation (2.2) tells us that the size of the islands or lakes should increase when approaching the mean position of the front. But this size, even at x_f, is bounded due to the finite gradient of $p(x)$. The maximum typical size of islands and lakes is then given by the width σ_f of the front, which represents the only characteristic length scale in the problem, and we can assume

$$\sigma_f \sim \xi(x_f \pm \sigma_f). \qquad (7.19)$$

This assumption expresses our observation that islands or lakes near the front have a size comparable to the width of the front. Using (7.18) and (2.2) and expanding $p(x)$ around x_f we obtain

$$\sigma_f \sim |p(x_f \pm \sigma_f) - p_c|^{-\nu} \cong \left| \sigma_f \frac{dp}{dx}(x_f) \right|^{-\nu},$$

which gives

$$\sigma_f \sim \left| \frac{dp}{dx}(x_f) \right|^{-\alpha_\sigma}, \quad \text{where} \quad \alpha_\sigma = \nu/(1+\nu). \qquad (7.20)$$

Let us next calculate the number of atoms of the front in a square of side σ_f. If the linear extent of the front is L, this number is simply $N_f \sigma_f / L$. As the front is fractal, this same number scales with σ_f as $\sigma_f^{d_{\mathrm{fr}}}$. From this, we deduce $N_f \sim L \sigma_f^{d_{\mathrm{fr}} - 1}$, and hence

$$N_f \sim \left| \frac{dp}{dx}(x_f) \right|^{-\alpha_N}, \quad \text{where} \quad \alpha_N = 1/(1+\nu). \qquad (7.21)$$

As percolation is a critical phenomenon (see Sects. 2.2, 2.3) one can expect that the above results are *universal*. The exponents depend only on the dimensionality of the system (here $d = 2$), and not on the particular lattice structure (square, triangular, etc.). Hence, d_{fr}, α_σ, and α_N should be the same on any $d = 2$ lattice, as well as in off-lattice systems situations [7.66].

7.3.2 Diffusion Fronts in $d = 3$

The study of $d = 3$ systems reveals interesting results, showing that the diffusion front structure is qualitatively different from that obtained in $d = 2$ [7.70,71]. In $d = 3$, the diffusion front extends over a large concentration range, proportional to the diffusion length l_D. Over that range, almost all occupied sites belong to the front. In Fig. 7.19, we show a schematic picture of a simulated sample, where the concentration p of occupied sites is decreasing from 1 at the bottom to 0 at the top. Only the sites which are connected to the $p = 1$ plane are represented.

Two different behaviors are observed. In the central part of the sample, the diffusion front is a compact object with dimension 3: from this point of view it behaves like an ordinary solid, but any point of that solid can be reached from the outside, i.e., belongs to its surface. In this sense, the diffusion front is an ideally porous material. At the top of the sample, which is the critical region around the percolation threshold, the front has a fractal dimension $d_{\rm fr} \cong 2.5$.

Up to this point, we have considered the diffusion of noninteracting particles, but obviously this is not the most common physical case. For example, even in the presence of screening effects, ions intercalated in a material have a repulsive interaction. Atoms or molecules also interact via van der Waals forces. Even rare gas atoms have nonnegligible effective attractive interactions through substrate deformation.

7.3.3 Diffusion Fronts of Interacting Particles

With attractive or repulsive interactions between the diffusing particles taken into account, several types of behavior have been predicted [7.67], depending on the temperature. Below a critical temperature T_c, an ordered phase should be observed, with important deviation of the variation of concentration with time and position from the noninteracting case. Well above T_c, only local precursors of this phase are expected and the concentration profile should be only slightly different from the noninteracting case. In this high-temperature region there exists a finite thermal correlation length related to the interaction. On length scales above this correlation length, we expect that the overall geometry is similar to that of noninteracting particles.

Even in the low-temperature situation ($T < T_c$), complex structures may be found in *transient* regimes, maintaining the memory of the fractal geometry of the high temperature situation. This is illustrated in Fig. 7.17 for an attractive interaction. The picture shows a diffusion front at low temperature, with different, fixed concentrations at the two ends of the sample. At time $t = 0$, one has a constant gradient across the sample and a random distribution of the particles. Then diffusion occurs, with attractive interaction between the particles. The situation shown in Fig. 7.17 is observed before equilibrium is attained; it simulates the situation created by *quenching* a high-temperature diffusion

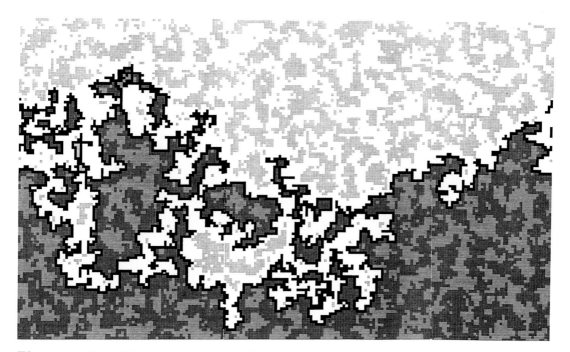

Fig. 7.17. This diffusion profile below T_c is obtained by freezing in, at time $t = 0$, a high-temperature steady state of diffusing particles with an attractive interaction (on a square lattice). The high-concentration region (land and lakes) is darkened for clarity. Spinodal decomposition leads to the growth of droplets of average size $R(t)$ (in the figure $R(t) \cong 7$ particles sizes); $R(t)$ now plays the role of the elementary length (intersite distance in the noninteracting case). The front has the same fractal geometry as in the noninteracting case

process. One observes that individual particles tend to aggregate, forming a droplet structure. This indicates a tendency for phase separation. This droplet structure has been observed experimentally by Mazur (Fig. 7.18) in a polyimide film in which diffusion and reduction of a silver salt have been obtained [7.68]. Wool has measured the fractal dimension of the diffusion front in this system and found $d_{\mathrm{fr}} \cong 1.74$ [7.69]. The same kind of structures should be observed in polymer–polymer interfaces [7.69].

7.3.4 Fluctuations in Diffusion Fronts

The study of the dynamical properties of the diffusion front has revealed a strongly fluctuating behavior [7.60]. Figure 7.20 shows the variation of the number of particles on the front (or length of the seashore) with time. There is something striking in this evolution. Although one generally expects a smooth behavior for diffusion, we observe here large, erratic fluctuations. Most of the time very little happens but from time to time there is a "catastrophic" event. This catastrophic event can be the connection or disconnection of an island; it can also be the transformation of a lake into a gulf, or, inversely, a gulf becomes obstructed. Then the shape of the seashore is drastically modified. There are giant fluctuations occurring on very small time scales [7.60].

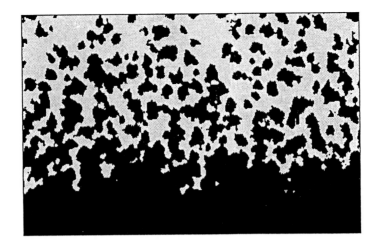

Fig. 7.18. Micrograph of a sample obtained by diffusion and reduction of silver chloride in a polyimide film [7.68] (the picture is digitized). The diffused metal silver is in black. The fractal dimension of the boundary of the largest cluster was found by Wool to be around 1.74 [7.69]

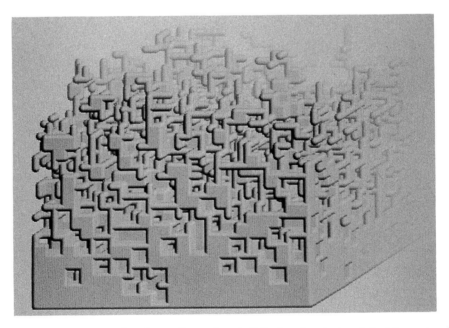

Fig. 7.19. Representation of a three-dimensional diffusion front simulated in a simple cubic lattice. In this representation, suggested to us by Mandelbrot, a single occupied site is shown as a sphere, two adjacent occupied sites are represented by a cylinder joining them, and if the four corners of a facet are occupied, all the facet is filled

Similar types of behavior appear in invasion of a porous medium by a fluid (see section 7.4), in growth of magnetic domains under the action of an external field [7.44], in thin-film corrosion [7.72], in some earthquake models [7.73], etc. These processes are intimately related to the notion of *self-organized-criticality* [7.74].

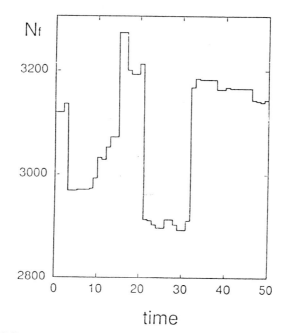

Fig. 7.20. During diffusion the front evolves in a very erratic manner. The figure shows the evolution of the number of particles in a 2d diffusion front in a given time interval. Most of the fluctuations are small, but from time to time large events appear [7.60]

The existence of these fluctuations may have many important physical consequences [7.75]: for example, in the case of diffusion of metallic atoms, they might be the origin of large impedance fluctuations occuring in electrical contacts. Even in a situation where diffusion is almost frozen, the observation of the fluctuations is in principle possible, because their characteristic time is much shorter than the characteristic diffusion time (the jump of one particle is sufficient to generate a large fluctuation, whereas (roughly) all the particles must jump to make the front diffuse by an intersite distance) [7.60,7.75].

Essentially two quantities have been studied in detail: the distribution of the "events" per unit time, and the time autocorrelation function of the front perimeter (or equivalently of the size of the region connected to the source). In principle they may be related to actual experiments such as impedance or conductivity measurements, they are also important in invasion experiments (see below). We cannot present here a detailed study of these fluctuations, and the interested reader is refered to [7.75,7.76]. The "size" of an event may be the number s of sites in a connected or disconnected cluster or its perimeter length h (or even any other physically relevant subset of a cluster). From percolation theory (see Chap. 2) we know that s scales with h according to

$$s \propto h^{d_f/d_{\mathrm{fr}}}. \tag{7.22}$$

where d_f is the fractal dimension of percolation clusters (note that the fractal dimension of the external perimeter of percolation clusters d_h is equal to d_{fr}, see Chap. 2). The number of events $N_{ev}(s)$ (respectively $N_{ev}(h)$) follows a power-law behavior as a function of s (resp. h):

$$N_{ev}(s) \propto s^{-\gamma_s}, \quad \text{where} \quad \gamma_s = 1 + (d_{fr} - d_{red})/d_f. \quad (7.23a)$$

$$N_{ev}(h) \propto h^{-\gamma_h}, \quad \text{where} \quad \gamma_h = 2 - d_{red}/d_{fr}. \quad (7.23b)$$

In these expressions, $d_{red} = 1/\nu$ is the fractal dimension of the red bonds (see Chap 2). These results have been shown to be valid in $d =1$, 2, and 6 dimensions [7.76], and are expected to be correct in any dimensions. Due to the existence of a concentration gradient, the diameter of the largest fluctuating clusters is limited to the order of the width σ_f of the diffusion front. As a consequence the above scaling applies in the limited range $s_m \leq s \leq s_M$, where the largest cluster size s_M is simply related to the front width (and then to the concentration gradient at x_f) by, $s_M \propto \sigma_f{}^{d_f} \propto |dp/dx|^{-\alpha_s d_f}$; s_m is the lower cut-off of the cluster size ($s_m = 1$).

Another important quantity is the *noise spectrum*: To find the noise generated by a fluctuating front, we need to know the autocorrelation function of these fluctuations: its Fourier transform. The power spectrum has the expression,

$$S(\omega) \propto A \int_{s_m}^{s_M} ds \, \frac{2 B s^{\varrho}}{4 B^2 s^{2\sigma} + \omega^2}. \quad (7.24)$$

where, $A \propto L^{d-1} |dp/dx|^{-\alpha_N}$, $B = \text{const}$, $\varrho = 1 - (d_{fr} - d_{red})/d_f$, $\sigma = d_{red}/d_f$, L is the size of the system.

It can be shown (see [7.76] for a discussion of this point) that in general two regimes can be observed: a white noise at low frequency $f < f_c$ and a Brownian noise at high frequency $f > f_c$, where $f_c \propto s_M{}^{d_{red}/d_f}$, is a crossover which depends via s_M on the concentration gradient at the interface.

7.4 Fractal Fluid–Fluid Interfaces

7.4.1 Viscous Fingering

Viscous fingering (see also Chaps. 1, 4, and 8) occurs when a low-viscosity fluid such as water displaces a high-viscosity fluid such as oil: the low-viscosity fluid forms "fingers", which extend into the high-viscosity fluid. This is because the interface between the low-viscosity fluid pushing against the high-viscosity fluid is basically unstable: protuberances in the interface cause an increase of the pressure gradient ahead of the protuberance in the high-viscosity fluid, which tends to further develop the protuberance (see for instance [7.77]). Often the fingers are not fractal.

However, under appropriate experimental conditions one can create fractal figures, resembling DLA patterns (see Fig. 7.21). Very probably, the key

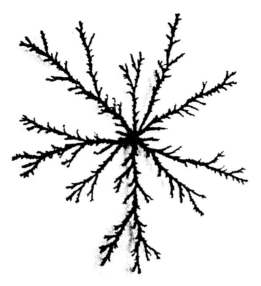

Fig. 7.21. Characteristic viscous fingers created by water injected into a radial Hele Shaw cell filled with a more viscous polymer solution (scleroglucan). (From [7.78])

to obtaining such fractal structures is that randomness must dominate over stabilizing processes (like interfacial tension) [7.78–80]. The detailed physical conditions responsible for this situation are not fully understood at present, and we refer to Chap. 1 for an extensive discussion.

7.4.2 Multiphase Flow in Porous Media

Real porous materials are irregular but in general porosity is not directly related to fractal geometry. The classical description considers porous media to be homogeneous on a macroscopic scale [7.81]. Nevertheless many porous media show fractal structures on some finite range of length scales. Three different classes of porous fractal structures may be distinguished [7.82]:

(1) fractal backbones (belonging to the mass fractals),
(2) fractal pore structures (belonging to the pore fractals) and
(3) cases where neither the solid medium nor the pore space is fractal but the pores have a fractal surface (surface fractals) (see Fig. 7.1).

The notion of fractal geometry has been used to study very different porous objects on very different scales, ranging from microscopic length scales in the case of aerogels [7.83] or organic polymer resins [7.84], to macroscopic length scales in the case of sponges, corals, sandstone or volcanic stones and even caves. For example, aerogels [7.83] may have a mass fractal geometry between 10 and 300Å, whereas sandstones have been claimed by Katz and Thomson to possess fractal structures from 10 nm to 0.1 mm [7.85]. The internal surface of many ultraporous solids seems to be fractal when one measures the quantity of adsorbed gas as a function of the molecular size [7.86]. Low permeability materials have been considered in relation to percolation theory [7.87].

The physical properties of porous media have been intensively studied [7.82], but the question of why and how they are formed has been little addressed in the literature. Among the possible mechanisms leading to fractal porosity are aggregation, spinodal decomposition and consecutive leaching [7.88], internal degasing, corrosion [7.88], and fracture.

Understanding fluid transport in porous media is of great industrial importance. Oil recovery, or preventing pollution of underground water by agriculture are some of the fields of application in this domain. Because of this practical interest, multiphase flow in porous media has been the subject of many theoretical and experimental studies.

The transport of a monophasic fluid across a well-connected porous medium is limited by its finite *permeability* k, which corresponds to the "conductivity" with regard to fluid transport. The permeability k, pressure difference ΔP across the sample, and volumetric flow rate Q, are related through Darcy's law,

$$Q = -\frac{kA}{\mu}\frac{\Delta P}{L}\,, \tag{7.25}$$

where μ is the viscosity of the fluid, A the normal cross-sectional area, and L the length of the sample in the direction of the flow. For polyphase flow the same equation holds when the permeability is large enough, μ then being an effective viscosity. The permeability can, however, be anomalous in many transport regimes, as we shall now describe. A polyphasic flow then becomes extremely complicated, due to voids (or pores and throats), and due to irregularities in the surface of these voids. In this case Darcy's law (7.25) is no longer applicable and fractal structures may appear in *nonfractal* porous materials; even a "narrow" random distribution of pore sizes may induce very complex, fractal figures. Because of its possible technological interest, we shall now discuss this point in detail. We shall mainly consider immiscible displacement of a liquid by a nonwetting one (drainage) in $2d$ systems.

a) Immiscible displacements: general case. A porous medium may be schematically represented by a set of pores and throats situated respectively at the sites and bonds of a regular lattice. Throats and pores are described by their radii (Fig. 7.22). The radii are randomly chosen, according to given distribution laws that may be different for pores and throats. The network is then characterized by five parameters: the mean radii of pores and throats, the widths of their distributions, and the distribution of the inter-pore distances. This allows an independent variation of the porosity (fraction of the bulk volume of the porous sample, which corresponds to pores and throats) and permeability.

First, we examine the invasion of individual pores. We consider a nonwetting fluid with viscosity μ_2 displacing a fluid with viscosity μ_1. Fluid 2 can invade a throat ij (with radius r) between two pores i and j only if the pressure

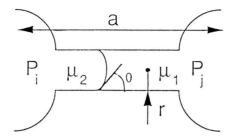

Fig. 7.22. Schematic representation of a throat of radius r linking two pores i and j. P_i and μ_2 are the pressure and viscosity of the invading fluid 2 in pore i, P_j and μ_1 are the pressure and viscosity of the displaced fluid 1 in pore j

difference $P_i - P_j$ is larger than the capillary pressure $\Pi = 2\gamma\cos\theta/r$ at the interface between the two fluids (γ is the surface tension and θ the contact angle, pore i is occupied by the nonwetting fluid and pore j by the wetting fluid). The flow rate in the throat is then given by

$$Q_s = \frac{\pi r^4}{8a\mu}(P_i - P_j - \Pi),\qquad(7.26)$$

with the condition $\Pi < P_i - P_j$. The effective viscosity μ is a weighted combination of the viscosities μ_1 and μ_2. Equation (7.26) shows that the transport process depends strongly on local parameters such as the throat radius or the surface quality (through θ).

In a porous network where all pores are connected, the pressure difference $P_i - P_j$ between neighboring pores depends on the flow rates from all other throats. This is a complicated many-body problem, and both experimental studies and computer simulations have been used to shed light on it.

Various experiments have been carried out in $d = 2$ and $d = 3$ lattices (random packing of glass marbles or crushed glass) or in $d = 2$ transparent etched networks [7.77,89]. To perform controlled experiments, artificial porous media have been built; an example of an experimental setup is presented in detail in [7.90]. Experiments and computer simulations have both shown that injection in porous media is governed by two dimensionless parameters: the viscosity ratio $M = \mu_2/\mu_1$ and the capillary number C, which is the ratio of viscous forces to capillary forces:

$$C = \frac{Q\mu_2}{A\gamma\cos\theta}.\qquad(7.27)$$

Lenormand and coworkers [7.91,92] have experimentally found three main types of interface morphologies, depending on the two parameters M and C (Fig. 7.23a–c). Figure 7.23a corresponds to the highest C and M values and shows a flat interface at the scale of a few pores. This regime is called the *stable displacement* regime: the dominant force comes from the viscosity of the *injected* fluid, and the capillary effects and pressure drop in the displaced fluid

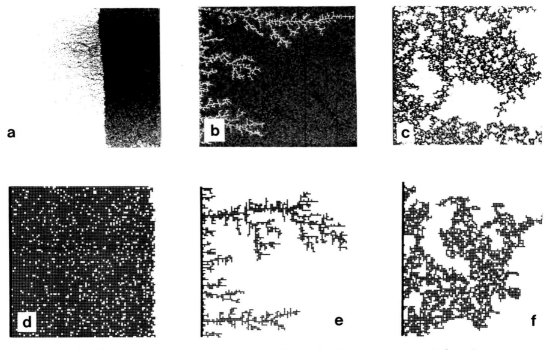

Fig. 7.23. (a–c) Results of experiments carried out by Lenormand *et al.* [7.92] for various values of M and C. (a) Air displacing oil: $\log M = 2.0$; $\log C = -1.0$. (b) Air displacing a very viscous oil: $\log M = -4.7$; $\log C = -6.3$. (c) Mercury displacing air: $\log M = 1.9$; $\log C = -5.9$. (d–f) Numerical simulations by Lenormand *et al.* of two-phase flow in porous media (25×25 and 100×100 networks) [7.92] for values of M and C very close to those in the experiments. (d) $\log M = 1.9$; $\log C = -0.9$. (e) $\log M = -4.7$; $\log C = -5.7$. (f) $\log M = 1.9$; $\log C = -5.9$

are negligible. Figure 7.23b represents the *viscous fingering* regime. Here, the dominant force comes from the viscosity of the *displaced* fluid and again the capillary effects and pressure drop in the displaced fluid are negligible. The interface has a tree structure without loops. Figure 7.23c is an example of the *capillary fingering* regime. The principal force is now due to capillarity. The injection is performed at very small C values corresponding to low flow rates Q. In the transition regions intermediate situations are observed.

Lenormand et al. [7.92] have combined these observations in a "phase" diagram in the (M, C) plane, which is shown in Fig. 7.24: region 'a' corresponds to the *stable displacement* regime, region 'b' corresponds to the *viscous fingering* regime, and region 'c' corresponds to the *capillary fingering* regime. Viscous displacement describes regions 'a' and 'b', in which viscosity is dominant, so that the growth is governed by the pressure field between the entrance and the exit (nonlocal effects).

Numerical calculations of the coupled hydrodynamic equations confirm this behavior [7.93]. Figure 7.23d–f can be favorably compared to the corresponding experimental patterns (Fig. 7.23a–c).

One can distinguish three basic mechanisms generating the different structures in regions 'a', 'b', and 'c':

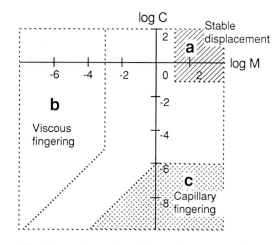

Fig. 7.24. "Phase" diagram for the various invasion regimes as a function of parameters M and C [7.92]

a) If M is large, the system can be described by *diffusion limited annihilation* ("anti-DLA")[7.93,94], which leads to nonfractal interfaces (region 'a'). In this model, a particle (representing the invading fluid) is released near a compact aggregate (corresponding to the pore network filled by the other fluid). The particle moves at random until it reaches an aggregate site, then both the particle and the aggregate site are removed and a new particle is released.

b) If the viscosity ratio M is small (viscous fingering in region 'b') the process is described by *diffusion limited aggregation* (DLA), see Chaps. 1, 4, 8, and 10.

c) If capillary forces are dominant, invasion percolation (see [7.95] and references therein) describes the low invasion rate of the nonwetting fluid (region 'c'). Pores are successively invaded; at each step, invasion proceeds through the throat of largest diameter connected to the already invaded region (see also Sects. 2.6.3, 4.3.3, and 8.2.5).

Finally, we note that if the invading fluid is more wetting than the displaced one, the porous medium is homogeneously invaded. In this case, a self-affine fractal interface between invading and displaced fluids has been predicted [7.96] and observed experimentally [7.97,99] in quasi-two-dimensional cells.

However, in various $2d$ experiments the roughness exponent α is found to be between 0.65 and 0.91: Horvath et al. [7.97] find $\alpha \cong 0.81$ in a fixed-rate invasion of a $2d$ Hele-Shaw cell filled with glass beads; Rubio et al.[7.98] obtain $\alpha = 0.73 \pm 0.03$ with a similar system; He et al. [7.99] find values of α ranging between 0.65 and 0.91 again in Hele-Shaw cells filled with glass beads; and Buldyrev et al [7.42] find for imbibition of blotting paper $\alpha = 0.65 \pm 0.05$. The rapture lines in paper sheets were found to have a similar value of the roughness exponent [7.43]. Bacterial colonies also show self-affine interface growth, with a roughness exponent around 0.78 [7.100,101]. The reason for this dispersion of values of α is at present not fully understood. The roughness seems to be sensitive to the flow conditions (situation in the phase diagram, see Fig. 7.24) and to the existence of nonlocal viscous pressure effects, and crossovers between various universality classes may be encountered.

In addition to the continuum models mentioned in Subsect. 7.2.3e, we should point out the studies on wetting-fluid invasion by Robbins and collaborators who find numerically a roughness exponent $\alpha \cong 0.81$ [7.102]. Another interesting model, the DPD model, which is related to directed percolation (see Sect. 2.6.4 for the definition of directed percolation), has been recently proposed [7.41,42,103,104]. The model yields $\alpha = \nu_\perp / \nu_\parallel \cong 0.63$ (where $\nu_\perp$ and $\nu_\parallel$ are the correlation length exponents normal and parallel to the rough surface), in reasonable agreement with the blotting paper imbibition experiment. The dynamical exponent z in $d + 1$ has been identified recently by Havlin et al [7.105] to be equal to $d_{\min}$ of isotropic percolation in d dimensions. Solving the KPZ equation in the presence of quenched noise lead to $\alpha \cong 0.75$ [7.45]. See also Amaral et al. [7.46] which argue that KPZ with quenched noise is in the same universality class of the DPD model. The existence of a nonuniform noise distribution can also lead to anomalous roughness exponents. This is obtained, for example, by using uncorrelated "noise" $\eta(x, t)$ that obeys a power-law distribution, $P(h) \propto \eta^{-\mu-1}$ [7.103,106,107]. Effects of the gradient of quenched noise, which is relevant to explain evaporation effects on the interface in the imbibition experiments, have been studied by Amaral et al [7.108].

b) Slow invasion by a nonwetting fluid under gravity. In most three-dimensional systems, the presence of gravity cannot be neglected. This has been shown in an experimental study of slow invasion of Wood's metal into homogeneous crushed glass [7.109]. In this experiment, Wood's metal, a low-melting-point liquid alloy, is injected at the bottom of a vertical evacuated crushed glass column of 10 cm inner diameter. The flow velocity is low (a few millimeters per hour) so that viscous pressure losses can be neglected, and the capillary number is much smaller than unity: this corresponds to region 'c' of the phase diagram in Fig. 7.24. When the front has reached a given height, the injection is stopped and the liquid is solidified. Horizontal sections of the front corresponding to various heights z are analyzed. The pictures are digitized into a square lattice of pixels and "invaded" or "empty" pixels are discriminated by a threshold procedure. Then the density-density correlation function $C(r, z)$ of the Wood's metal distribution in horizontal planes at height z is determined. The result is shown in Fig. 7.25. At low r, $C(r, z)$ has a power-law behavior, $C \sim r^{D(z)-3}$, whereas at large r, $C(r, z)$ becomes a constant, equal to the average saturation (concentration of invaded pores) at the height z. The z-dependent exponent $D(z)$ plays the role of an effective fractal dimension.

In these experiments, due to the high density of Wood's metal, the influence of gravity is important as it leads to a linear variation of the pressure in the liquid metal with height z. To model this behavior, one may think of a simple extension of the invasion percolation model (see Sect. 2.63), *gradient invasion percolation*, where in a lattice of size L^d random numbers between z/L and $1 + z/L$ are assigned to each bond. This way, the "resistance" to invasion increases linearly with z. The invasion starts at $z = 0$ and continues as in ordinary invasion percolation. The process is stopped before it reaches the top of the lattice. We expect that the front of the invaded region is similar to the front generated by gradient percolation [7.110], but contrary to gradient percolation, there are no islands of invaded bonds. The invaded region here corresponds to the infinite cluster in percolation.

Gouyet et al. have studied the scaling properties of gradient percolation [7.71,110], which allows for a quantitative understanding of the above experiments. For instance, the correlation function $C(r, z)$ has been fitted to its counterpart in the gradient percolation simulation (Fig. 7.25). From this fit one can determine the fractal dimension d_f of the growing pattern. One obtains $d_f \cong 2.4$, close to the value $d_f \cong 2.5$ found in the numerical calculations.

c) Fluctuations. As for diffusion fronts, one can observe *fluctuations* of all sizes during the progression of invasion fronts. We have shown above that *gradient invasion percolation* was appropriate to describe *slow invasion by a nonwetting fluid* in presence of a gravity field. As a consequence all the results of Sect. 7.3.4 and in particular (7.23), apply to the fluctuations of the interface created during the invasion process. The dynamics of the fronts then consists in sudden bursts [7.111,112] where large regions of the porous medium may be invaded. We have proposed to call it "interface bursting" in opposition to "interface depinning" which is observed in self-affine interfaces [7.76] and which has a different scaling behavior. These sudden burst have been clearly observed by Thompson and coworkers [7.113].

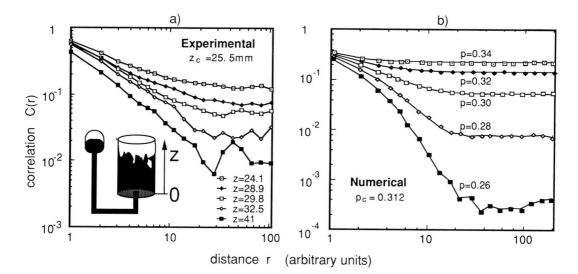

Fig. 7.25. Comparison of (a) experimental [7.109] and (b) calculated [7.110] correlation function $C(r)$ for various cuts corresponding to different heights z in the injection experiment (there is no one-to-one correspondence between these two sets of curves). A schematic representation of the experimental setup is shown in the inset

The slow invasion of a porous medium pertain to the general class of processes called *self-organized-criticality* (SOC) [7.74]. The above mentioned "bursts" are usually called "avalanches" (see [7.8], Chap. 2) in such processes. As in the case of diffusion fronts the noise generated during the injection of a fluid, which can be measured (see for instance [7.113]) presents two regimes: a white noise at low frequency, and a Brownian noise at high frequency, separated by a crossover depending on the pressure gradient.

It can be interesting to add here some words about the fluctuations in the *imbibition* regime. In this regime, self-affine interfaces are created, and the dynamical behavior belongs to a different universality class. We have already described above via its roughness exponent the static characteristics of the interfaces generated during imbibition. Cieplak et al. [7.96], in their studies of immiscible displacement in porous media, pointed out the existence of a critical contact angle θ_c, leading to a transition between drainage and imbibition. Martys et al. studied the fluctuations behavior in these systems [7.114], and independently they found the expression (7.23a) in the case of drainage: they proposed to extend this result to imbibition. In the latter case, the interface is pinned by the quenched disorder present in the system (here the porous medium), and the fluctuations of the front are due to depinning. The transition at θ_c between drainage and imbibition is accompanied by a "bursting-depinning" transition. Depinning processes are very general and can be found in various physical phenomena. Many aspects of the roughening and depinning behaviors are detailed by Kertész and Vicsek in [7.8] Chap. 4. See also [7.76] for a discussion on interface bursting and depinning.

d) Miscible displacements in porous media. The case of miscible displacements in porous media has been investigated in detail by Måløy et al. [7.115] (see also [7.79]) who have studied the structure of hydrodynamic dispersion in a $d = 2$ porous system. The experiment is done in a two-dimensional porous model 40 cm in diameter consisting of a monolayer of thin (1 mm) glass beads packed at random between two plastic sheets. This system is filled with a transparent glycerol–water mixture and the dispersion experiment is performed by injection of a black glycerol–water–Negrosine solution, in the center of the sample. Måløy et al. found that the concentration contours of the black agent are self-affine curves [7.115]. The result is shown in Fig. 7.26.

The self-affine contour, or dispersion front, reflects the probability distribution of where diffusing tracers end at time t. Its fractal dimension is $d_f = 1.42 \pm 0.05$. This value indicates clearly that this front belongs to a different universality class from that of diffusion fronts described in Sect. 7.3, and consequently has a different origin.

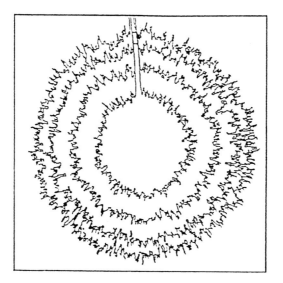

Fig. 7.26. Time development of the fractal contour during the dispersion of colored glycerol into glycerol of the same viscosity and density [7.115]

7.5 Membranes and Tethered Surfaces

We shall close this brief survey of fractal interfaces with a few words on fractal membranes. Indeed, not all membranes are fractals, but various physical processes may produce membranes with fractal geometries: we shall consider here only tethered surfaces [7.116].

A tethered or a polymeric surface is a surface, that preserves the connectivity between each pair of neighboring points on the surface. A piece of paper is a very simple example of a tethered surface: it conserves connectivity even when it is crumpled. A more general class of tethered surfaces is the surface consisting of a system of particles (atoms or monomers) bonded to form a regular $(d-1)$-dimensional array embedded in a d-dimensional space. The lattice structure and the precise shape of the interaction potential are not relevant, the crucial point is that the link between the particles cannot be broken.

In two dimensions, polymer chains represent a very simple case of a tethered "surface" (tethered contour). The attractive van der Waals interaction between the monomers tends to bend the joint between neighboring monomers. This leads to a compact structure $(d_f = d)$ at low temperature and to a fractal structure well represented by a self-avoiding walk with $d_f = 4/3$ at high temperature. The θ point separates both phases. At this temperature, the structure belongs to the same universality class as the diffusion front and the percolation hull, with $d_f = 7/4$ [7.117].

In three dimensions, tethered surfaces can be created by cross linking of monomers on liquid–liquid or liquid–gas or solid–gas interfaces (as the poly (methyl-methacrylate) extracted from the surface of sodium montmorillonite clays [7.118]). Polymeric tethered surfaces can also be obtained by cross polymerization of lipid bilayers (as done by Fendler and Tundo [7.119]).

These surfaces can exist in a large number of configurations, despite the strong constraint of connectivity. The configurations can be distinguished and classified by an order parameter, which can be defined as the correlation function between unit vectors normal to the surface. The (ordered) flat conformation (where the order parameter has a nonzero value) corresponds to the low-temperature situation. The (disordered) crumpled conformation (where the order parameter is zero) corresponds to the high-temperature situation. The transition is called the crumpling transition [7.116].

Tethered surfaces are in general surfaces with a repulsive interaction between monomers (resistance to bending of membranes with covalently bonded molecules). The transition between the flat and the crumpled geometry is a result of the competition between forces tending to flatten the surface (simulated by a coupling J between the normals of nearest neighbor plaquettes) and the entropy tending to crumple the surface. Figure 7.27 shows four equilibrium configurations of a tethered membrane for various values of the reduced interaction $k = J/k_B T$.

The tethered surfaces in the crumpled phase show interesting fractal behavior. Due to the presence of the excluded volume interactions (the usual physical case), the radius of gyration increases with the linear size L of the uncrumpled surface as $R_g \sim L^{\nu_t}$, where $\nu_t \cong 0.8$ in $d = 3$ space. Since L scales with the mass M of the uncrumpled membrane as $L \sim M^{1/2}$, the fractal dimension of the crumpled membrane is simply

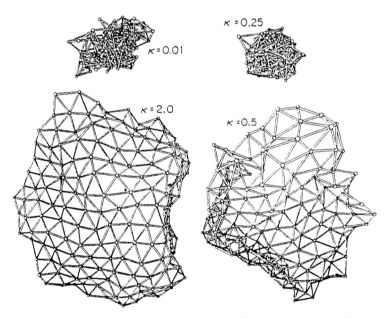

Fig. 7.27. Equilibrium conformations of hexagonal membranes constructed from triangular plaquettes. These are obtained for four different values of the interaction k between the normals of adjacent plaquettes. The values $k = 0.01$ and $k = 0.25$ correspond to a crumpled phase, $k = 0.5$ is close to the crumpling transition, and $k = 2.0$ corresponds to a flat phase [7.116]

$$d_f = \frac{2}{\nu_t} \cong 2.5 \,. \tag{7.28}$$

This is in agreement with a Flory calculation [7.116], which for $d = 3$ gives

$$d_f^{\text{Flory}} = \frac{5}{2} \,. \tag{7.29}$$

The exponent $\nu_t \cong 0.8$ can be very easily obtained in a table-top experiment by simply crumpling pieces of paper for various sizes L [7.120].

The structure of a tethered surface is very different from that of a diffusion front in a three-dimensional system (Sect. 7.3), despite their common character of being interfaces separating two and only two space regions, and of having approximately the same fractal dimension, $d_f \cong 2.5$. The diffusion front does not satisfy the tethered surface criterion (it is impossible to decrumple it). On the other hand, in $d = 2$, both diffusion fronts and polymer chains can be decrumpled. A comprehensive review on the statistical mechanics of fractal membranes and interfaces has been given by Nelson [7.121].

7.6 Conclusions

In this chapter, we have drawn attention to a variety of concrete and abstract problems connected with irregular surfaces. Most of the examples presented were devoted to the question of the formation of real fractal interfaces and we have discussed several models attempting to mimic their structure and the way they are created.

We have not examined the physical properties related to the fractal structure of these interfaces. Some of these aspects (adsorption on fractal surfaces, fractal electrodes, and scattering of light, X-rays, or neutrons by fractal interfaces) are discussed in Chaps. 3, 6, and 8.

Finally, we would like to conclude with several general comments about the importance of the concept of fractal geometry for irregular surfaces. Fractals are found at the confluence of observation and modeling [7.122]. Most of the surfaces existing in nature are irregular. The real geometry of a surface is the result of conflicting forces. Some forces tend to smooth surfaces, for example surface tension. Simultaneously some disorder may exist, for example due to thermal excitations. Disorder can also be found in situations where the time to reach an equilibrium is very large. Such situations are very common in nature. Think, for example, of the time of erosion for a mountain. In such circumstances the geometry is totally out of equilibrium.

It is noticeable that most far-from-equilibrium mechanisms produce fractals. Even very elementary processes, such as aggregation or diffusion, build fractals. It seems that the connection of random processes with fractality is ubiquitous

[7.122]. From this point of view, one should not be surprised to find fractal surfaces in very different situations. This field of study is only just in the beginning.

Acknowledgements. We wish to acknowledge A. Bunde, F. Family, H. Herrmann, M. Kolb, J.-P. Korb, P. Maaß, A. Margolina, R. Nossal, and T. Vicsek for many helpful discussions.

References

7.1   P. Pfeifer, M. Obert, in: *The Fractal Approach to Heterogeneous Chemistry*, ed. by D. Avnir (Wiley, New-York 1989), p. 11

7.2   B.B. Mandelbrot: *The Fractal Geometry of Nature* (W.H. Freeman, New York 1982)

7.3   R. Voss, in: *The Science of Fractal Images*, ed. by H.P. Peitgen, D. Saupe (Springer, Berlin, Heidelberg 1988), p. 21

7.4   D. Saupe, in: *The Science of Fractal Images,* ed. by H.P. Peitgen, D. Saupe (Springer, Berlin, Heidelberg 1988), p. 71

7.5   C. Roques-Carmes, private communication

7.6   M.W. Mitchell, D.A. Bonnell: J. Mater. Res. **5**, 2251 (1990)

7.7   A.-L. Barabási, H. E. Stanley: *Fractal Concepts in Surface Growth* (Cambridge University Press, Cambridge, 1995).

7.8   A. Bunde, S. Havlin, eds.: *Fractals in Science* (Springer, Berlin 1994)

7.9   P. Meakin: Phys. Rep. **235**, 189 (1993)

7.10   T. Halpin-Healey, Y.-C. Zhang: Phys. Rep. **254**, 215 (1995)

7.11   B.B. Mandelbrot, in: *Fractals in Physics*, ed. by L. Pietronero, E. Tossati (North-Holland, Amsterdam 1986), p. 3

7.12   J. Feder: *Fractals* (Plenum, New York 1988); J. F. Gouyet: *Physics and Fractal Structures* (Springer, New York,1995)

7.13   T. Vicsek: *Fractal Growth Phenomena,* 2nd ed. (World Scientific, Singapore 1991)

7.14   H.E. Hurst: Trans. Am. Soc. Civ. Eng. **116**, 770 (1951)

7.15   F. Family, in: *Universalities in Condensed Matter*, ed. by R. Jullien, L. Peliti, R. Rammal, N. Boccara (Springer, Berlin, Heidelberg 1988), p. 193;
       See also F. Family: Physica A **168**, 561 (1990) (Proc. of the 3rd Bar-Ilan Conf. on Frontiers in Condensed Matter, Tel-Aviv, January 1990)

7.16   R.F. Voss, in: *Fundamental Algorithms in Computer Graphics*, ed. by R.A. Earnshaw, (Springer, Berlin, Heidelberg 1985), p. 805

7.17   F. Family, T. Vicsek: J. Phys. A **18**, L75 (1985)

7.18   J. Kertész, D.E. Wolf: Physica D **38**, 221 (1989)

7.19   R. Messier, J.E. Yehoda: J. Appl. Phys. **58**, 3739 (1985)

7.20   T. Hashimoto, K. Okamoto, K. Hara, M. Kamiya, H. Fujiwara: Thin Solids Films **91**, 145 (1982)

7.21   H.J. Leamy, A.G. Dirks: J. Appl. Phys. **49**, 3430 (1978)

7.22   H.J. Leamy, G.H. Gilmer, A.G. Dirks: Curr. Topics Mater. Sci. **6**, 309 (1980)

7.23   O. Geszti, L. Gosztola, E. Seyfried: Thin Solid Films **136**, L35 (1986)

7.24   M.J. Brett: J. Mater. Sci. Lett. **8**, 415 (1989);
       see also *Deposition Processes* (parts I&II), MRS Bulletin **13**, Nos.11&12 (1988);
       for a comparison with experiments, see R.N. Tait, T. Smy, M.J. Brett: Thin Solid Films **187**, 375 (1990)

7.25   F. Family: J. Phys. A **19**, L441 (1986)

7.26   S.F. Edwards, D.R. Wilkinson: Proc. R. Soc. London, A **381**, 17 (1982)

7.27   M.J. Vold: J. Coll. Science **14**, 168 (1959)

7.28   P. Meakin, P. Ramanlal, L.M. Sander, R.C. Ball: Phys. Rev. A **34**, 5091 (1986)

7.29   J. Krug, P. Meakin: Phys. Rev. A **40**, 2064 (1989);
       P. Meakin, J. Krug: Europhys. Lett. **11**, 7 (1990)

7.30   N.G. Nakhodkin, A.I. Shadervan: Thin Solid Films **10**, 109 (1972)

7.31   P. Meakin: Physica D **38**, 252 (1989)

7.32   J.M. Kim, J.M. Kosterlitz: Phys. Rev. Lett. **62**, 2289 (1989)

7.33   D. Dhar: Phase Transitions **9**, 51 (1987)

7.34   A. Bunde, H.J. Herrmann, A. Margolina, H.E. Stanley: Phys. Rev. Lett. **55**, 653 (1985)

7.35   E. Medina, T. Hwa, M. Kardar, Y. Zhang: Phys. Rev. A **39**, 3053 (1989)

7.36   P. Meakin, R. Jullien: Europhys. Lett. **9**, 71 (1989)

7.37   A. Margolina, H.E. Warriner: J. Stat. Phys. **60**, 809 (1990)

7.38   J. Villain: J. Phys. I **1**, 19 (1991);
       D.E. Wolf, J. Villain: Europhys. Lett. **13**, 389 (1990);
       S. Das Sarma, P. Tamborenea: Phys. Rev. Lett. **66**, 325 (1991);
       Z.W. Lai, S. Das Sarma: Phys. Rev. Lett. **66**, 2348 (1991)

7.39   M. Kardar, G. Parisi, Y.C. Zhang: Phys. Rev. Lett. **56**, 889 (1986);
       Numerical solutions of the KPZ equation can be found in: J.G. Amar, F. Family: Phys. Rev. A **41**, 3399 (1990)

7.40   T. Sun, H. Guo, M. Grant: Physica A **40**, 6763 (1989)

7.41   S. Havlin, A.-L. Barabási, S. V. Buldyrev, C. K. Peng, M. Schwartz, H. E. Stanley, T. Vicsek, in: *Growth Patterns in Physical Sciences and Biology*, ed. by J. M. Garcia-Ruiz, E. Louis, P. Meakin, L. M. Sander (Plenum Press, New York 1993)

7.42   S.V. Buldyrev, A.L. Barabási, F. Caserta, S. Havlin, H.E. Stanley, T. Vicsek: Phys. Rev. A **45**, R8313 (1992); L.A.N. Amaral, A.L. Barabási, S.V. Buldyrev, S. Havlin, H.E. Stanley: Fractals **1**, 818 (1993); S.V. Buldyrev, A.L. Barabási, S. Havlin, J. Kertész, H.E. Stanley, H. Xenias: Physica A **191**, 220 (1992); S. V. Buldyrev, S. Havlin, H. E. Stanley: Physica A **200**, 200 (1993); S. V. Buldyrev, S. Havlin, J. Kertész, A. Shehter, H. E. Stanley: Fractals **1**, 827 (1993)

7.43   J. Kertész, V.K. Horváth, F. Veber: Fractals **1**, 67 (1993)

7.44   Y. Imry, S. Ma: Phys. Rev. Lett. **35**, 1399 (1975); G. Grinstein: J. Appl. Phys. **55**, 2371 (1984); A. Kirilyuk, A. Ferré, D. Renard: Europhys. Lett. **24**, 403 (1993)

7.45   Z. Csahók, K. Honda, T. Vicsek: J. Phys. A **26**, L171 (1993); Z. Csahók, K. Honda, E. Somfai, M. Vicsek, T. Vicsek: Physica A **200**, 136 (1993)

7.46   L. A. N. Amaral, A.-L. Barabási, H. E. Stanley: Phys. Rev. Lett. **73**, 62 (1994); L. A. N. Amaral, A.-L. Barabási, S. V. Buldyrev, S. T. Harrington, S. Havlin, R. Sadr-Lahijany, H. E. Stanley: Phys. Rev. E **51**, 4655 (1995); For theoretical considerations see also L.-H. Tang, M. Kardar, D. Dhar: Phys. Rev. Lett. **74**, 920 (1995)

7.47   W. Zdaniewski: J. Am. Ceram. Soc. **72**, 116 (1989)

7.48   R. Desmaison, in: *Protégeons la Montagne* (Nathan, Paris 1978)

7.49   B.B. Mandelbrot, D.E. Passoja, A.J. Paulay: Nature **308**, 721 (1984)

7.50   C.S. Pande, L.R. Richards, S. Smith: J. Mater. Sci. Lett. **6**, 295 (1987)

7.51   D.L. Davidson: J. Mater. Sci. **24**, 681 (1989)

7.52   E. Bouchaud, G. Lapasset, J. Planès: Europhys. Lett. **13**, 73 (1990)

7.53   A. Imre, T. Pajkossy, L. Nyikos: Acta Metall. Mater. **40**, 1819 (1992); V.Y. Milman, R. Blumenfeld, N.A. Stelmashenko, R.C. Ball: Phys. Rev. Lett. **71**, 204 (1993)

7.54   A. Chudnovsky, B. Kunin: J. Appl. Phys. **62**, 4124 (1987); M. Kardar, in *Disorder and Fracture,* ed. by J.C. Charmet, S. Roux, E. Guyon (Plenum, New York); S. Roux, D. François: Scripta Metall. **25**, 1092 (1991)

7.55   E. Bouchaud, S. Navéos: preprint

7.56   L. de Arcangelis, A. Hansen, H.J. Herrmann, S. Roux: Phys. Rev. B **40**, 877 (1989); For $d = 2$ models of crack propagation see: P. Meakin, G. Li, L.M. Sander, E. Louis, F. Guinea: J. Phys. A **22**, 1393 (1989); E.L. Hinrichsen, A. Hansen, S. Roux: Europhys. Lett. **8**, 1 (1989)

7.57   E. Louis, F. Guinea, F. Flores, in: *Fractals in Physics,* ed. by L. Pietronero, E. Tossati (North-Holland, Amsterdam 1986), p. 177

7.58   E.E. Underwood, K. Banerji: Mater. Sci. Eng. **80**, 1 (1986)

7.59   B. Sapoval, M. Rosso, J.F. Gouyet: J. Physique Lett. **46**, L149 (1985)

7.60   B. Sapoval, M. Rosso, J.F. Gouyet, J.F. Colonna: Solid State Ionics **18&19**, 21 (1986)

7.61   R.F. Voss: J. Phys. A **17**, L51 (1984)

7.62   R. Ziff, B. Sapoval: J. Phys. A **19**, L1169 (1986)

7.63   A. Bunde, J.F. Gouyet: J. Phys. A **18**, L285 (1985)

7.64   H. Saleur, B. Duplantier: Phys. Rev. Lett. **58**, 2325 (1987)

7.65   D. Stauffer: *Introduction to Percolation Theory* (Taylor & Francis, London 1985)

7.66   M. Rosso: J.Phys.A **22**, L131 (1989)

7.67   M Kolb, J.F. Gouyet, B. Sapoval: Europhys. Lett. **3**, 33 (1987);
       M. Kolb, T. Gobron, J.F. Gouyet, B. Sapoval: Europhys. Lett. **11**, 601 (1990)

7.68   S. Mazur, S. Reich: J. Phys. Chem. **90**, 1365 (1986)

7.69   R.P. Wool, in: *New Trends in Physics and Physical Chemistry of Polymers*, ed. by Lieng-Huan Lee (Plenum, New York 1989), p. 129

7.70   M. Rosso, B. Sapoval, J.F. Gouyet: Phys. Rev. Lett. **57**, 3195 (1986)

7.71   J.F. Gouyet, M. Rosso, B. Sapoval: Phys. Rev. B **37**, 1832 (1988)

7.72   J.L. Hudson, J. Tabora, K. Krisher, I.G. Krevrekidis: Phys. Lett. A **179**, 355 (1993);
       D.E. Williams, R.C. Newman, Q. Song, R.G. Kelly: Nature **350**, 216 (1991); L. Balázs, L. Nyikos, I. Szabó, R. Schiller, in press

7.73   D. Sornette: Phys. Rev. Lett. **69**, 1287 (1992); T.L. Chelidze: Phys. Earth. Planet. Inter. **28**, 93 (1982); G.A. Sobolev: Pure Appl. Geophys. **124**, 811 (1986)

7.74   P. Bak, C. Tang, K. Wiesenfeld: Phys. Rev. A **38**, 364 (1988)

7.75   J.F. Gouyet, Y. Boughaleb: Phys. Rev. B **40**, 4760 (1989);
       B. Sapoval, M. Rosso, J.F. Gouyet, Y. Boughaleb: Proc. of the 1988 Seminar on Fractals in Physics, Erice, Italy (Plenum, New York 1990) p.297

7.76   J.F. Gouyet, in *Soft Order in Physical Systems,* ed. by R. Bruinsma, Y. Rabin (Plenum Pub. Corp., New York, 1994);
       J.F. Gouyet, in Diffusion Processes: Experiment, Theory, Simulations Lecture Notes in Physics, vol. 438, A. Pekalski (ed.), (Springer Berlin, 1994), p. 115

7.77   H. Van Damme, in: *The Fractal Approach to Heterogeneous Chemistry,* ed. by D. Avnir (Wiley, New York 1989), p. 199

7.78   J. Nittmann, G. Daccord, H.E. Stanley: Nature **314**, 141 (1985);
       J. Nittmann, G. Daccord, H.E. Stanley, in: *Fractals in Physics*, ed. by L. Pietronero, E. Tossati (North-Holland, Amsterdam 1986), p. 193

7.79   J.D. Chen: Exp. Fluids, **5**, 363 (1987)

7.80   H.J.S. Hele Shaw: Nature **58**, 34 (1898)

7.81   F.A.L. Dullien: *Porous Media: Fluid Transport and Pore Structure* (Academic Press, London 1979)

7.82   P.M. Adler, in: *The Fractal Approach to Heterogeneous Chemistry*, ed. by D. Avnir (Wiley, New York 1989), p. 341, and references therein

7.83   R. Vacher, T. Woigner, J. Pelous, E. Courtens: Phys. Rev. B **37**, 6500 (1988);
       F. Chaput, J.P. Boilot, A. Dauger, F. Devreux, A. de Geyer: J. Non-Cryst. Solids **116**, 133 (1990)

7.84   C. Chachaty, J.P. Korb, J.R.C. Van der Maarel, W. Bras, P. Quinn: Phys. Rev.B **44**, 4778 (1991)

7.85   A.J. Katz, A.H. Thompson: Phys. Rev. Lett. **54**, 1325 (1985)

7.86   D. Farin, D. Avnir, in: *The Fractal Approach to Heterogeneous Chemistry,* ed. by D. Avnir (Wiley, New York 1989), p. 271, and references therein

7.87   E. Guyon, G.D. Mitescu, J.P. Hulin, S. Roux: Physica D **38**, 172 (1989)

7.88   B. Sapoval, M. Rosso, J.F. Gouyet, in: *The Fractal Approach to Heterogeneous Chemistry*, ed. by D. Avnir (Wiley, New York 1989), p. 227

7.89   T.M. Shaw: Phys. Rev. Lett. **59**, 1671 (1987)

7.90   R. Lenormand, C. Zarcone, A. Sarr: J. Fluid Mech. **135**, 337 (1983)

7.91   R. Lenormand: C.R. Acad. Sci. Paris II **301**, 247 (1985)

7.92   R. Lenormand, E. Touboul, C. Zarcone: J. Fluid Mech. **189**, 165 (1988)

7.93   L. Paterson: Phys. Rev. Lett. **52**, 1621 (1984)

7.94   P. Meakin, J.M. Deutch: J. Chem. Phys. **85**, 2320 (1986)

7.95   D. Wilkinson, in: *Physics of Finely Divided Matter*, ed. by N. Boccara, M. Daoud (Springer, Berlin, Heidelberg 1985), p. 280

7.96   M. Cieplak, M.O. Robbins: Phys. Rev. Lett. **60**, 2042 (1988);

7.97   V.K. Horvath, F. Family, T. Vicsek: J. Phys. A **24**, L25 (1991)

7.98   M.A. Rubio, C.A. Edwards, A. Dougherty, J.P. Gollub: Phys. Rev. Lett. **63**, 1685 (1989)

7.99   S. He, G. Kahanda, P.Z. Wong: Phys. Rev. Lett. **69**, 3731 (1992)

7.100 T. Vicsek, M. Cserzo, V.K. Horvath: Physica A **167**, 315 (1990)

7.101 Y.-C. Zhang: J. de Phys. **51**, 2129 (1990)

7.102 N. Martys, M. Cieplak, M.O. Robbins: Phys. Rev. Lett. **66**, 1058 (1991)

7.103 L.H. Tang, H. Leschhorn: Phys. Rev. **45**, R8309 (1992)

7.104 K. Sneppen: Phys. Rev. Lett. **69**, 3539 (1992); see also Z. Olami, I. Procaccia, R. Zeitak: Phys. Rev. E **49**, 1232 (1994)

7.105 S. Havlin, L. A. N. Amaral, S. V. Buldyrev, S. T. Harrington, H. E. Stanley: Phys. Rev. Lett. **74**, 4205 (1995)

7.106 Y.-C. Zhang: Physica A **170**, 1 (1990);
J. Krug: J. de Phys. I **1**, 9 (1991);
S. Havlin, S.V. Buldyrev, H.E. Stanley, G.H. Weiss: J. Phys. A **24**, L925 (1991)

7.107 J.G. Amar, F. Family: J. Phys. A **24**, L79 (1991);
S.V. Buldyrev, S. Havlin, J. Kertesz, H.E. Stanley, T. Vicsek: Phys. Rev. A **43**, R7113 (1991)

7.108 L. A. N. Amaral, A.-L. Barabási, S. V. Buldyrev, S. Havlin, H. E. Stanley: Phys. Rev. Lett. **72**, 641 (1994)

7.109 E. Clément, C. Baudet, E. Guyon, J.P. Hulin: J. Phys. D **20**, 608 (1987);
See also J.-D. Chen, M.M. Dias, S. Platz, L.M. Schwartz: Phys. Rev. Lett. **61**, 1489 (1988)

7.110 J.P. Hulin, E. Clément, C. Baudet, J.F. Gouyet, M. Rosso: Phys. Rev. Lett. **61**, 333 (1988)
J.F. Gouyet, M. Rosso, E. Clément, C. Baudet, J.P. Hulin, in: *Hydrodynamics of Dispersed Media*, ed. by J.P. Hulin, A.M. Cazabat, E. Guyon, F. Carmona (Elsevier, Amsterdam 1990), p. 179

7.111 S. Roux, E. Guyon: J. Phys. A **22**, 3693 (1993), in this paper the exponent τ of the number of events is only valid in $d = 2$

7.112 J.F. Gouyet: Physica A **168**, 581 (1990)

7.113 A.H. Thompson, A.J. Katz, C.E. Krohn: Adv. Phys. **36**, 625 (1987)

7.114 N. Martys, M.O. Robbins, M. Cieplak: Phys. Rev. B **44**, 12294 (1991)

7.115 K.J. Maløy, J. Feder, F. Boger, T. Jøssang: Phys. Rev. Lett. **61**, 2925 (1988)

7.116 Y. Kantor, in: *Statistical Mechanics of Membranes and Surfaces*, Vol. 5, ed. by D. Nelson, T. Piran, S. Weinberg (World Scientific, Singapore 1989), p. 115

7.117 R.M. Bradley: Phys. Rev. A **41**, 914 (1990), and references therein

7.118 A. Blumstein, R. Blumstein, T.H. Vanderspurt: J. Colloid Interface Sci. **31**, 236 (1969)

7.119 J.H. Fendler, P. Tundo: Acc. Chem. Res. **17**, 3 (1984)

7.120 Y. Kantor, M. Kardar, D.R. Nelson: Phys. Rev. Lett. **57**, 791 (1986)

7.121 D. Nelson, in: *Statistical Mechanics of Membranes and Surfaces*, Vol.5, ed. by D. Nelson, T. Piran, S. Weinberg (World Scientific, Singapore 1989), p. 1 and p. 137

7.122 B. Sapoval: *Fractals* (Aditech, Paris 1990)

8 Fractals and Experiments

Jørgen K. Kjems

8.1 Introduction

Real fractals are self-similar structures that result from physical, chemical, or biological growth processes. As pointed out by Benoit Mandelbrot in his pioneering book *Fractal Geometry of Nature* [8.1], such structures are abundant in our surroundings. Once our attention is attuned to recognizing patterns of self-similarity in an object we find it everywhere, for example in trees, bushes, plants such as the cauliflower, and in the cracks and grains of stones and rocks, as well as in polymers, aggregates, and flocculates. The natural abundance of such structures must mean that there are many processes that can give rise to fractal geometry. One of the goals of the current research in this field is to unravel the common features of these processes in order to find general principles that could be characterized as laws of nature. The objective of this chapter is less ambitious. Here we shall concentrate on the kind of fractal structures that can be produced in the laboratory under controlled circumstances and try to give answers to questions like: how do you make a fractal? How do you determine its structure, in particular the fractal dimension, d_f? What are the mechanical, electrical, and magnetic properties? It is through systematic variation of the control parameters during the preparation and the subsequent detailed characterization of the specimens that we can hope to make progress in the understanding of fractal systems.

The physical properties of normal matter with uniform density often stem from interactions on a particular length scale, for example, the bondings between the atoms in a crystalline lattice. With the knowledge of these interactions we can explain many of the static, dynamic, and transport properties that are observed in experiments on crystalline matter. Fractal objects are

---

◀ **Fig. 8.0.** Minaret cauliflower, courtesy of F. Grey

characterized by not having a characteristic length scale and they have a very inhomogeneous density distribution. We can therefore expect to find very different physical properties in materials with fractal structure compared to the ordinary solids. Furthermore, real fractals are disordered and highly irregular. In some sense they can be regarded as ideally disordered materials, and an understanding of their properties can therefore lead to a more general understanding of disordered materials such as polymers and glasses, which have more homogeneous density distributions.

8.2 Growth Experiments: How to Make a Fractal

8.2.1 The Generic DLA Model

Fractal structures can result from stochastic growth processes. It was Witten and Sander [8.2] who first realized that the simplest conceivable random aggregation process leads to such structures. They considered the process in which diffusing particles one by one make irreversible contact to a growing aggregate seeded by one particle. This is the diffusion limited aggregation model (DLA model, see also Chaps. 1, 4, 7, and 10). Figure 10.0 shows a large DLA cluster. In their pioneering computer experiments Witten and Sander were able to show that the DLA clusters had self-similar properties. The radial mass distribution measured from the center of the aggregate was shown to follow a power law and thus they were able to make the first estimates of the fractal dimension, d_f, and its dependence on the dimensionality of the embedding space.

We may ask ourselves the rhetorical question: why does this process result in aggregate clusters whose size increases faster than their weight? The answer is self-screening: the growth occurs predominantly at the outermost sites and the larger the cluster becomes the larger the size of the lacunae that can be embedded in the structure. In other words, the aggregate cluster has density fluctuations on all length scales up to its own size. This is equivalent to the statement that the density-density correlation function follows a power-law decay. It is by no means obvious that the same power law should be able to describe the density correlations over a wide range of length scales, but that is in fact borne out by extensive computer experiments [8.3,4]. More recent research suggests that the description of a DLA cluster in two dimensions requires an additional parameter such as the compactness and that fractal dimensions may vary with size [8.5]. The multifractal aspects of DLA clusters have also been explored [8.6].

It was soon realized that the DLA model is the generic example of a wide range of growth models that are governed by the diffusion equation or an equivalent mathematical form. Many of these models can be realized in real growth experiments. In the following the most prominent examples will be discussed.

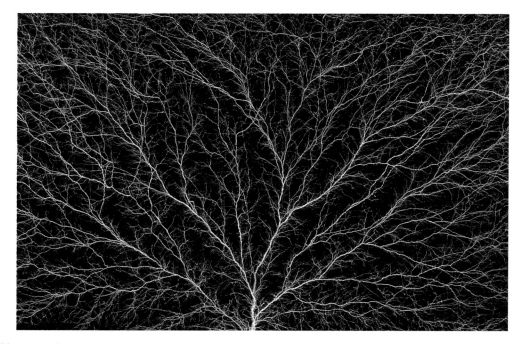

Fig. 8.1. Discharge pattern in a block of polymethyl-methacrylate (PMMA). The material is charged with 2 MeV electron beam and subsequently discharged through a point contact. The sample was prepared by A. Miller

8.2.2 Dielectric Breakdown

The fastest way to produce a fractal structure is dielectric breakdown. An elegant procedure for obtaining decorative "frozen lightning" is the electron beam charging of an insulating polymer material such as polymethyl-methacrylate (PMMA) followed by a discharge through a point connector [8.7]. The resulting tree-like patterns (Fig. 8.1) have a fractal character that is similar to that of DLA clusters. The shapes can be varied by the charging procedure and the discharge point and by applying mechanical stress during the discharge.

A two-dimensional geometry was used in the gas discharge experiments of Niemeyer et al. [8.8]. They observed the discharge patterns in a thin layer of SF_6 gas between a central point electrode and a circular counterelectrode at the perimeter. The resulting discharge patterns (Fig. 8.2), known as Lichtenberg figures [8.9], bare a striking resemblance to the two-dimensional (2d) DLA clusters. From a theoretical point of view this analogy stems from the equivalence of the differential equations that govern the two processes (see also Chaps. 1 and 4).

For the DLA, the diffusion equation governs the particle concentration, $c(x, y, t)$,

$$\frac{\partial c}{\partial t} = C \left(\frac{\partial^2 c}{\partial x^2} + \frac{\partial^2 c}{\partial y^2} \right), \qquad (8.1)$$

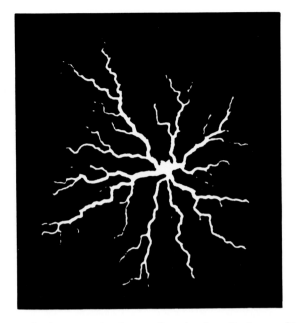

Fig. 8.2. Time-integrated photograph of a surface leader discharge (Lichtenberg figure) on a 2 mm glass plate in 0.3 MPa SF$_6$. Applied voltage pulse: 30 kV$\times$1μs. This experiment corresponds to an equipotential system growing in a plane with a radial electrode [8.8]

and for the discharge, the Laplace equation governs the electrostatic potential, $\Phi(x, y)$,

$$0 = \frac{\partial^2 \Phi}{\partial x^2} + \frac{\partial^2 \Phi}{\partial y^2}. \tag{8.2}$$

In the limit of very slow deposition rates, these equations become identical. The solutions depend on the boundary conditions. In the ideal case the growing discharge pattern is assigned an infinite conductivity and thus acquires a uniform potential. Similarly, the DLA cluster acts as a perfect sink for the diffusing particles, i.e., the sticking probability is unity. The stochastic mechanism in the latter case is associated with the random motion of the incoming particles. Each site on the cluster has a certain probability of encounter for the next particle. In the discharge problem the local growth probability is related to the local field. In the simplest theory it is just proportional to the field, but more generally it is expressed as the field raised to a power η.

The analysis of Lichtenberg figures from a series of two-dimensional discharges indicates that the value of the fractal dimension is 1.7, close to that found for the two-dimensional DLA clusters [8.10]. This indicates that the exponent η is close to 1. It has been shown that the branching is suppressed at higher values of η [8.11]. Similarly the discharge patterns seen in the electron beam loaded PMMA appears to have a strong resemblance to the 3d DLA structures, but in this case there is no quantitative analysis available yet.

8.2.3 Electrodeposition

The process of electrodeposition combines the physics of diffusion experiments and that of discharge experiments. It therefore provides a wide variety of control parameters such as geometry, concentration, temperature, viscosity, field strength and current values. Correspondingly, a rich variety of growth patterns can be synthesized. Electrodeposition is an important industrial process and it is well known that it can lead to rough surfaces and dendritic growth [8.12].

Brady and Ball were the first to point out the similarity between the fractal growth of copper electrodeposits and the diffusion limited aggregation process [8.13]. In a clever experiment they deposited copper on a central cathode in a dilute (0.01–0.02 M) sulfate solution where the viscosity was enhanced by addition of a high-molecular-weight polymer. In this manner they were able to suppress convection and obtain results that were dominated by diffusion. They carefully monitored the current at constant voltage and used Faraday's law to measure the mass of the deposited aggregate. If Smoluchowski kinetics [8.14] are assumed, the effective radius of the growing cluster is proportional to the current. Hence they were able to experimentally determine the relation between the linear size and the mass and in this manner obtain a value for the fractal dimension. They found $d_f = 2.43 \pm 0.03$, very close to the DLA computer model value of 2.495 ± 0.06.

Similar experiments have been carried out in a 2d geometry by Matsushita et al. [8.15]. They studied the growth of zinc metal leaves in the interface between a zinc sulfate aqueous solution with a molar concentration of 2 M and a layer of n-butyl acetate. The electrodeposition was initiated at a centrally placed graphite cathode. The anode was a circular zinc electrode with a diameter of 17 cm and the deposition proceeded with a constant current.

In these experiments as well, the resulting structures have fractal character and resemble those of the two-dimensional DLA model. In this case the mass distribution was analyzed on the basis of digitalized pictures. For applied voltages less than a critical value ($V_c = 8.2\,\mathrm{V}$ for the chosen geometry) the resulting fractal dimension was found to be constant at a value of 1.66 ± 0.03, which coincides with the mean field result 5/3 for the DLA model. At higher deposition voltages d_f was found to increase, i.e., the clusters became more dense. Similar experiments have been carried out by Kondoh and Sawada [8.16] with a wider variation of the growth conditions. They used an experimental setup with a thin layer of solution, 0.25 mm, between circular Plexiglass plates and found a rich variety of patterns from almost linear protuberances or dendrites at high voltages and high concentrations to very dense, ramified clusters at low voltages and low concentrations (Fig. 8.3). With high concentration and low voltage, the growth forms an open fractal, and with medium concentration and high voltage, the growth form is dendritic, with a straight backbone [8.17]. Ball has suggested that these crossovers are related to changes in the effective dimensionality of the process [8.18]. At very slow growth rates the

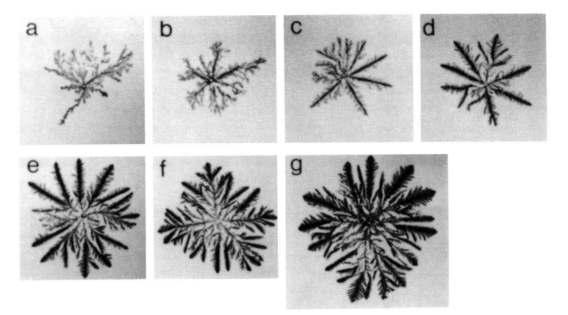

Fig. 8.3. Growth patterns for 0.05 M zinc sulphate solutions at (a) 2 V, (b) 3 V, (c) 4 V, (d) 6 V, (e) 8 V, (f) 10 V, and (g) 12 V. (From [8.16])

transport of material to the cluster becomes three-dimensional and results in denser clusters.

The deposition rate and the ionic concentration are also the key control parameters for the experiments where the growth is initiated along a line. With increasing concentrations, changes are observed from an initial fractal character to an initial dendritic character of the resulting structures [8.17]. These experiments are relevant for the ballistic growth models and ultimately they produce interface layers of constant density with varying degrees of openness depending on the initial ionic concentrations.

8.2.4 Viscous Fingering

The interface between immiscible fluids can have a very complex geometry. In this context we focus on the patterns that are formed when a low-viscosity fluid is used to displace a more viscous one. Experiments of this type are easiest to analyze when they are carried out in a two-dimensional geometry, such as the Hele Shaw cell [8.19]. The control parameters other than the viscosities are the interface tension, the injection rate, the wettability of the plates, and the geometric anisotropies.

Examples of such experiments are shown in Fig. 8.4, which illustrates the obvious similarity to the other growth processes. In this case it is the pressure field, P, that governs the growth. The pressure variation is very small in the intruding, low-viscosity fluid, i.e., it acts as an isobaric medium in analogy with the equipotential of the metallic electrodeposited clusters. As an aside

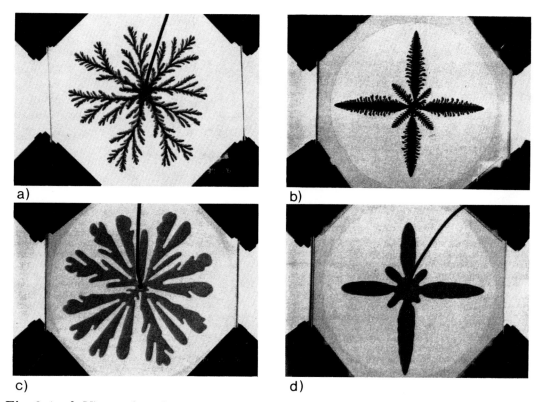

Fig. 8.4a-d. Viscous fingering patterns in a thin gap beween two glass plates. The less viscous fluid displaces a more viscous fluid at a high flow rate. (a) The miscible liquids, water and glycerine, interact in a Hele-Shaw cell. There is no surface tension between the two fluids and a DLA-like pattern results. (b) A square lattice of grooves etched into the glass plates imposes an anisotropy. (c) and (d) like (a) and (b) but with oil invading glycerine. The surface tension results in wider fingers. (Photographs courtesy of J.D. Chen, Mead Imaging)

we mention that the temperature field plays a similar role in the formation of snowflakes. In mathematical terms we have

$$\frac{\partial^2 p}{\partial x^2} + \frac{\partial^2 p}{\partial y^2} = \frac{\kappa \mu}{k} \boldsymbol{v} \cdot \nabla p, \qquad (8.3)$$

where κ is the compressibility and μ the viscosity of the defending fluid. The constant k is a geometry factor. These parameters combine to a length scale, $k/\kappa\mu v$, which is very long for a typical experiment. It is therefore a good approximation to assume zero compressibility and hence obtain an equation analogous to the DLA case [8.20], namely

$$\frac{\partial^2 p}{\partial x^2} + \frac{\partial^2 p}{\partial y^2} = 0. \qquad (8.4)$$

The surface tension γ tends to suppress the branching structure at short length scales. The relevant parameter is the capillary length l_a given by $\gamma/\delta p$. This is an example of the competition between the macroscopic field induced growth controlled by the Laplace equation or equivalent and the interface dynamics, which can lead to a variety of morphologies in nonlinear growth problems. Moving interfaces are subject to instabilities and the resulting morphology of the interface is the result of competition between the different possible solutions. Ben-Jacob has proposed that this selection is controlled by the interface dynamics and that as a rule the fastest growing morphology wins [8.21]. The most prominent instability is the tip splitting, which leads to regular dendritic growth in the presence of anisotropy, as illustrated in Fig. 8.4b. The morphologies that result from nonlinear growth experiments can be classified into the following essential shapes [8.21]:

1) fractal,
2) dense branching,
3) dendritic,
4) faceted;

all of which can be studied in the Hele-Shaw cell experiments. An active area of current research is the possible transitions between the different morphology classes, which may be induced and hence studied by the variation of the control parameters.

8.2.5 Invasion Percolation

The invasion experiments using Hele-Shaw cells with patterns engraved in the plates are examples of invasion percolation in anisotropic media (see also Chaps. 2, 4, and 7). They represent a simplified version of the more general problem of fluid invasion into a porous medium. This problem has obvious applications in the oil industry and has therefore attracted considerable attention, but it is also an interesting academic question in its own right.

Controlled experiments have been carried out using artificial porous media, such as vessels filled with glass beads [8.22]. The structures formed by the invading fluid depend on the interfacial pressure and the wettability of the porous medium. A wetting fluid will first invade the small pores, whereas a nonwetting fluid first enters the large pores.

If the pore size is smaller than the capillary length of the fluid–fluid interface then the viscous fingering and tip-splitting instabilities are unimportant and the structure becomes that of a percolation cluster for a nonwetting fluid. When the capillary number is reduced one can obtain a capillary length of a scale between the pore size, and the cell size and both viscous fingering patterns and invasion percolation patterns are observed [8.23]. For wetting fluids the patterns are best described as compact fingers whose widths, we believe, have no obvious theoretical explanation. The effects of gravity have also been studied experimentally by Meakin et al [8.24].

8.2.6 Colloidal Aggregation

A colloidal suspension is a fluid containing small charged particles, monomers, that are kept afloat by Brownian motion and kept apart by Coulomb repulsion. A change in the interaction can be induced by changing the chemical composition of the solution and in this manner an aggregation process can be initiated. The aggregation of the colloidal monomer particles into clusters of growing size is a classical problem in physical chemistry. The pioneering work was done by Smoluchovski [8.14], who formulated a kinetic theory for the irreversible aggregation of monomers into clusters and further clusters combining with clusters. The inclusion of cluster–cluster aggregation makes this process distinct from the DLA process, which in this terminology evolves by cluster–monomer contact only. However, for practical colloid concentrations and aggregation conditions the clustering of clusters will always occur. Recent work by Lin et al. [8.25] has indicated that the colloidal aggregation process has universal features, which are independent of the detailed nature of the colloidal systems. It appears that there are two distinct limiting regimes of the irreversible colloidal aggregation process, namely

1) a fast, diffusion limited regime (DLCA), in which the reaction rate is determined solely by the time neeeded for the clusters to encounter each other by diffusion, and

2) a slow, reaction limited regime (RLCA) in which the cluster–cluster repulsion has to be overcome by thermal activation, a process that may require numerous encounters.

In both cases the resulting clusters have fractal character with cluster masses that scale like $M \sim (R_g/a)^{d_f}$, where R_g is the radius of gyration of the cluster (see Sects. 2.5 and 4.2) and a is the monomer radius. The universal features have been demonstrated for colloidal aggregates of gold, silica and polystyrene and qualitatively it is apparent from the electron micrographs shown in Fig. 8.5.

a) Gold colloid aggregates. Colloidal gold suspensions can be produced by well-known chemical methods [8.26]. The monomer size distribution can be made very homogeneous with a spread in the diameters of about 10%. A typical diameter is 15 nm and a typical concentration is 10^{12} particles per cubic centimeter.

When formed, the surface is highly charged with adsorbed negative ions. The positive ions in the solution screen the surface charges with a Debye–Hückel screening length with a scale of the order of 10 nm. This means that there is a strong repulsion between the monomers at the scale of the screening length and the potential barrier is typically of order $20k_BT$. The barrier can be reduced in a controlled fashion either by reducing the surface charge on the monomers or by changing the screening length by adding more ions to the suspension.

In the case of gold colloids the surface charge can be reduced by adding a small concentration of pyridine, which is strongly adsorbed as a neutral molecule on the monomer surface and thus displaces the negative ions. Concentrations above 10^{-3}M virtually displace all the negative surface charge. Once the monomers can come into close contact they experience a strong attraction due to the van der Waal's interaction, and ultimately metallic bonds are formed. This means that the monomers once in contact stick together and that the aggregation process is irreversible. By proper selection of the monomer concentration, of the ionic strength of the solution, and of the neutralizing agent one can control the aggregation process and let it evolve in either the DLCA or the RLCA regime or in between the two.

b) Silica colloid aggregates. Colloidal suspensions of silica particles, often called sols, can be obtained commercially [8.27] and consist of SiO_2 monomer particles, charge stabilized with OH^- or SiO^{--} groups on the surfaces. A typical monomer diameter is 7.5 nm. The sol is stable in an alkaline medium, typically NaOH at a pH of about 10.

The aggregation can be induced in different ways and proceeds by the formation of $Si-O-Si$ bonds. The OH^- groups on the surface catalyze the process. An increase in the ionic strength of the solution will reduce the screening length as described above. Changing the pH by addition of an acid such as HCl will remove the OH^- from the surface and will initially enhance the aggregation, but if too many OH^- groups are removed the bonding between the monomers will no longer be catalyzed, and the sol becomes stable again.

The effects of temperature variations are more delicate. The screening length is essentially temperature independent but the siloxane catalysis depends strongly on the temperature and becomes more effective at lower temperatures. This is counteracted by the reduced dissociation and by the reduced kinetic energy of the diffusing monomers. The net effect at low concentrations (less than 1 wt% silica) is an increasing aggregation rate with decreasing temperature [8.28]. At high concentration the opposite effect is observed. With a suitable choice of aggregation conditions the process can evolve in either of the regimes, DLCA or RLCA, as illustrated in Fig. 8.5.

c) Aerogels. A fascinating new class of materials with fractal and many other interesting physical properties are the silica aerogels. They can be made in solid, transparent, monolithic form. They are extremely light, and have a very low thermal conductivity, a solid-like elasticity, and very large internal surfaces [8.29,30].

The initial step in the preparation of silica aerogels is the hydrolysis of an alkoxysilane $Si(OR)_4$, where R is CH_3 or C_2H_5 [8.31]. The hydrolysis produces SiOH groups that polycondense into $Si-O-Si$ bonds and small particles start to grow in the solution. The reaction is catalyzed by addition of either an acid or a base to the solution, probably in a manner similar to that described for the colloidal aggregation processes. Other control parameters are the relative concentrations of the reagents and of the alcohol. The aggregation clusters

DLCA   RLCA

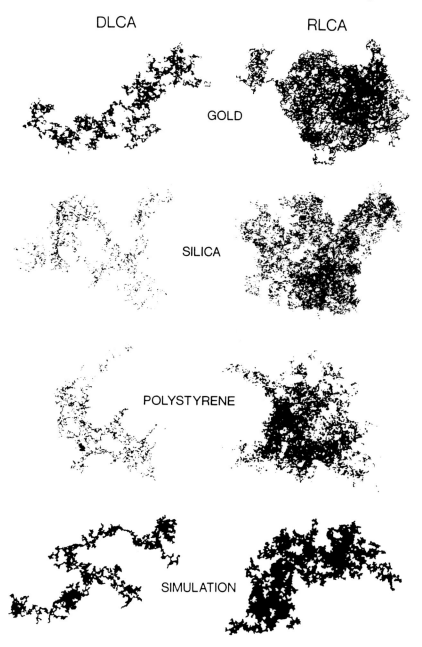

GOLD

SILICA

POLYSTYRENE

SIMULATION

Fig. 8.5. Electron micrograph of DLCA and RCLA clusters from [8.25]. Courtesy of D.A. Weitz

continue to grow until they eventually fill the reaction volume, at which point the solution gels.

The reactions are normally not complete at this point and the cluster networks continue to grow in the "alcogel phase". After suitable aging the solvent is removed by supercritical drying, i.e., at a temperature and pressure above the critical point of the solvent. In this manner the open-porous structure of the network is preserved and decimeter-size monolithic blocks with a range of densities from 50 to 500 $kg\,m^{-3}$ can be obtained. Intuitively one expects the short and intermediate range structure of this material to have the same fractal morphology as the colloidal aggregates. This is borne out in the structural experiments to be discussed below.

d) Smoke particle aggregates. An example of commercially available fractal materials is smoke particle aggregates, which are supplied under the trade names Alfasil and Cab-O-Sil [8.32]. They are produced by the process of flame hydrolysis in which $SiCl_4$ vapor is burned in a flame of hydrogen and oxygen. This results in a snow-like product in which the basic monomer particles are amorphous SiO_2 spheres of approximately 5 nm diameter. The production process has been studied in detail by Ulrich and Riel [8.33] and the control parameters are the reaction-chamber geometry, the flow rates, and the cooling rates. Scattering experiments and electron micrograph analysis indicate a crossover from DLA-like clustering at short length scales to a structure characterized by clustering of clusters at larger scales [8.28].

e) Other processes. There are many other curing and aging processes that can lead to fractal structures. One example is the solidification of cement, which evolves through quite a complicated gel process. Ordinary Portland cements contain four major components, namely CaO, SiO_2, Al_2O_3 and Fe_2O_3, in the form of sintered aggregates, typically with a mean size of about 40 μm. When it is mixed with water a complex series of hydration reactions takes place, a prominent product of which is an amorphous calcium-silicate hydrate with a gel-like structure. In aged cements this component accounts for 20% of the volume of the hydrated material. Neutron scattering studies during the aging process show the development of structures with fractal correlations in the range 2−50 nm and electron micrographs indicate self-similar structures at even larger length scales but the relation to the chemical and mechanical properties of cement has not yet been established [8.34].

Another example is the structure that results from the segregation of two phases far from equilibrium by the spinodal decomposition of binary mixtures. It can be observed in metals, fluids, and polymers and involves large fluctuations of the local order parameter, i.e., the concentration of a given component. The starting point is a rapidly quenched homogeneous state, which is thermodynamically unstable. The process evolves through the formation of clusters with dynamic interfaces. For a general review of this process the reader is referred to the article by Gunton, San Miguel, and Sahni [8.35]. Molecular dynamics experiments show that the cluster morphology strongly resembles that of fractal aggregates and suggest fractal dimensions similar to that of the DLA clusters [8.36].

So far, we have discussed solid structures that are kept together by chemical bonds. Another important class of fractal structures comprises the percolation networks (see Chaps. 2 and 3), and here we will mention the example of diluted random magnets. The percolation clusters are well-known fractal objects that can be created by the random substitution of atoms on a regular lattice. The introduction of magnetic ions into a nonmagnetic host will lead to the build-up of magnetic-exchange-coupled clusters. At the percolation threshold the largest cluster has the size of the sample and the system can undergo a phase transition to an ordered state when cooled sufficiently.

In most magnetic systems the dilution also introduces an element of frustration due to competing interactions between local fields and/or oscillatory exchange forces. In these cases even the static magnetic properties are governed by dynamical effects, which can lead to the occurrence of spin glass phases. However, there are simple systems without frustration effects in which both the static and the dynamic properties are determined purely by the geometry; an example being the two-dimensional Ising antiferromagnet $Rb_2(Co,Mg)F_4$ studied by Cowley, Birgeneau, and Uemura [8.37] and by Aeppli, Guggenheim, and Uemura [8.38]. Near the percolation threshold of 59.3 at% cobalt concentration the magnetic exchange bond network of this system represents a very good physical realization of a percolation system with fractal geometry, and important results for both the static and the dynamic properties have been obtained.

Having discussed some of the many ways in which fractal structures can be synthesized we now turn to the problem of the experimental characterization of the fractal systems and we begin with the structural aspects.

8.3 Structure Experiments: How to Determine the Fractal Dimension

The most convincing way to illustrate fractal geometry is the self-similarity under magnification. This is easily done for the regular mathematical fractal shapes such as the Koch island, but for real systems the self-similarity is a statistical property and we can only expect a qualitative similarity between different magnifications of the same object (see, for example, Figs. 2.4 and 7.16), although a cauliflower comes pretty close (a minaret cauliflower is shown in Fig. 8.0).

Instead we have to resort to a quantitative mathematical description of the objects in terms of the density $\rho(r)$ and correlation functions $G(r_1 - r_2)$ in order to judge their fractal character. This immediately raises the question of the limiting scales. The systems that have been described up until now in this chapter clearly have a lower bound for the fractal character of these distribution functions. For aggregates and polymers it is the monomer size, a, and for the magnetic percolation clusters it is the lattice parameter. The upper bound ξ corresponds either to the size of the aggregates/clusters or, in the case of gels and percolating systems, to the scale on which the density of the system in question becomes uniform. The presence of these bounds will limit the information that can be obtained from the analysis of the data from structural experiments and it constrains the ranges over which the notion of a fractal structure is justified.

8.3.1 Image Analysis

Microscope photographs and electron micrographs have been used as the basis for image analysis of aggregate structures. They represent projections of the three-dimensional structures and this may be a cause of systematic error. The pictures are digitized with a suitably chosen pixel scale, typically the scale of the monomer diameter, which then becomes the lower cutoff bound. The full pair correlation function

$$G(r) = \frac{1}{v\rho} \int_v \langle \rho(r')\rho(r'+r) \rangle dr', \qquad (8.5)$$

where $\rho = \langle \rho(r) \rangle$ is the mean density, $\rho(r)$ is the local density, and the brackets mean ensemble average, is then approximately given by

$$G(r) \cong \frac{1}{N} \sum_{r'} \theta(r')\theta(r+r'). \qquad (8.6)$$

The function $\theta(r)$ takes the value 0 or 1 depending on whether a given pixel is occupied or not, and the sum runs over all the pixels. The finite-size effects can be minimized by restricting the summations from a given site to the largest circle centered on that site that can be contained in the cluster circumference. The normalization has to be adjusted accordingly. Another technique to suppress the edge effects is Fourier analysis of the density distribution of the cluster. This involves the evaluation of the power spectrum and the construction of the density–density correlation function by another Fourier transformation. This procedure is extensively used to analyze computer-generated clusters [8.39].

For a fractal system the resulting correlation function should follow a power law of the form

$$G(r) \sim (r/a)^{-A}, \qquad (8.7)$$

where A is related to the fractal dimension d_f and the embedding space dimension d by $d_f = d - A$ (see also Chaps. 1, 2 and 4). This means that a double logarithmic plot of the measured results for $G(r)$ is the most convenient representation of the data. Fractal correlations manifest themselves as straight lines with the slope $-A$ in such plots. It is interesting to note that the image analysis of the projection of a three-dimensional aggregate onto a plane leads to the correct value for the exponent provided d_f is less than 2, i.e., A is greater than 1 [8.40]. The proof of this statement can be regarded as a pleasant exercise for the reader.

8.3.2 Scattering Experiments

Coherent scattering from fractal objects is a very direct method of determining
structural parameters and correlations. The same scattering formalism can be
used for probes such as X-rays, neutrons, and laser light, and data from different
experiments can be combined to cover many decades of length scales (see Table
8.1) The experiment consists of recording the distribution of scattered radiation
from a specimen that is illuminated with a collimated, monochromatic beam.

The scattering, expressed as the differential cross section $d\sigma/d\Omega$, measures
the Fourier components of the spatial fluctuations in the scattering length density. For light scattering it is the variations in the refractive index that give rise
to scattering, for X-rays it is the electron density, and for neutrons it is the nuclear scattering lengths. In cases where the monomer units are large compared
to the interatomic distances it is a good approximation to assume a uniform
average scattering length density within the monomer. This means that the
scattering at a given momentum transfer (see Fig. 8.6)

$$q = 4\pi/\lambda \, \sin\theta \tag{8.8}$$

Table 8.1. Typical ranges for the momentum transfers in small-angle scattering experiments.
The values are quoted for pinhole geometries. Better nominal values can be obtained in line
geometry but this requires more extensive deconvolution procedures

Probe	Momentum transfer range (nm^{-1})
Laser light	0.001–0.025
Cold neutrons	0.01–2.5
Synchrotron X-rays	0.03–2.5

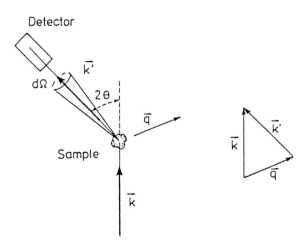

Fig. 8.6. Schematic diagram for an elastic scattering experiment with incoming wave vector
k, outgoing wave vector k', and momentum transfer $q = k - k'$

for the three probes gives the same scattering pattern in this regime, after correction for the trivial angular Lorentz factor for photon scattering. In typical X-ray and neutron experiments the length scales of interest are much larger than the wavelength of the probe. It is therefore the scattering at small angles that contains the relevant information.

For the resolution in the neutron scattering experiments this is an additional advantage because the resolution effects from the wavelength distribution of the incident beam are proportional to the scattering angle in the small-angle limit. This means that a good spatial resolution can be obtained by restriction of the angular spread, often in a pinhole geometry, and at the same time an adequate intensity is obtained by a relatively broad distribution in the wavelength, typically 5%–10%. This is also necessary because of the diffuse nature of neutron sources.

The use of large-area detectors is essential to obtain a high efficiency and reproducability in such experiments. It eliminates the setting errors that can be encountered for scanning detector systems and it ensures that the scattering at all angles of interest is recorded simultaneously. This is important for the normalization and for studies of temporal variations such as in situ aggregation. A schematic layout is shown in Fig. 8.7.

X-ray scattering at small angles can be carried out using both ordinary X-ray generators and synchrotron radiation sources. The latter have the advantages of being naturally collimated and providing a range of wavelengths. These advantages can be utilized to obtain excellent spatial resolution even with the choice of relatively short wavelengths that ensures good penetration of the X-ray radiation into the specimen.

Laser light is also a very adequate source for small-angle scattering experiments and it offers the possibilities of both good resolution and good intensity in the spatial range that can be covered. Examples of the typical ranges in momentum transfer that can be covered by the three probes are given in Table 8.1.

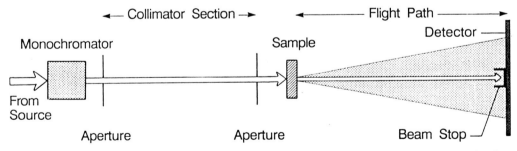

Fig. 8.7. Schematic layout of a small-angle neutron scattering experiment, SANS

8.3.3 Scattering Formalism

The scattering experiments measure the Fourier transform of the pair correlation function $g(r)$,

$$S(q) = 1 + \frac{N}{v} \int_v |g(r) - 1| e^{i\boldsymbol{q} \cdot \boldsymbol{r}} d\boldsymbol{r}, \qquad (8.9)$$

where $g(r) = G(r)/\rho$ for $r > 0$ and, by definition, $N/v = \rho$. For systems made up of monomer units it is convenient to define a form factor for the monomers,

$$F(q) = \int_{v_p} |\rho(\boldsymbol{r}) - \rho_0| e^{i\boldsymbol{q} \cdot \boldsymbol{r}} d\boldsymbol{r}, \qquad (8.10)$$

which describes the scattering from a single unit, and an average form factor, $P(q)$, that describes the scattering from a dilute uncorrelated ensemble of the monomers,

$$P(q) = \langle |F(q)|^2 \rangle. \qquad (8.11)$$

In this manner one can identify the inter-monomer scattering function

$$\bar{S}(q) = 1 + \frac{|\langle F(q) \rangle|^2}{\langle |F(q)|^2 \rangle} |S(q) - 1| \qquad (8.12)$$

expressed by the spatial correlations between the centers of the monomers. The observed intensity $I(q)$ is related to the inter-monomer scattering function through

$$I(q) = \frac{N}{v} P(q) \bar{S}(q). \qquad (8.13)$$

If the spatial correlations can be represented as $r^{-A} = r^{d_f - d}$, (8.5), we obtain the following simple result for the scattering function:

$$\bar{S}(q) \sim q^{-d_f}. \qquad (8.14)$$

In other words, the scattering corrected for the average form factor, $P(q)$, gives a very direct measure of the fractal dimension and, in the ideal case, the value of d_f can be deduced from the slope of the observed corrected intensity versus the momentum transfer in a double logarithmic plot. This is illustrated in Fig. 8.8, which shows the results of static light-scattering experiments on fully aggregated colloid systems of different types, formed under both RLCA and DLCA conditions. The difference in the slopes for the two classes of experiments and the similarity within a given class is apparent.

   The finite range of the fractal correlations can be seen in the scattering from a not yet fully aggregated system. This is illustrated by the experiments on LUDOX silica colloids during aggregation shown in Fig. 8.9.

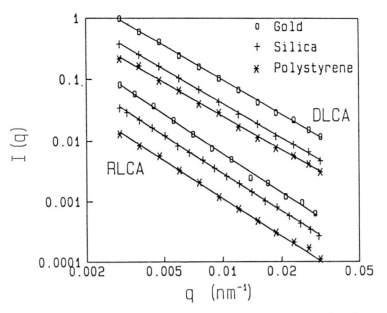

Fig. 8.8. Light scattering data in a double logarithmic presentation for samples of colloidal aggregates of particles of gold, silica, and polysterene (see Fig. 8.5). The slopes of the fitted lines give the fractal dimension $d_f \cong 1.85$ for DLCA and $d_f \cong 2.1$ for RLCA

In terms of the correlation function the finite-size effects can be included in the form of a scaling function that provides a cutoff at large scales. The most convenient but not unique choice is that of an exponential so that

$$g(r) = r^{-A} \exp(-r/\xi). \qquad (8.15)$$

With this choice one can obtain an analytic form for the resulting scattering function, namely

$$S(q) = 1 + \frac{1}{(qr_0)^{d_f}} \frac{d_f \Gamma(d_f - 1)}{(1 + 1/(q\xi)^2)^{(d_f-1)/2}} \sin[(d_f - 1)A \tan(q\xi)], \qquad (8.16)$$

which can be used in the detailed analysis of scattering data to extract the value of ξ [8.41]. An example is shown in Fig. 8.9. The effects of short-range correlations can also be treated more explicitly with specific assumptions of the form of the correlation function at near-neighbor distances. This has been demonstrated for silica smoke clusters [8.41] and for colloidal gold aggregates [8.42]. In the latter case the form for the pair correlation function shown in Fig. 8.10a was used, which provided fits to the observed synchrotron X-ray scattering data over a wide range of momentum transfer, as seen in Fig. 8.10b.

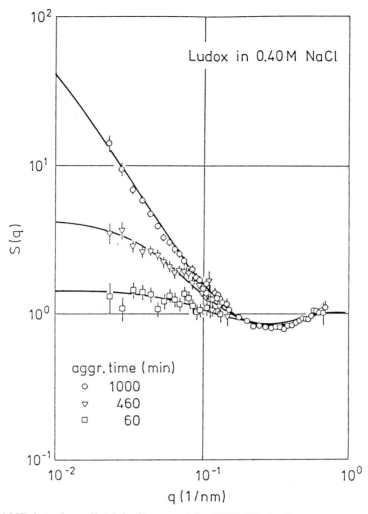

Fig. 8.9. SANS data for colloidal silica particles LUDOX during aggregation. The form factor for the unaggregated particles has been divided out to give the inter-particle $\bar{S}(q)$. From the fitted lines one can determine the cluster size ξ as a function of aggregation time

8.4 Physical Properties

The fractal geometry has a profound effect on the physical properties of the types of materials that have been discussed above. The randomness introduces strong scattering in the transport processes and in some cases it even causes localization (see Chap. 3). Similarly the morphology and in particular the connectivity of the random networks influences both the mechanical restoring forces and the transport properties as seen in the conduction of both charge and heat. This means that both the macroscopic and the microscopic physical properties are strongly dependent on the structural parameters such as density, correlation length, and connectivity. The fractal character of the materials implies that we can expect to express these dependences in terms of scaling relations and power

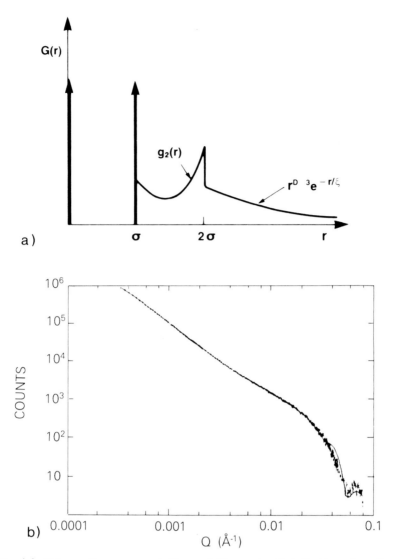

Fig. 8.10. (a) Schematic representation of the pair distribution function that includes nearest-neighbor and next-nearest-neighbor correlations as well as fractal long-range correlations. This was used to fit the data shown in (b). (b) Double logarithmic plot of intensity versus wave vector transfer for the scattering of synchrotron X-rays from a sample of gold colloid aggregates obtained by a RLCA process [8.42]

laws. The determination of these laws and relations is a main objective of the experimental efforts in this field. In this chapter we shall confine our interest to the mechanical and thermal properties of fractal networks.

8.4.1 Mechanical Properties

The concept of fractal geometry and its implications for the mechanical properties have been applied to a very wide range of disordered materials, for example, glasses, polymers, gels, and compressed powders [8.43,44]. These materials have inhomogeneities or fluctuations in the density distributions and in the bond networks on various scales, which affect the mechanical responses.

As an example let us discuss the silica aerogels, where these scales are most easily distinguished. Aerogels are produced in monolithic form with a range of densities. This means that they are mechanically stable with well-defined elastic properties and that they can support sound waves of suitable long wavelength. However, the sound waves are scattered by the disorder in the material and at a certain scale the waves become overdamped. Intuitively this scale is related to the upper length scale ξ for the density fluctuations but as we shall see these scales do not nescessarily coincide numerically. For wavelengths shorter than ξ, the vibrational excitations are those of the random network.

Alexander and Orbach [8.45] were the first to point out that these odes have a special character, and they coined the name *fractons* for them. Their theoretical analysis indicated that fractons are strongly localized and suggested a universal behavior of the density of states $z(\omega) \sim \omega^{d_s-1}$, where d_s is the spectral dimension.

Fracton modes are expected to exist for wavelengths in the range ξ_a to a, where a is the radius of the monomer particles. For wavelengths shorter than a, the vibrational excitations correspond to the phonon modes of bulk amorphous quartz with additional contributions from the surface modes of the monomer particles. Thus we have to consider three different regimes in wavelength and/or frequency in order to obtain a complete description of the mechanical and vibrational properties of these materials, here exemplified by the aerogels:

a) $\xi_a < \lambda$: long-wavelength phonons and elastic modes,
b) $a < \lambda < \xi_a$: localized fracton modes,
c) $\lambda < a$: short-wavelength vibrations within the monomers.

A corresponding separation into three regions can be made in the frequency regime, the regions being separated by the crossover frequencies ω_{co_1} and ω_{co_2}. (In Chap. 3, ω_{co_1} was denoted by $\omega_\times$.) We now discuss some of the experiments that have provided quantitative imformation about the different regimes in various materials.

a) Elasticity of silica networks. The elastic properties of silica network materials such as aerogels and compressed powders of Cab-O-Sil and Alfasil can be determined by three-point bending experiments on beam-shaped samples. Young's modulus, Y, is determined this way from the relation

$$Y = \frac{Fd^3}{4l\delta e^3},\tag{8.17}$$

where d is the span, F the applied force, e the beam thickness, l the beam width, and δ the observed deflection. Such measurements have been carried out for aerogels with a range of densities by Woignier et al. [8.46] and for prepared rods of compressed Cab-O-Sil by Forsman et al. [8.47]. The main results are shown in Fig. 8.11 as a function of the volume fractions.

References [8.46,47] have related the observed power laws to the predictions from theories pertaining to percolation but warn that the relations between the observed volume fraction and the relevant percolation parameters are uncertain. It is worth noting that these systems are intermediate between branched polymers and compressed powders in the sense that both the internal energy and the entropy may contribute to the elastic properties as discussed by Edwards and Oakeshott [8.44]. The relative role of this contribution depends on the system and there does not yet seem to be an adequate general method for calculation of the elastic properties.

b) Sound propagation. Sound propagation is related to the elastic properties of the medium. Highly disordered media such as liquids and gases can only support longitudinal sound waves and follow the rules of scalar elasticity. Most solids have tensorial elasticity and can support transverse as well as longitudinal sound waves. Systems with fractal geometry are special solids and one of the questions to be answered by the experiments is the nature of their elasticity. The propagation of sound has been extensively studied in the aerogels by both ultrasonic and Brillouin scattering techniques. The experiments give quantita-

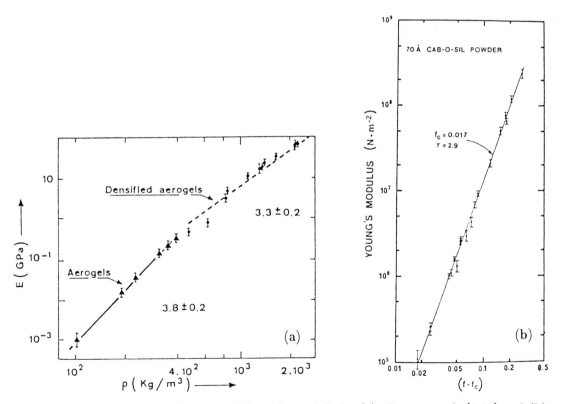

Fig. 8.11. Experimentally determined Young's moduli for (a) silica aerogels [8.46] and (b) compressed Cab-O-Sil powders [8.47] as a function of density and volume fraction, respectively

tive information on both the sound velocities and the damping as a function of the excitation frequency, and through studies with systematic variation of the sample preparation considerable insight has been gained concerning the effects of the fractal geometry of the random networks that support the sound propagation.

A particularly illuminating set of experiments have been carried out by Courtens et al. [8.48–50]. They have studied a series of neutrally reacted silica aerogels with densitites in the range 95–356 kg m^{-3} and determined the structutal parameters d_f, ξ, and a by neutron scattering as well as the dynamical parameters, from the Brillouin scattering spectra. The systematic preparation technique resulted in specimens that were found to be mutually self-similar. This means that all of the samples have the same generic structure for length scales up to ξ and that only the value of ξ varies for samples of different densities. This property is documented by the scaling result that $\xi \sim \rho^{1/(d_f - 3)}$ and by the consistency of the observed neutron scattering intensities. Using this observation it is possible to combine the experimental results from different samples in a meaningful way, as we shall see below.

First let us review the experiments on the individual samples. For each sample a set of Brillouin scattering spectra was recorded with very high resoloution using the 515.4 nm argon laser line and a six-pass tandem interferometer [8.50]. All of these spectra can be fitted with a theoretical form for the lineshape, which is derived on the basis of Green's functions [8.51]:

$$I(q,\omega) = A\frac{c^2 q^2}{\omega^2} \frac{\Gamma}{(\omega^2 + \Gamma^2 - c^2 q^2)^2 + 4\Gamma^2 c^2 q^2}, \qquad (8.18a)$$

where

$$c(\omega) = c_o \left[1 + \left(\frac{\omega}{\omega_{co_1}}\right)^m\right]^{z/m} \qquad (8.18b)$$

and

$$\Gamma(\omega) = \frac{\omega^4}{\omega_{co_1}^3} \frac{1}{\left[1 + \left(\frac{\omega}{\omega_{co_1}}\right)^m\right]^{3/m}} \qquad (8.18c)$$

are crossover expressions in which the parameters $m = 2$ and $z \cong 0.5$ can be fixed by continuity arguments in the long-wavelength limit or left as fitting parameters [8.52]. An example of a fitted spectra is shown in Fig. 8.12a; the values for the crossover frequency, ω_{co_1}, and the sound velocity c_o can be obtained from the analysis. The crossover length, $\xi_a = 1/q_{co_1}$, is given by c_o/ω_{co_1}. The crossovers indicate the points where the linear dispersion of the sound excitations meets the dispersion relation for the fracton excitations and thus they can be used to trace the fracton dispersion for a set of mutually self-similar samples. This is done in Fig. 8.12b. The result is consistent with the theoretical conjecture for the dispersion relation, namely $\omega \sim q^{d_f/d_s}$, and fits to the data give $d_f/d_s \cong 1.9$.

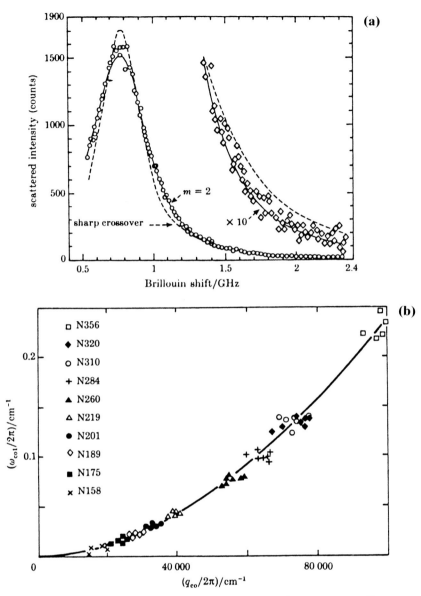

Fig. 8.12. (a) Example of Brillouin scattering spectrum from a neutrally reacted aerogel, $\rho = 200\,\mathrm{kg\,m^{-3}}$ fitted with the functional form given in (8.18) (*solid line*) [8.50]. (b) Effective fracton dispersion curve derived from the Brillouin determination of ω_{co_1} and q_{co} on a series of mutually self-similar aerogels [8.50]

The results for the acoustic correlation lengths ξ_a shown in Fig. 8.13 are approximately a factor of five larger in numerical value than the correlation lengths obtained from the neutron scattering measurement of the density–density correlations. Part of this difference is due to different definitions of the length scales, in that the neutron length refers to a radius of a correlated volume whereas the acoustic length refers to a diameter. However, there seems to be a genuine difference between the correlations seen in the density, ξ, and in the dynamics, ξ_a. This indicates that the frequency range over which one can

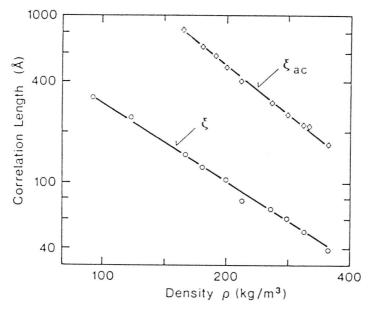

Fig. 8.13. Comparison of the correlation length ξ obtained from SANS experiments and the correlation length ξ_a obtained from Brillouin scattering experiments in a range of neutrally reacted aerogels with different densities [8.50]

observe fracton dynamics is larger than anticipated on the basis of the density fluctuations. This implies that fracton modes may be relevant also in systems with only short-range density fluctuations, such as polymers and glasses.

The Brillouin scattering experiments at the highest densities showed the presence of longitudinal as well as transverse sound modes. The transverse sound velocities amounted to approximately 60% of the longitudinal velocities and the intensity of the transverse signal was weak. This implies that the aerogels have tensorial elasticity at least in the long-wavelength region.

c) Density of states. A direct way to obtain information on the vibrational properties of systems with fractal geometry is the measurement of the density of states, $z(\omega)$. Alexander and Orbach first suggested the simple power law $z(\omega) \sim \omega^{d_s - 1}$, where d_s is the spectral dimension. Furthermore, their analysis suggested that the exponent d_s might have a universal value, at least for systems with scalar elasticity [8.45]. Subsequent work by Webman and Grest showed that percolation networks with tensorial elasticity gave $d_s \cong 0.9$, and, more recently, a range of values for d_s have been found for calculations pertaining to silica networks [8.53]. In this situation it is clearly important to obtain reliable experimental information, which can establish the value of d_s. A very direct method is incoherent inelastic neutron scattering, which measures the amplitude weighted density of states for a given system [8.54]:

$$I(q,\omega) = q^2 \frac{k'}{k} \frac{n(\omega)}{\omega} \sum_i e^{-2W_i} z_i(\omega), \qquad (8.19)$$

where $n(\omega)$ and W_i are the temperature-dependent Bose factor and Debye-Waller factor, respectively. The summation extends over the different sites for the atoms with an incoherent scattering cross section, each of which contributes proportionally to the amplitude of vibration at the frequency ω.

Protons have a particularly large cross section, which often dominates the incoherent scattering from systems containing even small amounts of hydrogen. Deuterons have a small incoherent cross section. This has been used to obtain $z(\omega)$ for silica networks through measurements of the difference in the scattering from samples in which small amounts of OH and OD, respectively, were bonded to the surfaces of the monomer particles [8.55]. In the low-frequency regime the motions of the protons (deuterons) follow the motions of the monomer particles due to the strong hydroxyl bonds. The observed scattering therefore reflects the density of states of the network [8.56,57].

A less direct method of obtaining the same information from the coherent scattering relies on the so-called incoherent approximation [8.58], which can be applied in situations where the probing wavelength is much smaller than the characteristic wavelength of the excitations. This situation occurs when the coherent scattering from fractons in silica networks is probed with momentum transfers of order $1\mathrm{\AA}^{-1}$ [8.58].

Measurements of the incoherent inelastic scattering have been carried out in both silica networks [8.56] and in aerogels [8.57] and the results for the former are shown in Fig. 8.14. A careful analysis has been applied to extract the intrinsic $z(\omega)$ by extrapolation to zero momentum transfer. In this manner possible systematic errors due to the frequency variation of the effective Debye-Waller factor and to the angular variation of the absorption in the sample is avoided. In both cases the result is a power law for $z(\omega)$ in the frequency region that corresponds to the fracton regime. Hence, values of d_s could be derived and were found to be in the range 1.9–2.1 for the fumed silica and 1.85 ± 0.15 for the aerogel. Both experiments cover the crossover from the fracton to the short-wavelength phonon regime, which is seen particularly clearly in the 136 K data for the fumed silica. On the other hand, the energy resolution was insufficient to observe the crossover to the long-wavelength phonon regime. Experiments in this region using the powerful neutron spin-echo technique have recently been published [8.59]. They indicate a somewhat lower value for d_s, closer to 4/3, in the fracton regime and the expected Debye law in the phonon regime.

Information on the density of states can also be obtained by means of Raman scattering. In this case it is also the incoherent contribution to the scattering that dominates and the observed intensity involves the product of $z(\omega)$ and the square of the polarization amplitude $P(\omega)$ produced by the strain of the fracton excitation. This means that the observed power law involves more than one exponent [8.60] and can be expressed as

$$I(\omega) \sim z(\omega)\omega^{-2-2d_\theta d_s/d_f}, \qquad (8.20)$$

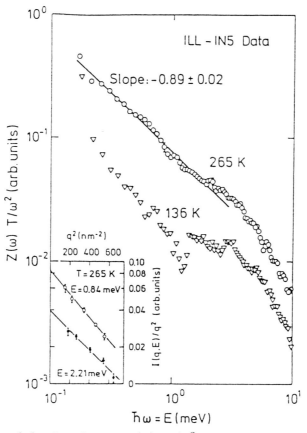

Fig. 8.14. Observed density of states $z(\omega) \cdot T/\omega^2$ for compressed powders of Cab-O-Sil at temperatures of 135 and 265 K, respectively. The observed slopes of the low-frequency part indicate that $d_s \cong 2$ for this material

where d_θ characterizes the scaling of the internal length of the fracton excitation. The experimental value is $d_\theta \cong 1.5$ for $d_s = 4/3$ [8.60].

Recently the density of states in silica aerogels has been studied in a wider frequency range by combining data from neutron time-of-flight and backscattering experiments [8.61]. These results indicate that different types of modes contribute to the scattering in the fracton range, and the authors suggest that the modes have a predominantly bending character at low frequencies and a stretching character at higher frequencies. Hence, these modes contribute with a changing weight across the frequency spectrum. Such a picture makes the interpretation of spectroscopic data less straightforward and consequently one has to be cautious in the interpretation of the observed power laws in scattering experiments and take possible variations in the structure factors into account.

8.4.2 Thermal Properties

The specific heat C_p and the thermal conductivity κ also reflect the fractal geometry. C_p is directly related to the density of states $z(\omega)$ through the relation

$$C_p = 3\frac{\delta}{\delta T} \int_0^{\omega_0} d\omega z(\omega)\hbar\omega \frac{1}{\exp(\hbar\omega/k_B T) - 1} , \qquad (8.21)$$

which means that $C_p \sim T^{d_s}$ in the fracton range. Measurements of the specific heat are therefore important to establish the validity of the fracton concept.

Such measurements are difficult to carry out at low temperatures in the fractal systems that have mostly been discussed in this chapter because of the extremely low thermal conductivity of the silica gels and aggregates. Nevertheless, the experiments are important in order to obtain a picture complementary to the spectroscopic studies, and clever procedures have beeen devised to overcome the difficulties. A very sensitive static measurement technique using a SQUID system null detector for thermoelectric currents has been developed by de Goer et al. [8.62]. An equally sensitive dynamic technique that allows simultaneous determinations of C_p and κ has been used by Bernasconi et al. [8.63]. Both groups have studied the thermal properties of silica aerogels in considerable detail with overall consistent results. An example of the specific heat measurements is shown in Fig. 8.15 in the usual representation as C_p/T^3.

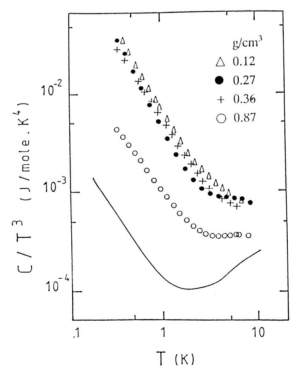

Fig. 8.15. Specific heat for three different samples of neutrally reacted aerogels with densities 0.87 g/cm^3 (a), 0.72 g/cm^3 (b), and 0.27 g/cm^3 (c) compared to bulk quartz (*solid line*) [8.62]

One notes immediately that the low-temperature specific heat of the aerogels below 3 K exceeds that of bulk α quartz by 1–2 orders of magnitude although the temperature dependence is similar, namely roughly linear in T. This linear temperature dependence is ubiquitous in amorphous materials and is normally ascribed to configurational fluctuations often referred to as two-level states. This physical picture is corroborated by the temperature dependence of the thermal conductivity, which typically follows a T^2 law in the same temperature ranges for many amorphous and/or disordered materials [8.64].

For several reasons the aerogels do not fit into this picture. First, the observed specific heat falls below the Debye values expected from the measured sound velocities. As we have seen, the sound velocities in the aerogels are quite small and correspondingly the Debye contribution from the propagating sound modes is much larger than in α quartz. In contrast, the low-temperature specific heat in α quartz exceeds the Debye limit. Second, the thermal conductivity of the aerogels has a plateau in the temperature dependence in the range where the specific heat is nearly linear in T, this is far from a T^2 law. It is therefore more appealing to interpret the thermal properties of the aerogels in terms of the fractal picture that has been used so far in this exposition. The temperature range where the specific heat is nearly linear corresponds to the fracton region in the dynamics. This gives a very direct interpretion of the observed power law, namely $C_p \sim T^{d_s}$, and from the quoted experiments one finds values for d_s in the range 0.9–1.2.

The crossover between the long-wavelength phonon Debye regime and the fracton regime is nontrivial. It has been argued on the basis of mode counting that a piling-up of modes is to be expected in this region [8.65]. So far this has not been borne out by computer simulations on percolation networks [8.53] (see also Sect. 3.4) but the most recent experiments on base-catalyzed aerogels support the notion of a "bump" in the density of states [8.66]. The results for C_p/T are shown in Fig. 8.16 for two different densities. For the higher density the data extend through the crossover region into the low-temperature phonon region. The density of states that can describe these data using (8.21) is shown in Fig. 8.17. This illustrates that one cannot necessarily expect a smooth crossover in the density of states from the phonon to the fracton region in a series of mutually self-similar aerogel samples since the limiting values from the two regions at the crossover point scale in a different manner with changes in the density.

The thermal conductivity in the fracton region has been the focus of extensive theoretical work [8.67–69] and the prediction is a linear temperature dependence that stems from the long-wavelength-phonon-assisted hopping of the localized fracton modes. The theoretical predictions for the temperature dependence of the thermal conductivity are confirmed by the experiment on base-catalyzed aerogels [8.70] and an example of the results is shown in the inset in Fig. 8.16. The magnitude of the coefficient to the linear term can be estimated on a theoretical basis and one finds reasonable agreement with the

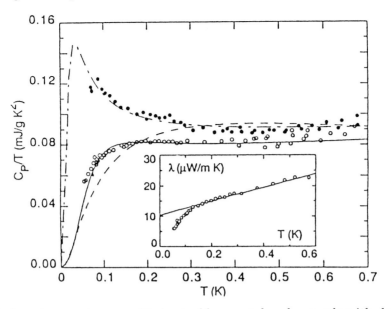

Fig. 8.16. Low-temperature specific heat of base catalyzed aerogels with densities of 0.145 g cm^{-3} (*full circles*) and 0.275 g cm^{-3} (*open circles*). The full lines are fits using the density of states shown in Fig. 8.17. The inset shows the thermal conductivity for the higher density sample [8.66]

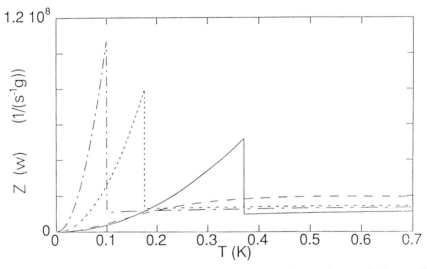

Fig. 8.17. Density of states that gives good fits to the observed specific heats shown in Fig. 8.16 using (8.21). For identification we note that the crossover temperature increases with increasing sample density [8.66]

observations. The extremely low level of the thermal conductivity paired with the relatively large specific heat is in itself a paradox and demands physical interpretations that invoke strongly localized modes, such as the fractons.

At higher temperatures the internal modes in the silica monomer particles become excited and hence contribute to the thermal properties. This means that the specific heat in these materials approaches the values for α quartz as the temperature increases beyond the fracton range. The thermal conductivities remain distinctly different both in magnitude and in the nature of the temperature dependences. This reflects the influence of the fractal morphology on the mean free paths of the heat-conducting phonons. In the aerogels the mean free path is restricted by the monomer particle size over the whole temperature range whereas the mean free path in α quartz is strongly temperature dependent. It can become very large at low temperatures. Furthermore, in the aerogels the heat has to migrate through the fractal network. The magnitude of the thermal conductivity will therefore also depend on the connectivity in much the same way as for conduction in a percolation network.

The thermal conductivity is given by the master formula

$$\kappa(T) = \frac{1}{3} \int z(\omega) C_p(\omega, T) v_s \ell(\omega, T) d\omega, \qquad (8.22)$$

where v_s is the sound velocity, ℓ is the mean free path, and $C_p(T)$ is the heat capacity, $C_p(T) = \int z(\omega) C_p(\omega, T) d\omega$.

Since the mean free path is kept constant by the monomer particle size in the aerogels we expect that κ and C_p will be proportional to one another in the high-temperature range. This is indeed borne out by experiments as illustrated in Fig. 8.18.

To conclude this section on the thermal properties, we note that a linear temperature dependence of the thermal conductivity has been observed over a wide range of temperatures in epoxy resins [8.71]. These results have also been interpreted in terms of the fracton model. In this case the fluctuations in the density distributions are confined to the atomic scale and in this sense the epoxy resins cannot be claimed to be fractal. On the other hand, the individual

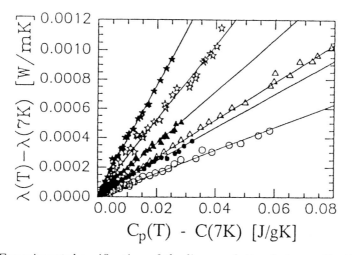

Fig. 8.18. Experimental verification of the linear relation between the thermal conductivity $\kappa(T)$ and the specific heat $C_p(T)$ for base catalyzed aerogels with different densities [8.70] for temperatures above 7 K

polymer chains between the cross-linking points in the resin are known to have a fractal morphology and this could lead to a dynamic behavior similar to that observed in the gels. This is an active field of research where many more materials have to be studied before a general understanding of the thermal properties of gels, polymers, and amorphous materials can be established.

8.5 Outlook

The aim of this chapter has been to provide a qualitative picture of the broad range of experiments that have been carried out on materials with fractal geometry. The selection of examples is by no means complete but hopefully it is large enough to give an impression of an active and diverse field of condensed matter science.

It is now well established that fractal structures can be synthesized under a range of circumstances both in the laboratory and by naturally occurring processes. The signature is the self-similarity or dilution symmetry that in the most prominent cases is found to extend over many orders of magnitude in length scale. Here, the colloidal aggregates are good examples. A particularly interesting class of materials that are synthesized by this route is the silica aerogels, and many of the examples of the physical properties in this chapter were taken from experiments with this kind of material.

It should be stressed that all the fractal materials are results of dynamic processes and they represent states of matter that are far from equilibrium. Each sample is in some sense unique and this makes it difficult to deduce universal properties from the measurements of the scaling exponents such as d_f and d_s, for example. The small selection presented here has illustrated the problem, which is also crucial for progress in the theoretical understanding of these systems. Each method of synthesis leads to different exponents although the values may cluster around generic values like those for the DLA model.

The structure determinations as well as the experiments to characterize the mechanical and thermal properties clearly show the importance of the limiting length scales. In the gels it is the monomer particle size that sets the lower length-scale cutoff and the correlation length ξ that determines the upper cutoff. These cutoffs are reflected in the frequency dependences of the dynamic reponses and give rise to characteristic crossover energies. Similarly one finds characteristic crossover temperatures in the thermal properties. The presence of these limits complicates the analysis of the experimental data and much of the controversy over the values of exponents such as d_s and d_f stems from such difficulties. It should therefore be one of the aims of the continuing work in this field to develop better overall models that allow for the inclusion of the cutoff effects in the numerical analysis of the experimental results. This will inevitably lead to more fitting parameters and hence require a broader base of complementary experiments. Again the experiments on the aerogels are good examples where the same materials can be studied by a variety of techniques. In this case a consensus description seems to be appearing, although some controversy still remains, mainly pertaining to the crossover regions.

The fractal materials can be regarded as generic examples of disordered materials, along with polymers, glasses, and amorphous solids. The fractals show scaling in most physical properties whereas the other types of materials typically have more restricted regions of scaling, if any. It is not yet clear whether a kind of unified description can be developed that encompasses all these materials, but clearly there are many common trends. Again this should serve as inspiration for further studies.

Finally, a comment on the practical relevance of the fractal materials by way of an example. If one regards a solitary tree in winter time when the branch structure is revealed without the foliage, one is struck by the fractal geometry. The absence of a particular length scale means that the tree also has no single resonance frequency, like many man-made mechanical constructions. This clearly protects the tree against disastrous agitations during storms. The example shows that there may be scope for incorporating fractal geometry into the synthesis of structural materials. An obvious example is composites, where fiber networks with fractal structure may be of advantage. The beauty of the fractal concept is that in principle the same idea applies to the design of structures on the scale of our common buildings, where there may be scope for better optimization of cost and performance using structures based on fractal geometry.

It is clear that the implications of the use of fractal concepts in many fields of experimental condensed matter science have not yet been fully explored. There is ample room for extensions of the experimental work along the paths that have been outlined in this chapter. A particularly interesting development has been the formulation of the concept of self-organized criticality by Per Bak and his co-workers [8.72], which appears to have considerable generality. This original work has made sand-pile a paradigm of the complexity of physics, and a recent review illustrates that theory and experiments on granular matter is far from being reconciled [8.73]. It should serve as a challenge to experimentalists to complement the current theoretical progress with carefully thought-out experiments under controlled circumstances. In this manner this field of science will continue to flourish.

References

8.1   B.B. Mandelbrot: *The Fractal Geometry of Nature* (Freeman, San Francisco 1983)
8.2   T. Witten, M. Sander: Phys. Rev. Lett. **47**, 1400 (1981)
8.3   P. Meakin, in: *Phase Transitions and Critical Phenomena*, ed. by C. Domb, J.L. Lebowitz (Academic Press, New York 1987)

8.4   D. Derrida and V. Hakim: Phys. Rev. A **45**, 8579 (1992)

8.5   T. Vicsek: Fractal Growth Phenomena (World Scientific, Singapore 1989)

8.6   S. Schwarzer, J. Lee, A. Bunde, S. Havlin, H.E. Roman and H.E. Stanley, Phys. Rev. Lett. **65**, 603 (1990); S. Schwarzer, S. Havlin, and H.E. Stanley: Physica A **191**, 117 (1992)

8.7   E.I. Knizhnik, A.D. Onisko, A.V. Gaydamaka: Radiat. Phys. Chem. **19**, 473 (1982)

8.8   L. Niemeyer, L. Pietronero, H.J. Wiesmann: Phys. Rev. Lett. **52**, 1033 (1984)

8.9   G.C. Lichtenberg discovered the Lichtenberg-figures (similar to Fig. 8.1) in 1777. He already noticed that the figures obey self-similarity.

8.10 S. Tolman, P. Meakin: Physica A **158**, 801 (1989); P. Ossadnik: Physica A **195**, 319 (1993)

8.11 J. Nittmann, H.E. Stanley: J. Phys. A **20**, L1185 (1987); F. Family, D. Platt, T. Vicsek: J. Phys. A **20**, L1177 (1987)

8.12 J. Langer: Rev. Mod. Phys. **50**, 1 (1980)

8.13 R.M. Brady, R.C. Ball: Nature **309**, 225 (1984)

8.14 M. v. Smoluchovski: Z. Phys. **17**, 557 and 585 (1916); Z. Phys. Chem. **92**, 129 (1917)

8.15 M. Matsushita, M. Sano, Y. Hayakawa, H. Honjo, Y. Sawada: Phys. Rev. Lett. **53**, 286 (1984)

8.16 H. Kondoh, Y. Sawada: *Science on Form 3* (KTK Scientific Publ., Tokyo 1988), p. 3

8.17 Y. Sawada, H. Hyosu: Physica D **38**, 299 (1989)

8.18 R.C. Ball, in: *On Growth and Form*, ed. by H.E. Stanley, N. Ostrowsky (Nijhoff, Dordrecht 1986), p. 69

8.19 J.D. Chen: Exp. Fluids **5**, 363 (1987)

8.20 L. Paterson: Phys. Rev. Lett. **52**, 1621 (1984)

8.21 E. Ben-Jacob, P. Garik: Physica D **38**, 16 (1989)

8.22 E. Guyon, L. Oger, T.J. Plona: J. Phys. D **20**, 1637 (1987)

8.23 D.A. Weitz, J.P. Stokes, R.C. Ball, A.P. Kushnich: Phys. Rev. Lett. **59**, 2967 (1987)

8.24 P. Meakin, A. Birovyev, V. Frette, J. Feder and T. Jøssang: Physica A **191**, 227 (1992)

8.25 M.Y. Lin, H.M. Lindsay, D.A. Weitz, R.C. Ball, R. Klein, P. Meakin: Proc. Roy. Soc. London, Ser. A, **423**, 71 (1989)

8.26 J. Turkevitch, A. Garton, P.C. Stevenson: J. Colloid. Sci. **9**, 26 (1954)

8.27 Ludox SM3 from E.I. Dupont de Nemours and Co., Wilmington, Delaware

8.28 T. Freltoft: Thesis, unpublished (Risø-M-2570, Risø National Laboratory 1986)

8.29 G. Henning, L. Svensson: Phys. Scripta **23**, 697 (1981)

8.30 M. Prassas, J. Phalippou, J. Zarzyki: J. Mater. Sci. **19**, 1656 (1984)

8.31 S.S. Kistler: J. Phys. Chem. **36**, 52 (1932)

8.32 Cab-O-Sil is a trade mark of Cabot Corporation, USA, and Alfasil is a trade mark of Alfa Products, UK

8.32 G.D. Ulrich, J.W. Riehl: J. Colloid. Interface Sci. **87**, 257 (1982)

8.34 D. Pearson, A.J. Allen: J. Mater. Sci. **20**, 303 (1985)

8.35 J.D. Gunton, M. San Miguel, P.S. Sahni, in: *Phase Transitions and Critical Phenomena, Vol. 8*, ed. by C. Domb, J.L. Lebowitz (Academic Press, New York 1983)

8.36 S.W. Koch, R.C. Desai, F.F. Abraham: Phys. Rev. A **27**, 2152 (1983)

8.37 R.A. Cowley, R.J. Birgeneau, Y.J. Uemura: Phys. Rev. B **21**, 4038 (1980)

8.38 G. Aeppli, H. Guggenheim, Y.J. Uemura: Phys. Rev. Lett. **52**, 942 (1984)

8.39 T.A. Witten, P. Meakin: Phys. Rev. B **28**, 5632 (1983)

8.40 D.A. Weitz, M. Oliveria: Phys. Rev. Lett. **52**, 1433 (1984)

8.41 T. Freltoft, J.K. Kjems, S.K. Sinha: Phys. Rev. B **33**, 269 (1986)

8.42 P. Dimon, S.K.Sinha, D.A. Weitz, C.R. Safinya, G.S. Smith, W.A. Varady, H.M. Lindsay: Phys. Rev. Lett. **57**, 598 (1985)

8.43 S. Alexander, C. Laermans, R. Orbach, H.M. Rosenberg: Phys. Rev. B **28**, 4615 (1983)

8.44 S.F. Edwards, R.B.S. Oakeshott: Physica D **38**, 88 (1989)

8.45 S. Alexander, R. Orbach: J. Physique Lett. **43**, 625 (1982)

8.46 T. Woignier, J. Phalippou, J. Pelous: J. Physique **49**, 289 (1988)

8.47 J. Forsman, J.P. Harrison, A. Rutenberg: Can. J. Phys. **65**, 767 (1987)

8.48 E. Courtens, R. Vacher, E. Stoll: Physica D **38**, 41 (1989)

8.49 R. Vacher, T. Woignier, J. Pelous, E. Courtens: Phys. Rev. B **37**, 6500 (1988)

8.50 E. Courtens, R. Vacher: Proc. Roy. Soc. London A **423**, 55 (1989)

8.51 G. Polatsek, O. Entin-Wohlman: Phys. Rev. B **37**, 7726 (1988)

8.52 E. Courtens, R. Vacher, J. Pelous, T. Woignier: Europhys. Lett. **6**, 245 (1988)

8.53 I. Webman, G. Grest: Phys. Rev. B **31**, 1689 (1985)

8.54 G.L. Squires: *Introduction to the Theory of Thermal Neutron Scattering* (Cambridge University Press, Cambridge 1978)

8.55 D. Richter, L. Passell: Phys. Rev. Lett. **44**, 1593 (1980)

8.56 T. Freltoft, J.K. Kjems, D. Richter: Phys. Rev. Lett. **59**, 1212 (1986)

8.57 R. Vacher, E. Courtens, J. Pelous, G. Coddens, T. Woignier: Phys. Rev. B **39**, 7384 (1989)

8.58 G. Reichenauer, J. Fricke, U. Buchenau: Europhys. Lett. **8**, 415 (1989)

8.59 D.W. Shaefer, C.J. Brinker, D. Richter, B. Farago, B. Frich: Phys. Rev. Lett. **64**, 2316 (1990)

8.60 Y. Tsujimi, E. Courtens, J. Pelous, R. Vachter: Phys. Rev. Lett. **60**, 2757 (1988)

8.61 R. Vacher, E. Courtens, G. Coddens, A. Heidemann, Y. Tsujimi, J. Pelons, M. Foret: Phys. Rev. Lett. **65**, 1008 (1990)

8.62 A.M. de Goer, R. Calumczuk, B. Salce, J. Bon, E. Bonjour, R. Maynard: Phys. Rev. B **40**, 8327 (1989)

8.63 A. Bernasconi, T. Sleator, D. Posselt, H.R. Ott: Rev. Sci. Instrum **61**, 2420 (1990)

8.64 W.A. Philips: J. Low Temp. Phys. **7**, 351 (1972); P.W. Andersson, B.I. Halperin, C.M. Varma: Phil. Mag. **25**, 1 (1972)

8.65 A. Aharony, S. Alexander, O. Entin-Wohlman, R. Orbach: Phys. Rev. B **31**, 2565 (1985)

8.66 T. Sleator, A. Bernasconi, D. Posselt, J.K. Kjems, H.R. Ott: Phys. Rev. Lett. **66**, 1070 (1991)

8.67 S. Alexander, O. Entin-Wohlman, R. Orbach: Phys. Rev B **34**, 2726 (1986)

8.68 T. Nakayama, K. Yakubo, R.L. Orbach: Rev. Mod. Phys. **66**, 381 (1994)

8.69 A. Jagannathan, R. Orbach, O. Entin-Wohlman: Phys. Rev. B **39**, 465 (1989)

8.70 D. Posselt, J.K. Kjems, A. Bernasconi, T. Sleator, H.R. Ott: preprint

8.71 H.M. Rosenberg: Phys. Rev. Lett. **62**, 780 (1989)

8.72 P. Bak, C. Tang, K. Wiesenfeld: Phys. Rev. Lett. **59**, 1398 (1987)

8.73 A. Mehta, G.C. Baker: Rep. Prog. Phys. **57**, 383 (1994)

9 Cellular Automata

Dietrich Stauffer

9.1 Introduction

Cellular automata [9.1] offer a very wide field of choices, and within the space given to me by the editors it will not be possible to write, say, 4 294 967 296 different chapters, each dealing with one particular automaton separately. Instead, I will concentrate on two more systematic approaches. The first approach mixes all possible cellular automata together by letting each site select at the beginning which cellular automata rule it wants to obey. This is the Kauffman model for genetics, and its fractal dimensions at the phase transition have been studied in detail by many people since 1986. The second approach tries to classify all possible cellular automata according to some unified simple criteria; such criteria could be the fractal dimensions, but since they are only beginning to be known we concentrate on the simpler Wolfram classification. Some readers may wish to look only at some restricted class of cellular automata, and not at all of them; for these readers I offer some appendices.

In this way the chapter starts with a definition of cellular automata, and a particular example in one dimension. Then we deal with fractality and multi-fractality in the Kauffman model for two to four dimensions, where all automata are mixed together randomly. This section is followed by one on the general classification of cellular automata. One appendix deals with the special case of Ising-like Q2R rules, another describes recent speculations on how to use more complicated cellular automata to model the immune system. (The second appendix is taken from an earlier review by this author.) A third appendix deals with hydrodynamics, such as flow through porous media.

Not everything I describe is fractal, but future research along these lines should find the still-hidden fractality. Simple computational techniques have already been described by Weisbuch in an appendix of his book [9.2]; our first appendix

---

◄ **Fig. 9.0.** Picture of a computer generated snow flake, courtesy of S. Wolfram

lists a multispin coding program which updates 10^9 sites per second on one Cray processor; for Kauffman models or general cellular automata the speed is about 300 sites per microsecond.

9.2 A Simple Example

Figure 9.1, kindly produced by R.W. Gerling, looks like the well-known Sierpinski gasket; but here it represents damage spreading (difference picture) of the one-dimensional exclusive-OR rule.

Imagine a long straight line of spins, e.g., of magnetic atoms which can have their magnetic dipole moment pointing either up or down. (Mathematically, such a spin is represented by a number which is either zero or unity, or by FALSE and TRUE.) Initially the spins are oriented randomly: half of them point up and the other half down. Then, at each time step $t = 1, 2, 3, \ldots$, each spin depends on the orientation of its two neighbors, to the right and to the left. For the exclusive-OR rule (sometimes abbreviated as XOR or as

Fig. 9.1. Damage spreading for the one-dimensional exclusive-OR rule. For a completely random initial spin distribution we check how flipping the center spin affects the later time development. The top line corresponds to zero time, then each time step corresponds to the next line

rule 90), the spin points up if at the previous time step its two neighbors had different orientations; it points down if the two neighbor spins were parallel. This finishes the definition of this particular cellular automat. We will ignore in this review any probabilistic automata where the orientation is determined only with a certain probability: all our cellular automata are deterministic and thus exclude the traditional Ising model simulations.

Figure 9.1 is, however, not a picture of one such XOR line for consecutive times. Instead we simulate two different XOR automata which follow the same exclusive-OR rule but have slightly different initial orientations of their spins. We talk about damage spreading if the set of spins which differ in their orientations in a spin-by-spin comparison of the two initial configurations is confined to a small region of the two lattices. In our one-dimensional example, initially only one spin is different; on two and more dimensions, simulations were made with one spin or with a whole line of initially different spins. We can imagine that the initially different spins are a computer error, and then a comparison of the later simulation of the two lattices shows how this computer error affects the whole configuration. Figure 9.1 shows the damage, i.e., the set of spins which differ in a comparison of the two configurations, when initially only the center spin differed. Each new line gives a new time step. In two dimensions, such a visualization would require three dimensions. Thus Fig. 9.2 shows instead one particular time step for the square-lattice XOR rule, again with initially only one spin damaged. Here a spin points up if an odd number of neighbors points up, whereas it points down for an even number of up neighbors.

We see from Figs. 9.1 and 9.2 that order and beauty can come out of disorder: the initial conditions were completely random, but nevertheless some

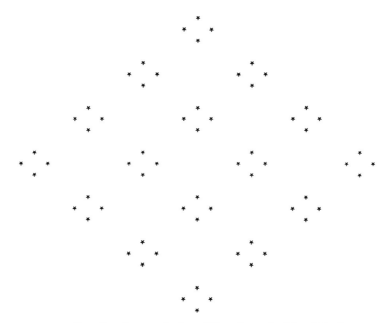

Fig. 9.2. Damage spreading for the exclusive-OR rule as in Fig. 9.1, but on square lattice, shown after several time steps. Again the initial spin distribution is random

regularity arises. This question of "order out of disorder" will accompany us throughout this chapter; we do not discuss here the opposite case of how a particularly shaped initial spin configuration decays into a less ordered state.

This XOR rule, all other one-dimensional rules, and some two-dimensional cases, based on nearest-neighbor interactions, have been studied by Wolfram [9.1,2]; many selected square-lattice rules are presented in work often based on a special-purpose computer [9.3] although the results are independent of the hardware used. In addition, the literature is full of special cases of interest for particular applications. Quite often these applications involve spins which have more choices than simply up or down, and which interact with more than just their nearest neighbors. Examples are the game of life [9.3] and self-organized criticality [9.4,5].

Rules with infinite interaction range allow each spin to depend on faraway sites, as in the original Kauffman model [9.6] invented for genetics. They can often be solved analytically by some sort of mean field approximation [9.7] though this identification does not always work [9.8]. The idea of damage spreading was introduced in this infinite-range Kauffman model; only later was it applied to nearest-neighbor models [9.9]. Nevertheless, the present review exclusively deals with nearest-neighbor interactions. The reader can get a more general impression from [9.1-3].

Thus our definition is: each spin in cellular automata points up or down depending on the orientation of its nearest neighbors.

9.3 The Kauffman Model

In 1969, Stuart Kauffman suggested [9.6] using cellular automata to describe cell differentiation in biology. All the cells of our body have the same genes but different cell types exist. For example, my fat cells cannot function as brain cells. Kauffman identified the genes with the spins, and the cell types with the limit cycles in which the whole system ends up. Genes can be on or off, just as spins point up or down, and are influenced in their status by other genes. Since N spins can be in 2^N different configurations and since the motion of our cellular automata is completely deterministic, after at most 2^N updates per spin the cellular automata must reach a configuration which they have had before. From then on the dynamics will be identical to the previously observed dynamics, and we have reached a limit cycle in configuration space. Depending on the initial configuration, one can end up in different limit cycles. (Fixed points are limit cycles of period 1.)

Through what rules do the genes interact? We do not know; and just as in statistical physics, we hide our lack of knowledge behind the assumption that the rules are randomly selected. (Similarly, we can today simulate only 4 million particles through Newton's law. A glass of beer contains about 10^{25} molecules.

Thus we assume the configurations to be random, with Boltzmann's probability proportional to $\exp(-E/k_BT)$.) Thus each site in the Kauffman model selects for itself at the beginning one of the possible cellular automata rules, and obeys this rule forever during its later life. In this sense, the Kauffman model also asks for order to come out of this disordered network of rules.

How many rules are there? Each site on a lattice has K neighbors, each of which can point up or down. This gives $C = 2^K$ neighbor configurations. For each of them, the rule can point the center spin either up or down, which results in $R = 2^C$ different rules. Thus for a square lattice without memory, if the center spin depends only on its four neighbors and not on itself, we have $K = 4$, $C = 16$, and $R = 65\,536$. A medium-sized square lattice can thus allow each of its spins to follow a different rule; for the largest lattices simulated so far, 6976×6976, about 743 usually well-separated spins follow the same rule. If we allow the center spin on a square lattice to depend also on itself, we have $K = 5$ and thus $R = 2^{32}$ different rules, on the triangular and cubic lattices their number is even larger, whereas for the honeycomb lattice ($K = 3$) with no dependence on the center spin, R is merely 256. Examples of rules are: the logical AND, where the spin points up if and only if (iff) all its K neighbors point up; the logical OR, where the spin points up iff at least one of its K neighbors points up; and the XOR where the spin points up iff an odd number of neighbors (for even K) points up.

Biologically, an infinite range of interaction seems appropriate [9.6] but also nearest-neighbor lattices have been simulated [9.9]. One can study whether flipping a single spin, or a small number of spins, changes the whole configuration. More precisely, if we initially damage only spins in a geometrically limited region of the lattice (by flipping them in one of two replicas of the system), does this damage spread with a finite probability to a finite fraction of all spins (in the thermodynamic limit), or does the number of spins damaged later remain limited, i.e., proportional to the number of initially damaged spins? (Similarly, Kauffman [9.6] asked whether flipping one gene moves the whole system from one limit cycle to another.) One can denote the case where the damage spreads as chaotic, and the case where it remains limited as frozen. However, to avoid ideological fights over the one and only true definition of chaos, this review talks about stability and instability in the case of limited and spreading damage, respectively. Is the dynamics stable against flipping a single spin? (For the stable case one can then distinguish between the damage remaining finite or dying out, as we will do in our general classification of cellular automata, in the next section.)

In one dimension, the Kauffman model is always stable. The damage cannot spread to infinity just as one-dimensional percolation (see Sect. 2.4) does not have an infinite cluster. For example, about every 16th spin obeys the rule "always point down" and thus does not transmit any damage information. For infinite dimension, the Kauffman model is unstable for K above 2. On a honeycomb lattice [9.9b] damage grows appreciably but does not spread to infinity:

the stability is close to its limit. On square, triangular, simple cubic, and hypercubic lattices in two to three dimensions, damage spreads easily and touches nearly all spins if we wait long enough. For example, on the square lattice in the stationary state about 43% of all spins are damaged at any moment (except if due to an unfortunate initial configuration the damage died out). An actual damage of 50% would correspond to two totally uncorrelated lattices, since half of the spins must agree by chance if they only have the choice of up or down. Thus the difference between 50% and 43% damage shows that there is still some degree of order in the system, due to the identical rules which the corresponding spins in the two lattices have to obey.

A transition to chaos, i.e., a phase transition between a stable and an unstable regime, can be found [9.9] if one selects up rules with probability p and down rules with probability $1 - p$. If we vary p from its past value $1/2$ down towards zero, or up towards unity, the stability is enhanced, and at some threshold p_c a transition from unstable to stable occurs. More precisely, when initially each site selects the rule it wants to obey, then for each site we go through all $C = 2^K$ neighbor configurations. For each configuration, we draw a random number between zero and one. If that random number is smaller than the probability p, we take the result for this particular neighbor configuration as up; otherwise it is down. In this way for the given neighbor configuration we select with probability p a rule giving the result "up", and with probability $1 - p$ a rule giving "down". Note that we are still dealing with deterministic cellular automata since random numbers are used only in the initial selection of rules; once selected, each spin follows its rule without ever deviating from it. Symmetry allows us to restrict p to the interval between 0 and $1/2$.

Now transitions show up, at $p_c = 0.31$, 0.16, 0.12, and 0.08 in the square ($K = 4$), triangular ($K = 6$), simple cubic ($K = 6$), and four-dimensional hypercubic ($K = 8$) lattices, respectively. For $p < p_c$ the system is stable against damage, whereas for $p > p_c$ the damage spreads over the lattice. These values are determined by Monte Carlo simulation since neither series expansions nor exact results are known to us. The history of the square lattice shows the inaccuracies involved in the determination of the threshold: we started at a value of 0.26 which then increased via 0.28, 0.29, and 0.298 to above 0.30 within three years [9.9–12]. Figure 1 in [9.12] shows nicely how earlier studies on smaller lattices could underestimate p_c: the finite-size effects are particularly strong here but could not be seen in earlier work. The efficient algorithm of da Silva and Herrmann [9.11], plus a lot of Cray time, made these improvements possible.

These effective thresholds are determined by the requirement that of the many independent samples studied, in half of the cases the damage should spread from the center of the lattice to its boundary. Initially, one lattice line in the center of the square was damaged and we asked if the damage later touched the border lattice line parallel to the initially damaged line. (Periodic boundary conditions were used. Most of the earlier work damaged initially only isolated sites. Then at the threshold we expect the fraction of unstable samples to decay with a power law of the system size, instead of being 50%. Thus the efficiency of that approach is lower.) Pictures of the damaged sites at the moment of touching the boundary show rather compact clusters. In contrast to percolation (see Chaps. 2 and 3), the sites damaged at a given moment can be separated by large gaps, and thus the damage constitutes a cloud consisting of, in general, several connected clusters. (On the square lattice with no dependence of the spin on itself, the damage is restricted to a checkerboard sublattice if initially only one site is damaged. Then the cluster connectivity works via next-nearest-neighbor interaction.)

This review would hardly be accepted in this book if it did not contain some fractal dimensions. Let M be the damage at the moment it touches the boundary of the lattice, and T the time (number of updates per spin) it needs to spread from the initial sites or line to that boundary. In a lattice of linear dimension L we then expect at the threshold:

$$M \sim L^{d_f} \tag{9.1}$$

and

$$T \sim L^{d'_f} \tag{9.2}$$

for sufficiently large L. For long times up to T we expect the damage to grow as

$$M \propto t^{d_f/d'_f}$$

from standard scaling arguments. Computer simulations [9.12] of large square lattices, Fig. 9.3, give a d_f of about 1.8 to 1.9, and d'_f of 1.3 to 1.4. Unfortunately these fractal dimensions are far less accurate than those of percolation clusters. We see that d_f is compatible with the percolation value 1.89, whereas d'_f seems to be larger than the value $d_{\min} = 1.1$ for the chemical distance (shortest path within the cluster) of percolation theory (see Sect. 2.3). The growth of damage as a function of time has a power-law exponent of about 1.2 in the simulations at $p = 0.298$, only roughly compatible with d_f/d'_f. For p above the threshold, a logarithmic dependence of the final damage as a function of initial damage was observed [9.13], which is still lacking an explanation. It thus seems that simple percolation ideas for the damage do not give a quantitative description of the Kauffman model.

Such percolation ideas have long been suggested [9.6] under the names of forcing functions and forcing arcs. The idea was that damage spreads or is blocked by special rules and bonds (pairs of neighboring rules). More recently, this percolation idea was put onto a better foundation by a generalization of percolation concepts [9.14]. Not only conducting sites and bonds (resistors) were used but also diodes (one-way streets), triodes (transistors), tetrodes, and pentodes. Then the flow of damage corresponded to the flow of electrical current in this complicated network. Unfortunately, the electrical network

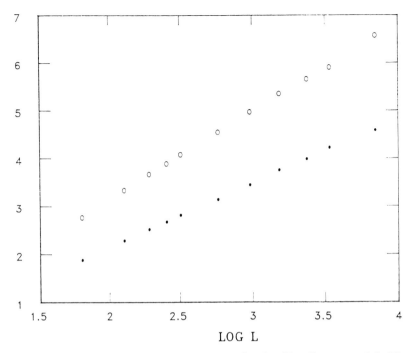

Fig. 9.3. Damage spreading for the $L \times L$ square-lattice Kauffman model. The decadic logarithm of the time needed for the damage to touch the boundary (*full lower symbols*) and for the amount of damage at this moment of touching (*open upper symbols*) is plotted versus the decadic logarithm of L, up to $L = 6976$, at the size-dependent thresholds near $p = 0.3$

corresponding to a given Kauffman model is so complicated that it also requires a difficult simulation and may not be numerically superior. The analogy is applicable, however, to general networks of disordered automata, not just to the square-lattice Kauffman model. A crucial element of the correpondence between Kauffman models and electrical networks is that the choice of the network elements depends not only on the rules but also on the given spin configuration.

The fractal dimensions are $d_f = 1.8$ and $d'_f = 2.1$ to 2.2 in three [9.15] and four [9.16] dimensions; perhaps four is the upper critical dimension [9.17] (see also Sect. 2.4).

Finally, instead of mixing all $65,536$ rules of the $K = 4$ square-lattice Kauffman model, one may randomly mix just two such rules, such as AND and OR [9.18]. In that case the fractal dimensions, but not the thresholds, seem percolation-like. Another subset of rules for the Kauffman model was shown by Kürten [9.18] to reproduce reasonably the full Kauffman behavior.

9.4 Classification of Cellular Automata

Many papers have discussed many different cellular automata and their particular applications or beauty. The reader is referred, for example, to the books on cellular automata machines [9.2,3] for an overview. The aim of this section is more systematic (more boring): we want to find the behavior of cellular automata in general, not just special cases. Thus in principle we want to go through all possible rules for cellular automata and find out how they behave. To do so requires, of course, some criteria for classifying the behavior of automata; these criteria should be simple enough to be evaluated automatically by a computer.

How many different automata do we need to analyze? As we discussed at the beginning of the section on Kauffman models, we have $C = 2^K$ different neighbor configurations if a site depends in its behavior on K neighbors; since each of these configurations can force the center spin either up or down, we have $N = 2^C$ different cellular automata. On a one-dimensional lattice, a honeycomb lattice, a square lattice, and a simple cubic lattice we have $K = 2$, 3, 4, and 6, respectively, if a spin depends only on its nearest neighbors and not on more distant neighbors; thus we have $N = 16$, 256, 65536, and more than 10^{19} different rules. Symmetry may reduce these numbers by about an order of magnitude, but even then it is clear that the simple cubic lattice cannot be investigated completely, nor the triangular lattice also with $K = 6$. If a spin depends on itself as well, then K is increased by 1, and even the square lattice would require four thousand million rules to be investigated, a task nobody has tackled so far. Thus only for K up to 4 do systematic investigations of all possible rules exist [9.19]; for larger K only a randomly selected subset of typically 10000 rules has been investigated [9.19].

This difference between complete (K up to 4) and randomly selective ($K = 6$) classification is not as fundamental as it first looks: even in the complete classification of all 65536 rules, the results are not exact. For the behavior of a rule depends also on the initial configuration. If, for example, initially all spins are up, we may get a special kind of behavior not seen with a random initial distribution. Comparing the results with different random initializations we still see some changes, though these fluctuations are very small for very large lattices. Therefore whatever results we have for the fraction of rules in certain classes, the numbers have some small error bars whether we evaluate all 2^C rules or only a large random subset of them. (Unless stated otherwise, initially our spins are random, half up and half down.)

The first systematic classification seems to be that of Wolfram [9.1] for one dimension, with the spin also depending on itself. Thus $K = 3$, and 256 rules had to be investigated. Wolfram grouped them into four classes depending on the final spin configuration: in the first class, finally all spins were parallel; in the second class, they formed finite clusters of up and down neighbors oscillating with finite periods; class 3 contained the chaotic rules with infinite periods; and propagating structures were found in class 4.

For an automatic computer evaluation, these distinctions are sometimes problematic. For example, any finite lattice always has a finite period, and thus classes 3 and 2 should be distinguished perhaps by median periods increasing exponentially or less than exponentially with system size. Such an investigation still needs to be done, however, and is nontrivial. Thus we make it simpler and check whether the period of the oscillations is 2 or greater than 2. The question of chaos will be addressed later by the different criterion of damage spreading. For the first Wolfram class, on the other hand, it is easy to distinguish between the case that finally all spins are up, and the case that finally all spins are down. Thus we classify the final outcome of cellular automata simulations as follows (mixed here means that some spins are up and some are down, as opposed to the case where all spins are parallel).

Class 0: fixed point with all spins down
Class 1: fixed point with all spins up
Class 2: fixed point with mixed configuration
Class 3: oscillations of period 2 between all up and all down
Class 4: oscillations of period 2 between mixed configurations
Class 5: translation of mixed configuration in one lattice direction

All other types of behavior then could be put into a sixth class. (A word of caution: classes 4 and 5 are not mutually exclusive. The translation of an antiferromagnetic structure with spins alternatively up and down means also an oscillation of period two.)

Table 9.1, from [9.19c], summarizes the numerical or exact results in one to three dimensions. We see that this Wolfram-type classification is quite useful in one dimension, for which it was invented. But the higher the dimension or the number K of neighbors, the smaller the fraction of rules belonging to one of

Table 9.1. Fractions, in percent, of cellular automata rules falling into various classes. The number K of neighbors varies from 2 to 6; for $K = 6$ we distinguish between triangular (tr) and simple cubic (sc) lattices

Class	$K = 2$	$K = 3$	$K = 4$	$K = 6$ (tr)	$K = 6$ (sc)
Class 0	12.5	5.5	2.8	0.34	0.39
Class 1	12.5	5.5	2.8	0.34	0.39
Class 2	–	–	0.1	0.02	0.00
Class 3	–	4.7	0.7	0.01	0.00
Class 4	12.5	4.7	2.1	0.17	0.14
Class 5	37.5	22.3	15.0	0.01	0.00
Class a	25.0	17.2	5.0	0.65	0.64
Class b	12.5	20.3	37.0	0.87	0.84
Class c	62.5	62.5	59.0	98.48	98.53

the above five classes: most rules cannot be characterized by these criteria. To some extent this result corresponds to experiences with the Kauffman model (preceding section) which is more chaotic for larger K.

(Numerically it is important to avoid linear lattice dimensions L which are a power of 2. For 128×128 and 256×256 square lattices, for example, the various classes get different weights [9.19]. This should be taken into account before using a special-purpose computer or program allowing only powers of two.)

An entirely different criterion arises from the question of damage spreading or chaos, used already for the Kauffman model in the preceding section. We check how sensitive the system is to small perturbations. Thus two lattices are simulated at the same time, with the same rule and the same initial spin configuration. However, in a small, geometrically localized region the spins of one lattice are initially (or after some equilibration time) reversed compared to the spins of the other lattice. The damage (or more generally, the Hamming distance) is the set and number of spins which differ later in a spin-by-spin comparison of the two lattice sites.

We thus distinguish three mutually exclusive cases for the spreading of the damage during the later time evolution.

Class a: the damage heals after some time (stable)
Class b: the damage neither heals nor spreads (marginal)
Class c: the damage spreads over the whole lattice (unstable)

To check for class c, one initially damages the center site or the center line of the lattice, and then checks whether the damage later touches a boundary of the lattice (parallel to the initially damaged line). Sometimes class c is also called chaotic. In class b, the number of damaged sites becomes neither zero nor infinite (where infinite means a finite fraction of an infinite lattice). Also for class c the number of damaged sites does not necessarily go to infinity in this sense. It is possible, for example in the bureaucratic rule where each spin obeys its top neighbor, for the damage to remain constant in size but to move with fixed velocity throughout the lattice. This happens quite often in one dimension [9.19c]. Table 9.1 also shows how the rules are distributed among these three classes. For two to four neighbors, this stability analysis seems useful since all three cases happen quite often; but with six neighbors in the triangular and cubic lattices, nearly all rules become unstable, and damage spreading ceases to be a practical classification scheme.

The fraction of lattice sites damaged at the moment the damage touches the boundary in the case of class c is normally between $1/4$ and $1/2$; thus the damage is not fractal. (If damage remains constant and touches the boundary, the fraction of damaged sites is very small.) The nonfractality of the damage is in full accord with the Kauffman model. Also in the Kauffman model fractality was not observed automatically (as in self-organized critical phenomena [9.4,5] such as DLA, (see Chaps. 1, 4, and 10), but only at some special critical point.

The two types of classifications (classes 1 to 6 for fixed points and short periods, classes a, b, c for damage spreading) can also be combined: how many class a rules belong to class 0? For the simple cubic case, in about 82% of class 0 or 1 the damage heals completely, whereas in 18% it touches the boundary before the fixed point is reached. Most of the class 4 rules belong to class b, and nearly all unclassified rules (class 6) are unstable [9.19c].

With classes $0 - 6$ as well as classes a, b, c losing their usefulness in higher dimensions, how else can we classify the various cellular automata? One way would be to look for phase transitions between the different classes: transition to chaos (from class b to class c), or transition to trivial fixed points (to class 0 or 1), etc. For the square lattice [9.19] such an analysis has been partly undertaken already, but for the other two-dimensional and three-dimensional lattices it is missing at present and thus we do not give details here. An analysis for the periods of limit cycles was useful for small one-dimensional chains [9.20] but not in large square lattices [9.19b].

In such a search for transition points one should vary the concentration of initial up spins (thus far always 50%) until the behavior jumps from one class to another. One has to be careful, however, to escape spurious transitions which vanish for large enough lattices. For example, let a spin be up if and only if three or four of its four neighbors on the square lattice are up. Then for low initial concentrations of up spins the rule belongs to class 0, i.e., finally all spins are down because of the rather stringent conditions for up orientations. If nearly all spins initially are up, then class 4 behavior (oscillations of period 2) is observed. However, the critical concentrations for this transition between trivial fixed point and short oscillation increases slightly (logarithmically) for increasing lattice size, and approaches unity in the thermodynamic limit. However, such a critical concentration approaching its limit of unity only logarithmically for $L \to \infty$ does not necessarily mean we have no true transition in a physics application: the logarithm of infinity in physics is not really infinite but somewhere between 10 and 100. Typical lengths L are 10^8 atomic spacings, and thus $\ln L$ is about 18. If, therefore, mathematically a law $1/(30 + \ln L)$ is proven for the difference between the size-dependent transition point and unity, then the correct expansion of that expression is not $1/\ln L$ but is $1/30 - \ln(L)/900 +$ quadratic terms. In this sense, a sharp transition would still exist in reality though not mathematically. Such difficulties are relevant for some bootstrap percolation cases, which are cellular automata like the above example except that a down spin always remains down, as reviewed recently [9.21].

9.5 Recent Biologically Motivated Developments

In Sect. 9.3 we discussed the possibility of "chaos" in cellular automata, where this type of instability meant that a small initial damage spreads over the whole lattice. Some cellular automata have a tunable parameter as a function of which the behavior changes from chaotic to nonchaotic. At the transition point to chaos, fractal dimensions and other critical exponents have been studied in several cases, e.g. for the Kauffman model on the nearest-neighbor square lattice (a disordered mixture of all possible cellular automata). These phase transitions may be biologically relevant: perhaps life happens at the edge of chaos, as reviewed in [9.24].

One can make this idea plausible as follows. The small initial perturbation or damage is identified with a biological mutation in the genes of the organism. If the effect of the mutation dies out quickly, it will hardly improve the chances of this individuum of surviving Darwinistic selection of the fittest. If, on the other hand, the small mutation changes half the genes of this individuum during the later development, then the complex coordination necessary for life will be destroyed. So, just like with whisky, to have nothing is bad, and to use too much of it is also bad. The effect of a mutation should be large but still finite. And that's what usually happens in the nonchaotic regime close to the transition to chaos.

We have to distinguish here between cellular automata like those discussed above, which as a function of an arbitrarily tunable parameter have a phase transition, and those which by themselves automatically move towards the transition point. In the first case, evolution may work best at or near the transition; in the second case we have evolution towards the edge of chaos, in the sense of Bak's self-organized criticality (for a recent review, see Chap. 2 in [9.25]). This second type of cellular automata is presently causing more excitement than the first, through the work of Packard, Langton, and Mitchell, as cited in [9.24], and other models of Bak et al [9.26].

Rapid progress has also been made with cellular automata in immunology. Lattice models with nearest-neighbor interactions between the antibodies were suggested by de Boer et al [9.27] and by Stewart and Varela [9.28], see also [9.29–34]. They are *not* supposed to model a living being as a three-dimensional lattice but rather describe the multi-dimensional space of different shapes of the antibodies. The automata values 1 and 0 correspond to the presence or absence of antibodies of a given shape, in a manner similar to the lattice gas interpretation of the Ising model. The dimensionality of this shape space should be about 10, according to an old estimate of Perelson and Oster. Now computers have become good enough to simulate such high dimensions. Barral et al. [9.35] looked at them more mathematically whereas Stauffer and Weisbuch [9.29] as well as Sahimi and Stauffer [9.33] had immunological applications in mind. It is now possible to study such cellular automata with thousands of millions of sites, about as many as the number of cells in a natural immune system. So in biology we have already reached the aim of simulating models as large as a natural system, whereas even a very healthy glass of alcohol contains much more water molecules than the world's best computer can handle at present. I find this a very satisfying feature of biologically motivated simulations.

Apart from the above exciting biological applications, there has also recently been progress in applying cellular automata to the following diverse systems: hydrodynamics [9.36,37], traffic jams [9.38,39], earthquakes [9.40], reaction-diffusion processes [9.41], quantum gravity [9.42], and calculations of Lyapunov exponents [9.43] and percolation thresholds [9.44]. A classification of the cellular automata rules has been described in [9.45,46].

Acknowledgements. I thank many colleagues, too numerous to be listed by name, for introducing me to this field of work, for researching important aspects, and for critical discussions.

9.A Appendix

9.A.1 Q2R Approximation for Ising Models

The most-studied cellular automata (if we ignore neural networks as reviewed by Weisbuch [9.2]) are presumably for two-dimensional hydrodynamics. These, however, have at least seven spins per site and are shifted to Appendix 9.A.3. The second-most studied case is probably the Q2R approximation for the Ising model [9.3,46].

Imagine the spins interact ferromagnetically, i.e., each has an energy proportional to the number of its antiparallel neighbors. This Ising model is usually simulated at a fixed temperature T with thermal probabilities proportional to $\exp(-\text{energy}/k_B T)$. In the microcanonical ensemble, on the other hand, one simulates this Ising model at fixed energy. Then a spin can be flipped only if it has equally many up and down neighbors. Q2R cellular automata on a square lattice do exactly this: a spin is flipped if and only if it has two up and two down neighbors. It is easy to program this model by going through the lattice like through a typewriter and by sequentially updating the spins.

However, in cellular automata all spins are traditionally updated simultaneously, and if one does this with the above Q2R rule and without further precautions, the energy is no longer constant. If a spin is flipped and its neighbor, too, niether yet know that the neighbor spin will be flipped, and thus base their flipping decision on a wrong assumption. We can flip simultaneously only those spins which do not directly interact with each other. (The same trick applies generally to vectorization and parallelization of algorithms.) Thus we separate the system into two sublattices as on a checkerboard, and update first one sublattice simultaneously, and then the other sublattice, using for the second sublattice the already flipped orientations of the first sublattice. A complete simple Fortran program is given, for example, in Ref. [9.47].

```
      DATA JMIN/1,2,2,1/,IMIN/1,2,1,2/
...
...
...
C     Simplification of updating: Idea of Sebastian Kremer
      DO 6 ITIME=1,MAX
      DO 6 LATT=1,4
      DO 1 J=JMIN(LATT),LL,2
      DO 1 I=IMIN(LATT), L,2
1     N(I,J)=   IEOR(N(I,J),IOR(
     1 IAND(IEOR(N(I-1,J),N(I+1,J  )),IEOR(N(I,  J-1),N(I,J+1))),
     2 IAND(IEOR(N(I-1,J),N(I  ,J-1)),IEOR(N(I+1,J  ),N(I,J+1)))))
      IF(LATT.EQ.2.OR.LATT.EQ.3) THEN
         DO 2  I=1,L
2        N(I,0)=SHIFT(N(I,LL),63)
      ELSE
         DO 3 I=1,L
3        N(I,LL1)=SHIFT(N(I,1),1)
      END IF
      IF(LATT.EQ.2.OR.LATT.EQ.4) THEN
         DO 4 J=JMIN(LATT),LL ,2
4        N(0  ,J)=N(L,J)
      ELSE
         DO 5 J=JMIN(LATT),LL ,2
5        N(LP1,J)=N(1,J)
      END IF
6     CONTINUE
```

Fig. 9.4. Compact multispin coding program for Q2R automata

To save computer memory and time, spins should be stored in single bits and treated in parallel by bit-per-bit manipulation (multispin coding). A vectorized multispin coding program is given by Herrmann [9.46] using four sublattices. The same four sublattices are treated in the Cray CFT77 Fortran program of Fig. 9.4 in a more compact form. The lattice is divided into even and odd lines, $N(I,J)$, where line index I runs from 1 to L, and column index J from 1 to LL=$L/6$ for a $L \times L$ square lattice (with 64 sites stored in one computer word). For $L=256$, the words with $J=1$ contain columns 1,5,9,...,253; the words with $J=2$ correspond to columns 2,6,10,...,254; $J=3$ contains sites 3,7,11,...,255; and $J=4$ corresponds to 4,8,12,...,256. We first treat odd J and odd I, then even J and even I, then even J and odd I, and finally odd J and even I. The boundary conditions are periodic, which is achieved by suitable circular shifts if $J=1$ or $J=$LL, and by buffer lines $I=0$ and $I=$LP1 (=$L+1$) for $I=1$ and $I=L$. A NEC-SX3 processor updates more than three thousand sites per microsecond for large lattices with such a program. (L must be a multiple of 128.) A multispin coding program for general cellular automata, following da Silva and Herrmann [9.11] is given in Refs. [9.13,53b] .

This algorithm is not an exact Ising model simulation since, contrary to the related but more complicated demon method [9.48], it is not [9.49] ergodic: the period of these (reversible) cellular automata increases exponentially with the number $N = L \times L$ of spins, but at most as $2^{N/4}$ and not as 2^N. An asymptotic increase like 2^N is needed, however, if the algorithm is supposed to reach all allowed configurations, as required for ergodicity. Nevertheless, the spontaneous magnetization agrees well with the exact solution if we start with a random distribution of spins (up with probability p, giving a thermal energy $8Jp(1-p)$ with exchange energy J on the square lattice; $12Jp(1-p)$ on the simple cubic lattice). One may have to wait, for however, millions of sweeps through the lattice before the proper equilibrium is established. The Curie point is reached for an initial concentration of up spins near 8%; for lower concentrations, Q2R is ferromagnetic, and for higher ones it is paramagnetic, with the maximum concentration $1/2$ corresponding to infinite temperature. The behavior in a magnetic field cannot be studied; however, the susceptibility comes out correctly from the magnetization fluctuations, if the thermodynamic formulas properly take into account that now the energy, not the temperature, is fixed [9.50]. The nearest-neighbor correlations allow an experimental determination of the temperature: if N_k counts how often a spin is surrounded by k parallel neighbors, then in equilibrium we have $N_4/N_0 = (N_3/N_1)^2 = 8J/k_BT$. However, again we may need millions of iterations in the ferromagnetic regime to get consistent temperatures from this formula.

If we determine local periods, the clusters of sites having the same period have a tendency to form rectangles, and may form an infinite connected network for initial concentrations above about 3.5% [9.51]. Damage spreading is easy for high temperatures and nearly impossible for low temperatures, with a threshold concentration again of several percent [9.52]. For low temperatures only a few spins change their orientation, whereas for high temperatures (concentrations above several percent) the set of changing spins spreads over the whole lattice [9.53a]. However, some of these thresholds may vanish logarithmically with increasing system sizes and observation time, up to $t = 10^9$. These difficulties may be related to very long relaxation times at low initial concentrations p [9.53b], which diverge exponentially for low p.

Thus the Q2R approximation is not just a cheap simulation method to get accurate Ising properties; it is a model. If one does not require precise data, then a simple sequential updating program is a good way to introduce ferromagnetism and collective behavior in teaching, before quantum mechanics and statistical physics are introduced.

9.A.2 Immunologically Motivated Cellular Automata

Reading papers by real (i.e., experimental) immunologists [9.54], one sees a wide difference between their style and that of the cellular automata or Ising community. Does it make sense to describe the complications of life by such simple models as spins being either up or down? Of course, reality is different and one can easily prove each of the models below to be unrealistic. However, studying complex reality alone does not guarantee success, as can be seen from cancer research. And sometimes success in science has come from unrealistic models. For example, Kepler's laws approximated the Earth by a point, even though everybody knew this to be wrong. And the Ising model ignores the well-known fact that magnets can be bent and set vibrating. Nevertheless, Kepler's laws started theoretical physics, and the Ising model is still a standard research subject of statistical physics. Of course, one should not expect Kepler's laws to describe continental drift, or the Ising model to describe the phonon spectrum of iron.

The real problem of these immunological cellular automata is that after only a dozen papers it is unlikely that the best model, corresponding to Kepler's laws or the Ising magnet, has already been found. But even then there is some faint hope: perhaps these models instead describe something else, just as the Ising model describes liquid–gas critical points. Columbus did not find what he was looking for, a sea-lane to India; nevertheless his travels had some impact.

The immune system consists of many different types of cells and molecules [9.54]. We need here B cells (bone-marrow derived lymphocytes) producing (indirectly) antibodies which attack and neutralize foreign antigens such as some virus. They are helped by helper cells and suppressed by suppressor cells. Macrophages eat the antigens and present parts of that food on their cell surface in order to alarm helper cells. Infection by some virus can stimulate the immune system in such a way that a second infection is dealt with much better. Vaccination against smallpox made so many people immune that this disease seems to have been eradicated. A more recent application of immunological knowledge is T-cell vaccination [9.54a]. The immune system can go wrong either by attacking its own body (auto-immune diseases such as multiple sclerosis or diabetes mellitus), or by being weakened, e.g., through the AIDS virus. These three cases, normal, too strong, and too weak immune responses, will be treated here; see Table 9.2.

Atlan [9.55] outlined the general philosophy of this cellular automata approach. We assume the concentrations of each cell type to be either very high (spin up, 1, true), or to be very low (spin down, 0, false). We assume that these cell variables interact by some deterministic rules, which we express as Boolean relations. Each cell variable at time $t + 1$ is determined by itself and the other cell variables at the previous time step. In this way one has a simple mean field

Table 9.2. Summary of immunological models. We give the number of spins used for each kind of cell type: A antibodies; B bone-marrow derived lymphocytes; D AIDS virus or other destructive agent; H helper cells; O self-antigen (own body attacked by killer cells); K killer cells attacking own body O, L: lymphokines; M macrophages; S suppressor cells; V antigen (e.g., virus)

Model	A	B	D	H	K	L	M	O	S	V
Modified KUT model	1	1	0	1	0	0	0	0	1	1
3-cell auto-immune	0	0	0	1	1	0	0	0	1	0
5-cell auto-immune	0	0	0	2	1	0	0	0	2	0
3-cell AIDS model	0	0	1	1	0	0	0	0	1	0
4-cell AIDS model	0	0	1	1	0	0	1	0	1	0
8-cell AIDS model	1	0	2	1	0	1	1	0	2	0
Unified model	1	1	1	1	1	0	0	1	1	1

theory, the whole human body being invaded by a homogeneous concentration of cells, with one variable only for each cell type. (In response to one specific antigen, we deal only with the part of the immune system reacting to that antigen, not the other antibodies and so on.)

Such a mean field approximation of spatially homogenous concentrations is presumably good in most cases since fluids spread through the body in minutes and hours, whereas the immune response takes days. In special cases, one might also have a slowly growing cancer-like immune response, which could be modelled on a lattice. Each lattice site now carries one set of spin variables, as used in the mean field model, and spins on different neighboring lattice sites interact.

a) Normal immune response. The modified KUT model [9.56,57] for the normal immune response deals with antibodies A, suppressors S, helpers H, B-cells B, and the external antigen V (for virus; denoted by E in [9.56]). Each of these five spins, A, S, H, B, and V, can be up (true, concentration 1) or down (false, concentration 0). The dynamics is governed by

```
A = V.AND.B.AND.H
S = H.OR.S
H = H.OR.(V.AND..NOT.S)
B = H.OR.(V.AND.B)
V = V.AND..NOT.A
```

Here the left-hand side gives the result at time $t+1$, as determined by the status at time t of the variables on the right-hand side (rhs). Thus the crucial last equation means that the virus survives only if there is a virus and no antibody present. As an alternative to the last equation, Neumann [9.59] suggests

```
V = V.AND..NOT.(A.and.H)
```

Starting from any of the $2^5 = 32$ possible configurations, we end up after some iterations in one of five fixed points: (ASHBV) = (00000), (01100), (01110), (01000), and (01001). The first three are healthy and correspond to the virgin, the vaccinated, and the immune state; if we add antigen V to any of these three fixed points, we return to the immune state (01110). The two other fixed points correspond to paralysis of the immune system, with (01001) or without (01000) antigen; addition of a virus results in paralysis with antigen, leading to permanent sickness because of this failure of the immune system.

If we now put this model onto a square lattice with nearest-neighbor interactions, we first have to clarify how different lattice neighbors interact with each other. High-speed multispin coding simulations are easiest if one first takes for each cell type the logical OR of the five lattice neighbors involved, and regards the outcome as the variable to be used on the rhs of the above equations. For example, one regards the virus as present at a lattice site if at least one of the five lattice sites contained a virus at the previous time step; the same rules apply to the antibodies; and finally, from `V(t+1) = V.AND..NOT.A`, the new value of V is determined.

The result of these lattice simulations, if we start with a small antigen concentration, is the immune fixed point (01110) everywhere, which is healthy but boring. The time needed to reach this fixed point increases as the square root of the logarithm of the system size, since the process has to fill the largest hole between two initial virus [9.58]. More interesting results were found [9.59] when positive thresholds were used, i.e., when just one true out of five neighbors no longer makes the corresponding variable true. Majority rules are helpul in this and other cases in order to get stable clusters or localized fluctuations. In the AIDS models to be discussed below [9.60], nontrivial end states can be found on lattices even without such tricks.

Simulations in the physics of condensed matter usually deal with only a small fraction of a real system, and often only with very short times. An immune system, on the other hand, consists of about 10^{10} cells. Thus a 9600×9600 lattice, with five cell variables on each lattice site, is already nearly of the size of a natural system. Then, a Cray processor each second makes a time step, corresponding to 10^2 hours in real immune systems. Thus if our models were realistic, we could simulate nature quite nicely.

b) Immune diseases: autoimmunity and AIDS. First we deal with the case where the immune system attacks its own body (autoimmune disease). The simplest case involves killer or effector cells K which attack, for example, the own nerves, helper cells H, and suppressor cells S. Between each type of cells one can have a supporting interaction, a diminishing interaction, or no interaction, giving 3^{n^2} possible models with n cells; thus for $n = 3$ we get 19 683 possibilities. Going through all of them, omitting unrealistic cases where helpers suppress and so on, omitting cases where two cell types have the same behavior, and requiring exactly three fixed points – healthy (KHS = 000), immune (KHS = 011), and sick (KHS = 111), Chowdhury [9.57] arrived at exactly one model:

```
H = H.OR.K
K = K + H > S
S = H
```

It would be somewhat difficult to repeat such a systematic search for systems of differential equations or other more complicated methods. (Again, the rhs is valid at time $t + 1$ if the rhs is the configuration at time t.) In the equation for K, the arithmetic on the rhs means that we treat the Boolean variables K, H, and S also as integers 0 and 1. (K is set true or 1 if the sum K+H is larger than S; otherwise K is false, K=0.) On a lattice, starting with a small concentration of helper cells, the whole lattice fills up with immune lattice sites. The time needed to reach this fixed point increases as the square root of the logarithm of the system size, since the process has to fill the largest hole between two initial virus [9.5]. When we change the threshold conditions on a lattice, we may get the clusters and oscillations shown in Fig. 9.5.

More realistic are five-cell models, with two types of helpers, two types of suppressors, and the active killers. The first model, of Weisbuch and Atlan

[9.61], had great influence on the later developments of [9.57–60]; Cohen and Atlan [9.62] suggested several modifications. Going through all possible five-cell models again and imposing many biologically motivated restrictions, Chowdhury [9.57] found one model giving only the three fixed points of healthy (0000), immune, and sick (11111):

```
K  = K + H1 + H2 > S1 + S2
H1 = H1 + H2 > S1
H2 = K.OR.H1
S1 = H1.OR.H2
S2 = H1
```

On a square lattice, all sites become permanently sick after a short time.

```
7 7 7 7 7 0 0 0 0 0 0 0 0 0 0 0 0 0 0 0 0 0 0 0 0 0 0 0 0 0 0 0 0 0 0 0 0 0 0 7 7 7 7 7
7 7 7 7 7 0 0 0 0 0 0 0 0 0 0 0 0 0 0 0 0 0 0 0 7 7 7 7 7 0 0 0 0 0 7 7 7 7 7
7 7 7 7 7 0 0 0 0 0 0 0 0 0 0 0 0 0 0 0 0 0 0 0 7 7 7 7 7 0 0 0 0 0 7 7 7 7 7
7 7 7 7 7 0 0 0 0 0 0 010 910 0 0 0 0 0 0 0 7 7 7 7 7 0 0 0 0 0 7 7 7 7 7
7 7 7 7 7 0 0 0 0 7 7 0 0 9 8 9 0 0 0 0 0 0 0 7 7 7 7 7 0 0 0 0 0 7 7 7 7 7
7 7 7 7 7 0 0 0 0 7 7 0 0 8 9 8 0 0 0 0 0 0 0 7 7 7 7 7 0 0 0 0 0 7 7 7 7 7
7 7 7 7 7 0 0 0 0 0 0 0 0 8 9 0 0 0 0 0 0 0 0 7 7 7 7 7 0 0 0 0 0 7 7 7 7 7
7 7 7 7 7 0 0 0 0 0 0 0 0 010 0 0 0 0 0 0 0 0 7 7 7 7 7 0 0 0 0 0 7 7 7 7 7
7 7 7 7 7 0 0 0 0 0 0 0 0 0 0 0 0 0 0 0 0 0 0 7 7 7 7 7 0 0 0 0 0 7 7 7 7 7
7 7 7 7 7 0 0 0 7 7 7 7 0 0 0 0 0 0 0 0 0 0 0 7 7 7 7 7 0 0 0 0 0 7 7 7 7 7
7 7 7 7 7 0 0 0 7 7 7 7 0 0 0 0 0 0 0 0 0 0 0 7 7 7 7 7 0 0 0 0 0 0 0 0 0 0
0 0 0 0 0 0 0 0 7 7 7 7 0 0 0 0 0 0 0 0 0 0 0 7 7 7 7 7 0 0 0 0 0 0 0 0 0 0
0 0 0 0 0 0 0 0 0 0 0 0 0 0 0 0 0 0 0 0 0 0 0 7 7 7 7 7 0 0 0 0 0 0 0 0 0 0
0 0 0 0 0 0 0 0 0 0 0 0 0 0 0 0 0 0 0 0 0 0 0 7 7 7 7 7 0 0 0 0 0 0 0 0 0 0
0 0 0 0 0 0 0 0 0 0 0 0 0 0 0 0 0 0 0 0 0 0 0 7 7 7 7 7 0 0 0 0 0 0 0 0 0 0
0 0 0 0 0 0 0 0 7 7 7 7 0 0 0 0 0 0 0 0 0 0 0 0 0 0 0 0 0 0 0 0 0 0 0 0 0 0
0 0 0 0 0 0 0 0 7 7 7 7 0 0 0 0 0 0 0 0 0 0 0 0 0 0 7 7 7 0 0 0 0 0 0 0 0 0
0 0 0 0 0 0 0 0 7 7 7 7 0 0 0 0 0 0 0 0 0 0 0 0 0 0 7 7 7 0 0 0 0 0 0 0 0 0
0 0 0 0 7 7 7 7 0 0 0 0 0 0 0 0 0 0 0 0 0 0 0 0 0 0 7 7 7 0 0 0 0 0 0 0 0 0
0 0 0 0 7 7 7 7 0 0 0 0 0 0 0 0 0 0 0 0 0 0 7 7 7 0 0 7 7 7 0 0 0 0 0 0 0 0
0 0 0 0 0 0 0 0 0 0 0 0 0 0 0 0 0 0 0 0 0 0 7 7 7 0 0 0 0 0 0 0 0 0 0 0 0 0
0 0 0 0 0 0 0 0 0 0 0 0 0 0 0 0 0 0 0 0 0 0 7 7 7 0 0 0 0 0 0 0 0 0 0 0 0 0
0 0 0 0 0 0 0 0 0 0 0 0 0 0 0 0 0 0 0 0 0 0 0 0 0 7 7 7 7 7 0 0 0 0 0 0 0 0
0 0 0 0 0 0 0 0 0 0 0 0 0 0 0 0 0 0 0 0 0 0 0 0 0 7 7 7 7 7 0 0 0 0 0 0 0 0
0 0 0 0 7 7 0 0 0 0 7 7 7 7 7 0 0 0 0 0 0 0 0 0 0 7 7 7 7 7 0 0 0 0 0 0 0 0
0 0 0 0 7 7 0 0 0 0 7 7 7 7 7 0 0 0 0 0 0 7 7 7 0 0 7 7 7 7 7 0 0 0 0 0 0 0
0 0 0 0 7 7 0 0 0 0 7 7 7 7 7 0 0 0 0 0 0 7 7 7 0 0 0 0 0 0 0 0 0 0 0 0 0 0
7 7 0 0 7 7 0 0 0 0 7 7 7 7 7 0 0 0 0 0 0 7 7 7 0 0 0 0 0 0 0 0 0 0 0 0 0 0
7 7 0 0 0 0 0 0 0 0 7 7 7 7 7 0 0 0 0 0 0 7 7 7 0 0 0 0 0 0 0 0 0 0 0 0 0 0
0 0 0 0 0 0 0 0 0 0 7 7 7 7 7 0 0 0 0 0 0 0 0 0 0 0 0 0 0 0 0 0 0 0 0 0 0 0
0 0 0 0 0 0 0 0 0 0 7 7 7 7 7 0 0 0 0 0 0 0 0 0 0 0 0 0 0 0 0 0 0 0 0 0 0 0
0 0 0 0 0 0 0 0 0 0 7 7 7 7 7 0 0 0 0 0 0 0 0 0 0 0 0 0 0 0 0 0 7 7 7 7 7
7 7 7 7 7 0 0 0 0 0 0 0 0 0 0 0 0 0 0 0 0 0 0 0 0 0 0 0 0 0 0 0 7 7 7 7 7
7 7 7 7 7 0 0 0 0 0 0 0 0 0 0 0 0 0 0 0 0 0 0 0 0 0 0 0 0 0 0 0 7 7 7 7 7
7 7 7 7 7 0 0 0 0 0 0 0 0 7 7 7 0 0 0 0 0 0 0 0 0 0 0 0 0 0 0 0 7 7 7 7 7
7 7 7 7 7 0 0 0 0 0 0 0 0 7 7 7 0 0 0 0 0 0 0 0 0 0 0 0 0 0 0 0 7 7 7 7 7
7 7 7 7 7 0 0 0 0 0 0 0 0 0 0 0 0 0 0 0 0 0 0 0 0 0 0 0 0 0 0 0 7 7 7 7 7
7 7 7 7 7 0 0 0 0 0 0 0 0 0 0 0 0 0 0 0 0 0 0 0 0 0 0 0 0 0 0 0 7 7 7 7 7
7 7 7 7 7 0 0 0 0 0 0 0 0 0 0 0 0 0 0 0 0 0 0 0 0 0 0 0 0 0 0 0 7 7 7 7 7
```

Fig. 9.5. Clusters of sick sites (7) in a sea of healthy sites (0) for a modified three-cell model of autoimmune disease. Numbers above 7 refer to oscillating sites. While such lattice models connect us to usual cellular automata they may be less relevant biologically if the immune response is slower than the spreading of diseases etc within the body

The AIDS disease is due to virus HIV and weakens the helper cell populations until the immune system is no longer working; the body may then no longer defend itself against normally nonlethal diseases such as pneumonia. Pandey and others [9.60] studied several cellular automata models for the beginning stages of AIDS; we review here only the one of intermediate complexity with four cells: virus D, helpers H, suppressors S, and macrophages M. Macrophages are important since the AIDS virus can hide there; in the model, M refers only to those macrophages which have the virus inside:

```
D = (D.OR.H.OR.M).AND..NOT.S
H = (H.OR.M).AND..NOT.D
S = D.AND.H.AND.M
M = M.AND.D
```

It leads, depending on the initial configuration, to one of five final states for the four concentrations (DHSM): fixed point healthy (0000), fixed point sick (1001), and a cycle of period 3: from susceptible (0001) to infected (1101) to strongly infected (1011), and back to susceptible. On a simple cubic lattice, all sites acquire macrophages and suppressors after some time, whereas the number of virus and helpers oscillate and depends strongly on the initial concentration.

One can also mix two different interactions (also for autoimmune diseases) in which case each site either selects at the beginning which rule it wants to follow (quenched disorder), or makes the choice randomly again at each different iteration (annealed disorder). The two types of disorder give qualitatively similar results. So far, the percolation threshold has not shown up as a sharp phase transition in these mixture models.

Finally, one may try a unified model, giving both the normal immune response and the above immune diseases. Enlarging [9.63] the extended KUT model, described above for the healthy case, by two fixed parameters, D and O, corresponding to the AIDS virus and the self-antigen (e.g., nerves attacked by the activated killers K of autoimmune diseases), adding for the killer cells

```
K = (O.AND.H).AND..NOT.S
```

to the equations, and modifying the helper equation to

```
H = (H.OR.((O.OR.V).AND..NOT.S)).AND..NOT.D,
```

we recover the normal immune response without self-antigen and without AIDS (O = D = false). The presence of the AIDS virus, on the other hand, destroys the helper cells and makes the body vulnerable to subsequent virus attack. (If for the helpers, .AND..NOT.D is replaced by .AND..NOT.(D.AND.V), then the AIDS virus spreads only when the immune system has to respond to another virus, in agreement with some observations (Chowdhury, private communication).) On a lattice, one can simulate how an initial infection (V = true) leads to immunity, which can then be destroyed by AIDS. A following infection (V = true) will then no longer be suppressed by the immune system.

9.A.3 Hydrodynamic Cellular Automata

Much more widely accepted than immonological automata are the so-called "lattice gas" cellular automata for hydrodynamics. They represent a simplified molecular dynamics simulation of a two-dimensional fluid. Whole conferences [9.64] have been devoted to these cellular automata approximations. This appendix, in view of the general area of the book, is restricted itself to disordered systems, that means to the simulation of fluid flow through a random porous medium.

In 1986, Frisch, Hasslacher, and Pomeau [9.45,65] considered particles moving with unit velocity on the nearest-neighbor bonds of a triangular lattice. At integer times these particles reach the lattice sites, and if several particles at one time reach the same site, they are scattered there with conserved total kinetic energy, momentum, and angular momentum. In some models, particles can also be at rest on the lattice sites. Each of the six bonds leading to one site can have at any time either one or no particle moving along its direction. Thus the status of the bonds leading to this site are defined by six bits; a seventh bit may be used for the angular momentum and an eighth for a rest particle. Because of this complication, these cellular automata are hidden in this appendix. Nevertheless, their simulation is orders of magnitude faster than real molecular dynamics based on Newton's law of motion. Thus also much larger systems (200 million lattice sites versus 4 million molecules) have been studied with these hydrodynamic cellular automata (discrete time, discrete space) than with molecular dynamics (continuous time, continuous space).

Although earlier methods of studying hydrodynamic flow exist, the cellular automata approximation seems to be particularly suited [9.66] for flow through complicated nonperiodic geometries such as sand. Flow simulations around a single obstacle [9.67] or in one given geometry [9.68] gave quantitative agreement with laboratory experiments for the same geometry. Thus we allow now for randomly placed obstacles of arbitrary shapes, on which the particles are reflected by 180° to take into account the nonslip condition for the average fluid velocity at a solid boundary.

Kohring's program seems to be the fastest FORTRAN program published [9.69], with 500 sites updated per microsecond on one NEC-SX3 processor (190 such updates on one Cray-YMP processor, 1.3 on a SUN Sparc station); programming the Connection Machine in a language closer to the machine yielded about 1000 such updates on all 65 536 processors together [9.70]. The updating subroutine, listed here, determines from the incoming particle bits X1 to X6 the scattered particle bits Y1 to Y6 in the six lattice directions (numbered consecutively), and then shifts the information Y1 to Y6 to those neighbors (in the form of updated variables X1 to X6) to which the velocity direction is pointing. If no collision happens, we simply have Y1=X1, ..., Y6=X6; if a solid boundary is encountered, reflection by 180° is realized through Y1=X4, Y2=X5, Y3=X6, Y4=X1, Y5=X2, Y6=X3. If two or more particles meet away from an obstacle, then Fig. 9.6 gives the scattering rules.

```fortran
      SUBROUTINE UPDATE(X1,X2,X3,X4,X5,X6,Y1,Y2,Y3,Y4,Y5,Y6,
     1                          ANG,AANG,OB,NOTOB)
      IMPLICIT NONE
      INTEGER L,H,LL,N,B,BM1
      PARAMETER (B=64,L=8192,H=2344,LL=L/B,N=(L*H-1)/B,BM1=B-1)
      INTEGER X1(0:N),X2(0:N),X3(0:N),X4(0:N),X5(0:N),X6(0:N),
     1        Y1(0:N),Y2(0:N),Y3(0:N),Y4(0:N),Y5(0:N),Y6(0:N),
     2        ANG(0:N),AANG(0:N),OB(0:N),NOTOB(0:N),
     3        COL,NCOL,RULE,POS,NEG,SAME,OPP,W
      RULE(POS,NEG,SAME,OPP) = (POS.AND.COL.AND.AANG(W)) +
     1 (NEG.AND.COL.AND.ANG(W)) + (SAME.AND.NCOL) + (OPP.AND.OB(W))
C     The following loop looks for collisions; sets outgoing bits
      DO 10 W=0,N
      NCOL=NOTOB(W).AND.((X1(W).XOR.X4(W)).OR.(X2(W).XOR.X5(W))
     1               .OR.(X3(W).XOR.X6(W)))
     2               .AND.((X1(W).XOR.X3(W)).OR.(X3(W).XOR.X5(W))
     3               .OR.(X2(W).XOR.X4(W)).OR.(X4(W).XOR.X6(W)))
      COL=(.NOT.NCOL).AND.NOTOB(W)
      Y1(W)=RULE(X2(W),X6(W),X1(W),X4(W))
      Y2(W)=RULE(X3(W),X1(W),X2(W),X5(W))
      Y3(W)=RULE(X4(W),X2(W),X3(W),X6(W))
      Y4(W)=RULE(X5(W),X3(W),X4(W),X1(W))
      Y5(W)=RULE(X6(W),X4(W),X5(W),X2(W))
      Y6(W)=RULE(X1(W),X5(W),X6(W),X3(W))
      ANG(W)=COL.XOR.ANG(W)
   10 AANG(W)=.NOT.ANG(W)
C     The next loop propagates the particles; the pipe is assumed
C     to have solid walls, OB=1, at the lattice top and bottom
      DO 20 W=LL,N-LL
      X1(W)=Y1(W+LL-1)
      X2(W)=Y2(W-1)
      X3(W)=Y3(W-LL)
      X4(W)=Y4(W-LL+1)
      X5(W)=Y5(W+1)
   20 X6(W)=Y6(W+LL)
      DO 30 W=0,LL-1
      X1(W)=Y1(W+LL-1)
      X3(W+N-LL+1)=Y3(W+N-2*LL+1)
      X4(W+N-LL+1)=Y4(W+N-2*LL+2)
   30 X6(W)=Y6(W+LL)
C     Then corrections (p.b.c.) are made for right and left edges
      DO 40 W=0,N-LL,LL
      X1(W)=SHIFT(Y1(W+2*LL-1),1)
      X2(W)=SHIFT(Y2(W+ LL-1),1)
      X4(W+2*LL-1)=SHIFT(Y4(W   ),BM1)
   40 X5(W+2*LL-1)=SHIFT(Y5(W+LL),BM1)
      RETURN
      END
```

Fig. 9.6. Program of [9.69] to update hydrodynamic cellular automata. The bottom shows schematically the collision rules

When two particles meet they can either be reflected by 180°, which is equivalent to no collision at all as long as the identity of the particles is ignored, or the line of motion can be rotated by 60°. Whether that rotation is clockwise or counterclockwise is determined by an angular momentum bit for each site; this bit is switched after each such collision. Thus on average angular momentum is conserved at each site. A head-on collision of two particles requires that each of the six bonds has the same status (occupied or empty) as the bond in the opposite direction, i.e., X1=X4, X2=X5, and X3=X6. If these three conditions are all fulfilled, it may also correspond to a four-body collision or a six-body collision, which is thus taken into account automatically. (The six-body collision does not change anything as long as we do not keep track of individual particles; thus it is ignored in Fig. 9.6.) For a three-body collision the two conditions X1=X3=X5 and X2=X4=X6 must be fulfilled simultaneously. In the program, the variable COL is true if one of these collisions takes place; otherwise it is false; the variable NCOL has the opposite meaning. They are both determined in the first two statements of loop 10. Of course, multispin coding with 64 sites per 64-bit word is employed; thus the exclusive-OR operation, XOR, marks those bits which differ in the two arguments.

The change of direction of a particle is zero (e.g., Y2=X2) if no collision takes place. The velocity vector changes in one direction (e.g., Y2=X1) if the angular momentum bit ANG is true, and in the opposite direction (e.g., Y2=X3) if the angular momentum bit ANG is false, i.e., if the bit AANG for the opposite angular momentum is true. The statement

Y2=(X1.AND.COL.AND.ANG).OR.(X3.AND.COL.AND.AANG).OR.(X2.AND.NCOL)

gives this described example. Since we need analogous statements for the five other directions, with cyclic permutations of direction indices, it is simpler to use the FORTRAN statement function RULE for this purpose. The six calls to RULE still belong to the main loop 10, which is finished with the updating of the angular momentum bits if a collision has changed the local angular momentum.

So far obstacle boundaries were ignored, NOTOB was true and OB was false. If a site is part of an obstacle, OB is true and NOTOB false. With

Y2=(Y2.AND.NOTOB).OR.(X5.AND.OB)

we would get the desired effect, where Y2 at first is our earlier expression without obstacles. The actual program of Fig. 9.6 does the same thing more efficiently by incorporating into COL and NCOL the condition that we have no obstacle. To create a porous medium or other more or less disordered geometry, we have to set the obstacle bits OB suitably outside the updating subroutine; these "obstacles" include the solid walls which may limit the sample on some sides. Taking into account the obstacles in this way reduces the speed to 190 sites or 1140 bonds per microsecond.

Loop 20 shifts the information Y1,...,Y6 gained from these collisions to the incoming velocity bits X1,...,X6 of the neighbors corresponding to the velocity directions. Loop 30 deals with the upper and lower solid walls, loop 40 with the left and right periodic boundary conditions. These three loops depend in their details on how the directions are numbered and how the lattice is oriented.

To apply such algorithm to reproducible geometries one may randomly distribute suitable obstacles (e.g., circles) on the triangular lattice. A hydrodynamic flow is then created by the boundary conditions at one end of the lattice: slightly more particles are set there to move in the direction of the lattice than in the opposite directions. By averaging over a region near the other end of the lattice we find the average velocity created by this boundary condition. The pressure can be found from the forces exerted by, for example, the particles reflected on the obstacles or the solid walls. In this way, the permeability, i.e., the ratio of flux to pressure gradient in suitable units, can be found by these simulations.

This permeability goes to infinity (in an infinite system) if there are no obstacles present, i.e., for porosity p equal to 1. (This porosity is the fraction of lattice sites not belonging to obstacles.) For p less than the percolation threshold, the permeability is zero. For p far above the percolation threshold and below unity the permeability has been approximated empirically by $\exp(6.6p)/(1-p)$ in such simulations [9.69]. However, finite-size effects are much stronger here than, for example, in magnetism. The obstacle size, not only the lattice size, has to be much larger than one lattice constant, and the results for obstacle sizes of the order of 10 can differ by a factor of 2 from those for intermediate sizes. Presumably this effect is due to a rather large mean free path [9.71] of the particles. For even bigger lattices containing up to 10^8 sites, a crossover to a different behavior may take place [9.69].

References

9.1   S. Wolfram: Rev. Mod. Phys. **55**, 601 (1983);
      N.H. Packard, S. Wolfram: J. Stat. Phys. **38**, 901 (1986)
9.2   S. Wolfram: *Theory and Applications of Cellular Automata* (World Scientific, Singapore 1986);
      G. Weisbuch: *Dynamique des systèmes complexes* (Editions du CNRS, Paris 1989);
      English translation: *Complex System Dynamics* (Addison-Wesley, New York 1991);
      for programming techniques see also D. Stauffer: Computers in Physics, Jan/Feb 1991, p. 62, Ref. [9.51b], and R.W. Gerling, D. Stauffer: in preparation
9.3   G.Y. Vichniac: Physica D **10**, 96 (1984);
      T. Toffoli, M. Margolus: *Cellular Automata Machines* (MIT Press, Cambridge MA 1987)
9.4   P. Bak, C. Tang, K. Wiesenfeld: Phys. Rev. Lett. **59**, 381 (1987);
      P. Bak: Physica D **38**, 5 (1989)
9.5   P. Grassberger, S.S. Manna: J. Physique **51**, 1077 (1990), with further references
9.6   S.A. Kauffman: J. Theor. Biol. **22**, 437 (1969);
      U. Keller, B. Thomas, H.J. Pohley: J. Stat. Phys. **52**, 1129 (1988)
9.7   B. Derrida, Y. Pomeau: Europhys. Lett. **1**, 59 (1986);
      B. Derrida, G. Weisbuch: J. Physique **47**, 1297 (1986)
9.8   J.G. Zabolitzky: Physica A **163**, 447 (1990)
9.9   B. Derrida, D. Stauffer: Europhys. Lett. **2**, 739 (1986);
      D. Stauffer: Phil. Mag. B **56**, 510 (1987);

for Ising models see:
U.M.S. Costa: J. Phys. A **20**, L 583 (1987);
B. Derrida, G. Weisbuch: Europhys. Lett. **4**, 657 (1987);
H.E. Stanley, D. Stauffer, J. Kertesz, H.J. Herrmann: Phys. Rev. Lett. **59**, 2326 (1987);
A. Coniglio, L. de Arcangelis, H.J. Herrmann, N. Jan: Europhys. Lett. **8**, 315 (1989);
G. Le Caer: Physica A **159**, 329 (1989);
L. de Arcangelis, H.J. Herrmann, A. Coniglio: J. Phys. A **23**, L265 (1990)

9.10   A. Coniglio, N. Jan, D. Stauffer: J. Phys. A **20**, L1103 (1987)

9.11   L.R. da Silva, H.J. Herrmann: J. Stat. Phys. **52**, 463 (1988)

9.12   D. Stauffer: Physica D **38**, 341 (1989)

9.13   D. Stauffer, in: *Computer Simulation Studies in Condensed Matter Physics II*, ed. by D.P. Landau, K.K. Mon, H.B. Schüttler (Springer, Berlin, Heidelberg 1989)

9.14   A. Hansen, S. Roux: Physica A **160**, 275 (1989)

9.15   L. de Arcangelis: J. Phys. A **20**, L369 (1987)

9.16   A. Hansen: J. Phys. A **21**, 2481 (1988)

9.17   S.P. Obukhov, D. Stauffer: J. Phys. A **22**, 1715 (1989)

9.18   G. Y.Vichniac, H. Hartmann, in: *Disordered Systems and Biological Organization*, ed. by E. Bienenstock, F. Fogelman-Soulie, G. Weisbuch (Springer, Berlin, Heidelberg 1986);
L.R. da Silva: J. Stat. Phys. **53**, 985 (1988);
K.E. Kürten, in: *Models of Brain Functions* (Cambridge University Press, Cambridge 1990)

9.19   D. Stauffer: Physica A **157**, 645 (1989);
S.S. Manna, D. Stauffer: Physica A **162**, 176 (1990);
R.W. Gerling: Physica A **162**, 187 and 196 (1990);
D. Stauffer: J. Phys. A **23**, 5933 (1990);
see also H.A. Gutowitz, J.D. Victor, B.W. Knight: Physica D **28**, 18 (1987)

9.20   P.M. Binder: J. Phys. A **24**, L31 (1991)

9.21   J. Adler: Physica A **171**, 453 (1991)

9.22   S.A. Kauffman, *The Origin of Order*, Oxford University Press, New York 1993. For later evolution simulations of Kauffman models see D. Stauffer, N. Jan: J. Theor. Biol. **168**, 211 (1994); J. Stat. Phys. **74**, 1293 (1994)

9.23   P. Bak, M. Creutz, in: *Fractals in Science*, ed. by A. Bunde and S. Havlin (Springer, Heidelberg, New York 1994)

9.24   P. Bak, K. Sneppen: Phys. Rev. Lett. **71**, 4087 (1993); T.S. Ray, N. Jan: Phys. Rev. Lett. **72**, 4048 (1994); for game of life see P. Bak: Physica A **191**, 41 (1992)

9.25   R.J. de Boer, L.A. Segel, A.S. Perelson: J. Theor. Biol. **155**, 295 (1992); R.J. de Boer, J.D. van der Laan, P. Hogeweg, in: *Thinking about Biology*, ed. by F.J. Varela and W.D. Stein (Addison-Wesley, New York 1992)

9.26   J. Stewart, F.J. Varela: J. Theor. Biol. **153**, 477 (1991)

9.27   D. Stauffer, G. Weisbuch: Physica A **180**, 42 (1992)

9.28   S. Dasgupta: Physica A **189**, 403 (1992)

9.29   M. Deffner: Physica A **195**, 279 (1993)

9.30   R.M. Zorzenon dos Santos: Physica A **196**, 12 (1993)

9.31   M. Sahimi, D.Stauffer: Phys. Rev. Lett. **71**, 4271 (1993)

9.32   D. Chowdhury, V. Deshpande, D. Stauffer: Int. J. Mod. Phys. C **5**, 1049 (1994)

9.33   B. Barral, H. Chate, P. Manneville: Phys. Lett. A **163**, 279 (1992), citing earlier papers of this group; J. Hemmingson, G.W. Peng: J. Phys. A **27**, 2735 (1994)

9.34   G.A. Kohring: Physica A **186**, 97 (1992); see also D.W. Grunau, T. Lookman, S.Y. Chen, A.S. Lapedes: Phys. Rev. Lett. **71**, 4198 (1993); J.F. McCarthy: Phys. Fluids **6**, 435 (1994)

9.35   R. Benzi, S. Succi, M. Vergassola: Phys. Rep. **222**, 3 (1992)

9.36   K. Nagel, M. Schreckenberg: J. Physique I **2**, 2221 (1992)

9.37   T. Nagatani: J. Phys. Soc. Jpn. **63**, 1228 (1994) and his earlier papers

9.38   J.B. Rundle, W. Klein: J. Stat. Phys. **72**, 405 (1993)

9.39   B. Chopard, P. Luthi, M. Droz: Phys. Rev. Lett. **72**, 1384 (1994); J.R. Weimar, J.P. Boon: Phys. Rev. E **49**, 1749 (1994)

9.40   C.F. Baillie, D.A. Johnston: Phys. Lett. B **326**, 51 (1994)

9.41   F. Bagnoli, R. Rechtman, S. Ruffo: Phys.Lett. A **172**, 34 (1992)

9.42   Ph. Blanchard, D. Gandolfo: J. Stat. Phys. **73**, 399 (1993)

9.43   G.A. Kohring: Physica A **182**, 320 (1992); see also B. Voorhees, S. Bradshaw: Physica D **27**, 152 (1994)

9.44   D. Makowiec: Z. Physik B **95**, 519 (1994)

9.45   U. Frisch, B. Hasslacher, Y. Pomeau: Phys. Rev. Lett. **56**, 1505 (1986)

9.46   H.J. Herrmann: J. Stat. Phys. **45**, 145 (1986);
J.G. Zabolitzky, H.J. Herrmann: J. Comp. Phys. **76**, 426 (1988)

9.47   D. Stauffer, F.W. Hehl, V. Winkelmann, J.G. Zabolitzky: *Computer Simulation and Computer Algebra*, 2nd ed. (Springer, Berlin, Heidelberg 1989)

9.48   M. Creutz: Phys. Rev. Lett. **50**, 1411 (1983); Ann. Physics **167** 62 (1986)

9.49   M. Schulte, W. Stiefelhagen, E.S. Demme: J. Phys. A **20**, L1023 (1987)

9.50   C. Moukarzel, N. Parga: J. Phys. A **22**, 943 (1989);
C. Moukarzel: J. Phys. A **22**, 4487 (1989)

9.51   H.J. Herrmann, H.O. Carmesin, D. Stauffer: J. Phys. A **20**, 4939 (1987)

9.52   S.C. Glotzer, D. Stauffer, S. Sastry: Physica A **164**, 1 (1990), and Ref. [9.9]

9.53   D. Stauffer: J. Phys. A **23**, 1847 (1990); J. Phys. A **24**, 909 (1991), (review, see in particular Fig. 4 there)

9.54   I.R. Cohen, in: *Theories of Immune Networks*, ed. by H. Atlan, I.R. Cohen, Springer Ser. Syn., Vol.46 (Springer, Berlin, Heidelberg 1989);
W.E. Paul, in: *Fundamental Immunology*, ed. by W.E. Paul (Raven, New York 1989);
A.S. Perelson, ed.: *Theoretical Immunology* (Addison Wesley, Reading, MA 1988)

9.55   H. Atlan: Bull. Math. Biol. **51**, 247 (1989)

9.56   M. Kauffman, J. Urbain, R. Thomas: J. Theor. Biol. **114**, 527 (1985);
G.W. Hoffman: J. Theor. Biol. **129**, 355 (1987);
for a recent differential equation approach see U. Behn, J.L. van Hemmen: J. Stat. Phys. **56**, 533 (1989)

9.57   D. Chowdhury, D. Stauffer: J. Stat. Phys. **59**, 1019 (1990)

9.58   I. Dayan, D. Stauffer, S. Havlin: J. Phys. A **21**, 2473 (1988);
U. Wiesner: J. Undergrad. Res. Phys. **7**, 15 (1988)

9.59   A.U. Neumann: Physica A **162**, 1 (1989)

9.60   R.B. Pandey: J. Stat. Phys. **54**, 997 (1989); ibid. **61**, 235 (1990), and J. Phys. A **23**, 3421 (1990);
Ch.F. Kougias, J. Schulte: J. Stat. Phys. **60**, 263;
for a differential equation approach see A.R. McLean, T.B.L. Kirkwood: J. Theor. Biol. **147**, 177 (1990)

9.61   G. Weisbuch, H. Atlan: J. Phys. A **21**, L 189 (1988);
see also K. E. Kürten: J. Stat. Phys. **52**, 489 (1988)

9.62   I.R. Cohen, H. Atlan: J. Autoimmunity **2**, 613 (1989)

9.63   D. Chowdhury, D. Stauffer, P.V. Choudary: J. Theor. Biol. **145**, 207 (1990);
D. Chowdhury, M. Sahimi, D. Stauffer: J. Theor. Biol. **152**, 263 (1991)

9.64   See e.g., the Proceedings of the NATO Advanced Workshop on Lattice Gas Methods for PDE's, ed. by G.D. Doolen, Physica D **47**, 1 (1991)

9.65   J. Hardy, O. de Pazzis, Y. Pomeau: Phys. Rev A **13**, 1949 (1976);
S. Succi: Comp. Phys. Comm. **47**, 173 (1987)

9.66   G.D. Doolen: First Topical Conference on Computational Physics, APS, Boston 1989 (unpublished invited talk);
U. Brosa: J. Physique **51**, 1051 (1990)

9.67   J.A.M.S. Duarte, U. Brosa: J. Stat. Phys. **59**, 501 (1990)

9.68   S. Chen, K. Diemer, G.D. Doolen, K. Eggert, C. Fu, S. Gutman, B. Travis, in Ref. [9.64]

9.69   G.A. Kohring: J. Stat. Phys. **63**, 411 (1991); J. Physique I **1**, 87 (1991) and preprints

9.70   B.M. Bhogosian: Computers in Physics (APS), Jan/Feb. 1990

9.71   D.H. Rothman: Geophys. **53**, 509 (1988);
see also S. Succi, A. Cancelliere, C. Chang, E. Foti, M. Gramignani, D. Rothman, in: *Computational Methods in Subsurface Hydrology*, ed. by G. Gambolati, A. Rinaldo, C.A. Brebbia, W.G. Gray, G.F. Pinder (Springer, Berlin, Heidelberg 1990);
J.P. Boon, in: *Correlations and Connectivity — Geometric Aspects of Physics, Chemistry and Biology* ed. by H.E. Stanley, N. Ostrowsky (Kluwer, Dordrecht 1990), p. 56 and D. Stauffer, p. 18

10 Exactly Self-similar Left-sided Multifractals

Benoit B. Mandelbrot and Carl J.G. Evertsz
with new Appendices B and C by
Rudolf H. Riedi and Benoit B. Mandelbrot

10.1 Introduction

10.1.1 Two Distinct Meanings of Multifractality

The term "multifractal" may have at least two meanings, depending upon the role given to *multi*. At one time, this distinction seemed to lack practical bite, but we feel that recent work on fully developed turbulence and on DLA has made it essential.

The earlier and more general meaning comes from the notion of *"multiplicative cascade* that generates nonrandom or random measures", and describes "measures that are *multiplicatively generated*". The virtue of *all* multiplicatively generated measures is that they are exactly renormalizable (just like the simplest self-similar fractals, such as the Sierpinski gasket). This general meaning of multifractality is investigated in early [10.1–3] and recent papers by one of us and co-workers. For self-similar fractals, multifractals, and DLA, see also Chaps. 1–4.

In addition, a second meaning has since been introduced by Frisch and Parisi [10.5] and by Halsey et al. [10.6]. Let space be subdivided into small boxes of side ϵ and let $\mu_j(\epsilon)$ denote the measure within the jth box. A multifractal can be defined as "a nonrandom measure for which it is true for all $-\infty < q < \infty$ that the *partition function* $\chi(q, \epsilon) = \sum_{j=1}^{N(\epsilon)} \mu_i^q(\epsilon)$ scales like a power of the form $\epsilon^{\tau(q)}$." An alternative form of this definition involves the Hölder exponent $\alpha = \log \mu_j / \log \epsilon$. Then, it is useful to represent such a multifractal by a *multiplicity*

◄ **Fig. 10.0.** A 50 000 particle cluster formed by diffusion limited aggregation. Particles are colored according to when they joined the cluster. White shading denotes the strength of the surrounding Laplacian (electrostatic) potential, which, as discussed by Mandelbrot and Evertsz in [10.15], is low (bright) not only at the cluster's center, but also at points throughout the cluster. This feature may provide clues to the way clusters grow. Craig Kolb has assisted in the preparation of this illustration

of intertwined measures, each of which is supported by a Cantor set and is characterized by a uniform α. The proportions of different αs in this mixture are characterized by a function $f(\alpha) \geq 0$ whose graph is shaped like the sign $\cap$, perhaps asymmetric and leaning to the side, but always standing above $f = 0$. Applying the method of steepest descents, one shows that this function $f(\alpha)$ is related to the function $\tau(q)$ by the Legendre transform. The basic facts about this approach are restated in [10.7].

We now come to one of the central points of this chapter. In *restricted cases* of our multiplicative multifractals, the scaling relation that defines $\tau(q)$ is indeed valid for all q. As a matter of fact, this was precisely the point of departure of [10.5]. On the other hand, we shall give explicit examples of multiplicative multifractals with the property that the scaling relation *fails to hold*, either for small enough negative qs ($q < q_{\text{bottom}}$) or for high enough positive qs ($q > q_{\text{top}}$). Insofar as they differ *qualitatively* from the familiar restricted class, these multiplicative measures are "anomalous". Nevertheless, they preserve the desirable property of being exactly renormalizable. As already implied, some of the examples we have constructed fit fully developed turbulence. We propose to provide further examples that we believe fit DLA, the diffusion limited aggregates [10.8].

The reader may be reassured to know that, in our more general multiplicative multifractals, counterparts of the functions $f(\alpha)$ and $\tau(q)$ continue to be needed and continue to be cap-convex (like $-x^2$). They also continue to be linked by Legendre transforms, but only at the cost of introducing (here again) a fine distinction. It happens that $f(\alpha(q)) = \min_q[q\alpha(q) - \tau(q)]$ continues to be true, which implies that the graph of $f(\alpha)$ is straight for $q > q_{\text{top}}$, and for $q < q_{\text{bottom}}$. But the alternative Legendre transforms that express both α and f as functions of q only apply in the restricted range $q_{\text{bottom}} < q < q_{\text{top}}$.

The first consequence of using the notion of multiplicative multifractal is that the function $f(\alpha)$ of a multiplicative multifractal need not, in general, be ≥ 0 and shaped like $\cap$. Instead, it can take one of several alternative overall shapes, each of which corresponds to a *qualitatively* distinct behavior for the corresponding multiplicative process.

The new multiplicative multifractals introduced in this chapter also raise a second important new issue. We shall show in Sect. 10.5.2, that even a fairly detailed empirical knowledge of $f(\alpha)$ gives unexpectedly incomplete information about such a measure, and we shall claim that an acceptable description of DLA requires information that goes beyond the $f(\alpha)$ that one can hope to obtain empirically. In the simplest case, yet another form of scaling comes in, characterized by a new scaling exponent λ.

Could one further generalize our multiplicative multifractals, without abandoning renormalizability? We prefer not to try until the need arises. However, the end of Sect. 10.6 refers to a small step in that direction.

10.1.2 "Anomalies"

Let us elaborate on our claim that, from the viewpoint of the steepest-descents formalism of Frisch-Parisi and Halsey et al., the important application to DLA turns out to present deep "anomalies". In the example of DLA, the growth probabilities are known to be ruled by the harmonic measure [10.9]. One observes [10.10–13] the experimental "anomaly" that, for $q < 0$, the partition function $\chi(q, \epsilon)$ *fails to scale* like $\epsilon^{\tau(q)}$. Therefore, a blind application of existing computer programs yields wildly disagreeing $f(\alpha)$s and even $f(\alpha)$s with cusps. The restricted theory of multifractals *does not* allow for these baffling and contradictory results, which is why the harmonic measure has been described as being "nonmultifractal" [10.14]. We believe strongly that $q_{bottom} = 0$ for DLA, and our experimental evidence is discussed in [10.15–17] (see also Chaps. 1 and 4). Similar anomalies have also been observed in resistor networks [10.18], see also Sect. 3.7.

Two challenges arise. An experimental challenge is to establish the exact nature of the anomalies. A methodological challenge is to provide a conceptual framework for the above-mentioned behavior. This chapter's goal is purely methodological. One may be tempted to elaborate on this "nonmultifractality" and conclude that the harmonic measure on DLA is not renormalizable. However, the multiplicative cascades described in this chapter are exactly self-similar, even though $\tau(q)$ is not defined for all q. The fact that they are self-similar makes them multifractals in the sense advocated in the introduction and shows that exact renormalizability or self-similarity can be compatible with a failure of the partition function to scale. This is very reassuring for theoretical approaches to DLA and DBM, because it means that existing methods [10.19,20] for the analytical estimation of the multifractal spectrum of DLA and DBM, which assume the self-similarity of the harmonic measure, may be extended to include the "anomalies".

The price to pay is that, in the examples that follow, the steepest descents method is altogether invalid for $q < q_{bottom}$, and the validity of the broader Legendre transform is something of a coincidence: this is an issue we hope to discuss in detail in the near future.

Combining these conclusions with those of several recent publications [10.15,16,21–24] we venture to claim that the known multifractal anomalies basically disappear if one follows the open-ended approach of our general multifractals, and if one accepts the need for a separate discussion for each of several qualitatively distinct basic categories, each illuminated by well-chosen special cases.

For a summary and overview of this chapter, we refer the reader to Sect. 10.7.

10.2 Nonrandom Multifractals with an Infinite Base

One reason why many measures are expected to be multifractal (e.g., the harmonic measure on DLA) is because they are supported by a fractal set (e.g., a Julia set, or DLA boundary). However, the basic ideas behind multiplicative fractals are best understood if this complication is postponed, and if one first examines measures supported by the interval $[0, 1]$ (e.g., linear cuts through developed turbulence). Also randomness brings in genuine complications, which would here serve no purpose, and will thus be avoided by considering *nonrandom* measures. This section describes the general idea. Sections 10.3 and 10.4 provide two basic illustrative examples.

A construction. We start with the interval $[0, 1]$ carrying a mass equal to 1, and generate a fractal measure μ by means of a conservative deterministic multiplicative cascade. Hence, the measure μ we shall obtain will be usable as a probability measure. At each stage of the cascade, $[0, 1]$ is divided into an infinity of (necessarily unequal) subintervals, which we index from right to left by the unbounded integer β. The βth subinterval I_β of $[0, 1]$ is taken to be of length r_β with $\sum_{\beta=1}^{\infty} r_\beta = 1$, and to contain the mass m_β, with $\sum_{\beta=1}^{\infty} m_\beta = 1$. The subintervals are thus

$$I_\beta = \left[1 - \sum_{j=0}^{\beta} r_j, 1 - \sum_{j=0}^{\beta-1} r_j \right], \tag{10.1}$$

with $r_0 = 0$ and $\beta = 1, 2, \ldots, \infty$. Next, the subintervals of lengths r_β are subdivided into sub-subintervals of lengths $r_\beta r_{\beta'}$ for all combinations of β and β' and these sub-subintervals are made to contain the respective masses $m_\beta m_{\beta'}$. The process is allowed to continue ad infinitum. It is clear that the measure μ it generates is *self-similar,* in the sense that the *relative* distribution of mass is exactly the same in all $(\text{sub})^k$-intervals.

When $m_\beta \equiv r_\beta$, this measure is, of course, uniform, but in all other cases, it is a multiplicative multifractal.

To describe a multifractal, it is the custom to evaluate its function $\tau(q)$. We must postpone to a later occasion a critical discussion of the meaning of $\tau(q)$ when the subdivision is infinite. We shall be content to explore a blind generalization of a formula that is familiar in the restricted theory. There, $\tau(q)$ is known to be given implicitly by the following relation due to Hentschel and Procaccia [10.25], which we call the *generating equation*:

$$Z(q, \tau) = \sum m_\beta^q r_\beta^{-\tau(q)} = 1. \tag{10.2}$$

A special case. Take $m_\beta = m^{\beta-1} - m^\beta$, with $0 < m < 1$, and $r_\beta = r^{\beta-1} - r^\beta$ with $0 < r < 1$. A moment's thought shows that the resulting multiplicative measure reduces to the "skew binomial" multifractal, which is defined by the finite base cascade, assigning the masses m and $1 - m$ to subintervals of $[0,1]$ having the lengths r and $1 - r$, and so on. The measures at finite stages of these two cascades are different. Namely, the infinite base cascade is a mixture of an infinity of stages of the finite base cascade. But their resulting $f(\alpha)$s are the same. The $f(\alpha)$ of the "skew binomial" multifractal is well known to be shaped like $\cap$. The $\alpha_{\min}$ and $\alpha_{\max}$ are, respectively, the smaller and the larger of the two quantities $\log(m)/\log(r)$ and $\log(1 - m)/\log(1 - r)$.

Generalization. The reason for introducing the new construction is, of course, that it also allows a variety of more interesting outcomes. Since β is unbounded and we imposed $\sum_{\beta=1}^{\infty} r_\beta = 1$, the right-most interval $[1 - \delta, 1]$ $(0 < \delta \ll 1)$ is subdivided into a finite number of pieces at the successive stages of the cascade, and no surprises are expected there.

The situation at the left-most interval is different. The nature of the anomalies there will be seen to depend on the following functions:

$$\begin{aligned} M^*(\beta) &= -\log_2 M(\beta), \\ R^*(\beta) &= -\log_2 R(\beta), \\ \alpha^*(\beta) &= M^*(\beta)/R^*(\beta), \end{aligned} \tag{10.3}$$

where we define

$$M(\beta) = \sum_{u=\beta}^{\infty} m_u \quad \text{and} \quad R(\beta) = \sum_{u=\beta}^{\infty} r_u. \tag{10.4}$$

Clearly, $M(\beta)$ is the measure of the left-most subinterval $[0, R(\beta)]$ of $[0, 1]$, i.e., $M(\beta) = \mu([0, R(\beta)])$. The size of this interval goes to 0 in the limit $\beta \to \infty$. If $\alpha^*(\beta)$ exists in this limit, then it is the Hölder (or singularity) strength associated with the measure in the neighborhood of the left-most point 0 of $[0, 1]$. The choice of the base 2 for the logarithm is for later convenience and we come back to it at the end of this section. Note that the intervals I_β defined in (10.1) can now be rewritten as

$$I_\beta = [R(\beta + 1), R(\beta)]. \tag{10.5}$$

The condition that the limit $\alpha^*(\infty) = \lim_{\beta \to \infty} \alpha^*(\beta)$ exists and satisfies $0 < \alpha^*(\infty) < \infty$ is sufficient to obtain a "usual" restricted multifractal, in which $0 < \alpha_{\min} < \alpha_{\max} < \infty$, hence $q_{\text{top}} = \infty$ and $q_{\text{bottom}} = -\infty$. The case $\alpha^*(\infty) = 0$, which is related to $\alpha_{\min} = 0$, is postponed to the Appendix, Sect. 10.A.2.

The case $\alpha_{\max} = \infty$. A sufficient condition for $\alpha_{\max} = \infty$ is that one can identify at least one point where $\alpha = \infty$. When $\alpha^*(\beta) \to \infty$, this is indeed the case at the left-most point of $[0, 1]$. In that case $\alpha_{\min} < \alpha < \infty$.

A consequence of $\alpha_{\max} = \infty$ is that $q_{\text{bottom}} = 0$, meaning that for $q < 0$ the function $\chi(q, \epsilon)$ *fails* to scale like $\epsilon^{\tau(q)}$ and the function $\tau(q)$ fails to be defined. To prove this, observe that for $q < 0$ the largest addend in the sum $\sum \mu_j^q(\epsilon)$ always comes from the interval where $\mu(\epsilon)$ is smallest. In this instance, this interval goes from 0 to ϵ. Let $\beta(\epsilon)$ be defined by $\epsilon = R(\beta(\epsilon))$. Then

$$\mu([0, \epsilon]) = M(\beta(\epsilon))$$

and thus

$$\min_j \mu_j(\epsilon) \sim \epsilon^{\alpha^*(\beta(\epsilon))}. \tag{10.6}$$

So, $\sum \mu_j^q(\epsilon) \geq \epsilon^{q\alpha^*(\beta)}$. Now suppose there exists a finite $\tau(q)$ such that $\sum \mu_j^q(\epsilon) = \epsilon^{\tau(q)}$ for $\epsilon \to 0$. The last inequality would imply that for every $q < 0$, $|\tau(q)/q| \geq \alpha^*(\beta)$. Since $\beta \to \infty$ as $\epsilon \to 0$, and $\lim_{\beta \to \infty} \alpha^*(\beta) = \infty$ is true by hypothesis, it thus follows that $\tau(q)/q$ would be infinite for all $q < 0$ and thus that $\tau(q) \to -\infty$ for all $q < 0$. The function $\tau(q)$ is therefore undefined for $q < q_{\text{bottom}} = 0$.

Conclusion. When $\alpha^*(\beta) \to \infty$ our infinite base multiplicative multifractal satisfies $\alpha_{\max} = \infty$, and the function $\tau(q)$ fails to be defined for $q < q_{\text{bottom}} = 0$, hence the measure is not a restricted multifractal. The shape of the $f(\alpha)$ corresponding to the function $\tau(q)$ for $q > 0$ will very much depend on the (singular) behavior of $\tau(q)$ for small positive values of q. This is studied in great detail in the next sections.

The very special role of $m_\beta = 2^{-\beta}$. We now come back to the choice of base 2 in the definition of M^* and R^* in (10.3). Note that, in the preceding discussion, the algebra can be greatly simplified by simplifying the dependence of either $M^*(\beta)$ or $R^*(\beta)$ upon the parameter β. Given that $M^*(1) = R^*(1) = 0$, this dependence may simply be set to be $\beta - 1$. We take $M^*(\beta) = \beta - 1$, i.e., $M(\beta) = 2^{1-\beta}$, and the expressions for m_β and for the criterion function $\alpha^*(\beta)$ become

$$m_\beta = 2^{-\beta}$$

and

$$\alpha^*(\beta) = (\beta - 1)/R^*(\beta). \tag{10.7}$$

It is easy to see that one may redefine $M^*(\beta)$ and $R^*(\beta)$ with logarithms of base $1/m \neq 2$. This would suggest selecting $m_\beta = m^{\beta-1} - m^\beta$ with $m \neq 2^{-1}$, and $0 < m < 1$. However, the change would bring no significant generalization, only the replacement of $\log 2$ by $-\log m$ in most formulas. In the rest of this chapter we set $m_\beta = 2^{-\beta}$.

Section 10.A.2 (Appendix) extends the above considerations to the case $\alpha_{\min} = 0$, which is not needed in Sects. 10.3 to 10.6.

10.3 Left-sided Multifractality with Exponential Decay of Smallest Probability

The one-parameter ($\lambda > 0$) family of exactly self-similar multiplicative measures studied here is defined by $m_\beta = 2^{-\beta}$ and $r_\beta = \beta^{-\lambda} - (\beta + 1)^{-\lambda}$. Thus, $M(\beta) = 2^{1-\beta}$ and $R(\beta) = \beta^{-\lambda}$. Equations (10.3) and (10.7) give

$$\alpha^*(\beta) = (\beta - 1)/\lambda \log_2 \beta, \tag{10.8}$$

and as $\beta \to \infty$, this quantity tends (rapidly) to ∞ so that $\alpha_{max} = \infty$.

The interval of size ϵ with the smallest measure is again $[0, \epsilon]$. From (10.6) and the fact that for the present choice of r_β one finds $\beta(\epsilon) = \epsilon^{-1/\lambda}$, the smallest probability behaves as

$$\min_j \mu_j(\epsilon) = 2^{1-\beta(\epsilon)} = \exp(-c\epsilon^{-1/\lambda}). \tag{10.9}$$

The constant c is nonintrinsic: here, $c = \log 2$, but taking $m_\beta = m^{\beta-1} - m^\beta$ with $m \neq 1/2$ would yield $c = -\log m$. Thus the smallest ϵ-coarse grained probability in this family of measures has *stretched exponential* decay, with exponent $1/\lambda$. Such a decay of the minimal probability was postulated for DLA in [10.14], and has been found in resistor networks [10.18]; see also Chaps. 1 and 4.

From the generating equation $Z(q, \tau(q)) = 1$, see (10.2), it is easy to show that for all λ

$$\tau(q) = \begin{cases} \text{undefined,} & q < 0, \\ -1, & q = 0, \\ \tau_\lambda(q), & q > 0. \end{cases} \tag{10.10}$$

To find out about $\alpha(0)$, we observe that $Z(q, \tau) = 1$ for $q = 0$ and $\tau(0) = -1$. For q to the right of q_{bottom}, it follows that

$$Z(q, \tau(q)) \simeq Z(0, -1) + q \frac{\partial}{\partial q} Z(q, \tau(q))|_{q=0} = 1 + q \frac{\partial}{\partial q} Z(q, \tau(q))|_{q=0}.$$

To insure that for small $q > 0$, $Z(q, \tau(q)) = 1$, we impose the condition

$$0 = \frac{\partial}{\partial q} Z(q, \tau(q))|_{q=0} = -\sum_{\beta=1}^{\infty} 2^{-q\beta} r_\beta^{-\tau(q)} \ln(2^\beta r_\beta^{-\alpha(q)})|_{q=0}$$

$$= -\ln 2 \sum_{\beta=1}^{\infty} \beta r_\beta - \alpha(0) \sum_{\beta=1}^{\infty} r_\beta \ln r_\beta, \tag{10.11}$$

so $\alpha(0) = (-\ln 2 \sum_{\beta=1}^{\infty} \beta r_\beta)/(\sum_{\beta=1}^{\infty} r_\beta \ln r_\beta)$. Using the approximation $r_\beta \simeq \lambda \beta^{-\lambda-1}$ for $\beta \gg 1$, and replacing the sums by integrals, we find that

$-\sum_{\beta=1}^{\infty} r_\beta \ln r_\beta$ is finite for all $\lambda > 0$, whereas the numerator behaves like $\sum_{\beta=1}^{K} \beta r_\beta \sim K^{1-\lambda}$. We therefore find that

$$\alpha(0) = \begin{cases} \infty, & \lambda \le 1, \\ \alpha_\lambda(0) < \infty, & \lambda > 1. \end{cases} \tag{10.12}$$

In order to determine the precise behavior of the function $f(\alpha)$ near $\alpha = \alpha(0)$, we solve (10.2) near $q = q_{\text{bottom}} = 0$. The analysis, which is rather technical, as seen in Sect. 10.A.1, yields

$$\tau(q) = -1 + \begin{cases} c_1 q + c_2 q^2 + \dots, & \lambda > 2, \\ c_1 q + c_2' q^2 \log q, & \lambda = 2, \\ c_1 q + c_\lambda q^\lambda + \dots, & 1 < \lambda < 2, \\ c_1' q \log q + c_1 q, & \lambda = 1, \\ c_\lambda q^\lambda, & 0 < \lambda < 1, \end{cases} \tag{10.13}$$

where the constants c_1, c_2, etc. depend on λ. This establishes the singular behavior of $\tau(q)$ near $q = q_{\text{bottom}}$. From (10.13) it is easy to compute $\alpha(q)$ near $q = q_{\text{bottom}}$ and recover the results in (10.12). Inverting this $\alpha(q)$ and eliminating q from the Legendre transform $f(\alpha(q)) = q\alpha(q) - \tau(q)$, gives the following asymptotical behavior

$$f(\alpha) \simeq 1 - \begin{cases} c(\alpha_\lambda(0) - \alpha)^\kappa, & \lambda > 1, & \alpha \uparrow \alpha_\lambda(0), \\ ce^{-c'\alpha}, & \lambda = 1, & \alpha \to \infty, \\ c\alpha^\kappa, & 0 < \lambda < 1, & \alpha \to \infty, \end{cases} \tag{10.14}$$

where c and c' are positive constants which depend on λ, and $\kappa = \max\{2, \lambda/(\lambda-1)\}$.

The order of the transition in the $f(\alpha)$ at $q_{\text{bottom}} = 0$ is given by the smallest value of n such that $\partial_\alpha^n f(\alpha)|_{\alpha=\alpha_\lambda(0)} \ne 0$. It thus follows that the transition is smooth, i.e., of infinite order, for $0 < \lambda \le 1$. For $\lambda > 1$, the order is $n = \kappa$ when κ is integer, and $n = \lfloor \kappa \rfloor + 1$ for κ noninteger. Here $\lfloor \kappa \rfloor$ is the integer part of κ. Hence, the transition is of order 2 for all $\lambda \ge 2$. Then, as $\lambda \searrow 1$, the order of the transition increases continuously and diverges as $(\lambda - 1)^{-1}$. It stays infinite for all $0 < \lambda \le 1$.

For $\lambda > 1$, which means that $\alpha(0)$ is finite, the left half of $f(\alpha)$ is shaped like the left half of $\cap$ for $\alpha < \alpha(0)$ (see Sect. 10.3.1). The right-hand side of $f(\alpha)$ is the horizontal $f = 1$, and the order of the discontinuity increases from 2 to ∞, as λ approaches 1 from above. For $\lambda \le 1$, $\alpha(0) = \infty$, and the right side of $f(\alpha)$ is nonexistent. These behaviors of $f(\alpha)$ are schematically depicted in Fig. 10.1.

Numerical solution of the generating equation. The above analysis says nothing of the shape of $f(\alpha)$ between $\alpha_{\min}$ and the asymptotic range. In alternative models, an analytical solution may even be lacking for small q. It is therefore of importance to test numerical solutions against the analytical solutions when

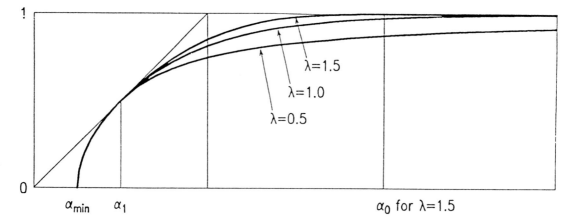

Fig. 10.1. The left-sided $f(\alpha)$ of the family introduced here, (10.14), can have a finite or infinite $\alpha(0)$. In the finite case, the discontinuity in the $f(\alpha)$ at $\alpha(0)$ can be of any finite order larger than 2. In the infinite case, the transition is smooth and f converges either exponentially or as a power law to its asymptotic value $f = D_0 = 1$

the latter are known. To this end, we have solved (10.2) for $\tau(q)$ by Newton's method, that is, using the iteration $\tau^n(q) = \tau^{n-1}(q) + \Delta\tau^{n-1}$, where $\Delta\tau^{n-1} = [1 - Z(q, \tau)]/[\partial_q Z(q, \tau)]|_{\tau^{n-1}}$.

In Fig. 10.2 we plot $\ln[\tau(q) + 1]$ versus $\ln q$, for $q = 10^{-k}, k = 1, \ldots, 7$ and $\lambda = 0.1, 0.5, 1, 1.5, 2, 3$. The slopes of these curves are estimates of the exponent x in $\tau(q) = -1 + cq^x$, for $0 < q \ll 1$. This exponent is related to the exponent λ in (10.13), in an obvious way. For $\lambda = 0.1$ the above two smallest values

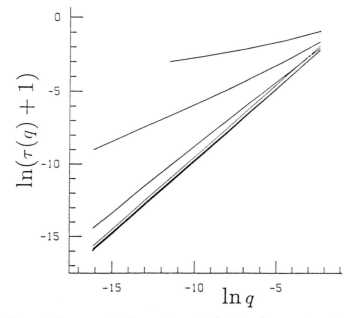

Fig. 10.2. Plots of the numerically evaluated $\ln[\tau(q)+1]$ versus $\ln q$, for $q = 10^{-k}, k = 1 \ldots 7$, obtained with Newton's method from (10.2). For $\ln q = -10$, the values of λ are from top to bottom $0.1, 0.5, 1, 1.5, 2$, and 3

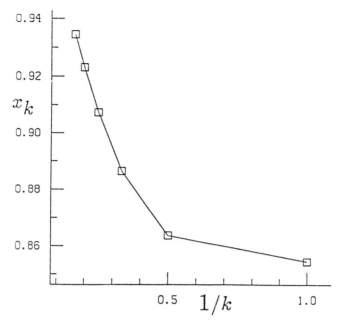

Fig. 10.3. To study the curvature in the $\lambda = 1$ curve in Fig. 10.2, we plot the local slope $x_k = \log(\tau_k) - \log(\tau_{k-1})$ versus $1/k$, where $\tau_k = \tau(10^{-k})$. A simple linear extrapolation yields $x_\infty(\lambda = 1) = 0.99 \pm 0.01$, which is in close agreement with the exact result

of q have not been computed, because of the slow convergence of the sums involved in $Z(q,\tau)$ and $\partial_q Z(q,\tau)$. From the slopes of the lines in Fig. 10.2, one finds that $x \approx 1$ for $\lambda = 1.5, 2, 3$. The case $\lambda = 1$ is more difficult to handle: the plot is not quite straight and its curvature becomes apparent in the plot (Fig. 10.3) of the local slope $\log[\tau(10^{-k})] - \log[\tau(10^{-k+1})]$ versus $1/k$. A simple linear extrapolation yields $x(\lambda = 1) = 0.99 \pm 0.01$. The above results, and those obtained from a similar analysis for $\lambda = 0.1$ and 0.5, agree very well with the exact results in (10.13).

Taking the Legendre transform of the $\tau(q)$ obtained by the above method for $0.0005 < q < 102$, we find the $f(\alpha)$ shown in Fig. 10.4 for $\lambda = 1$ and 0.5. As expected, the convergence of $f(\alpha)$ to $f = 1$ is much faster for $\lambda = 1$ than for $\lambda = 0.5$, while the convergence of $\alpha(q)$ to infinity, as $q \to 0$, is faster for $\lambda = 0.5$. An analytical solution of $\alpha_{\min}$ is discussed in [10.24].

It is clear from these examples that, in certain cases, the asymptotics of $f(\alpha)$, or the value of the critical exponent x, depend on values of q that are either extremely close to 0, or extremely large.

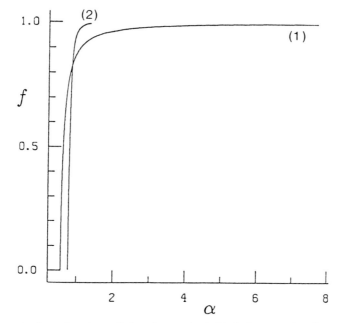

Fig. 10.4. Numerical evaluation of the theoretical $f(\alpha)$ for $\lambda = 0.5$ (*curve 1*) and for $\lambda = 1$ (*curve 2*). Both were obtained by Legendre transforms of numerical solutions of (10.2), using Newton's method

10.4 A Gradual Crossover from Restricted to Left-sided Multifractals

The input $r_\beta \sim \beta^{(-\lambda-1)}$ used in Sect. 10.3 was knowingly "tailored" for very specific goals. Recalling the very special example with which we had started in Sect. 10.2, we also know already that our new infinite base construction allows $\cap$-shaped $f(\alpha)$. It remains to substitute alternative sequences r_β in order to span the wide gap between the skew binomial and the family $r_\beta \sim \beta^{(-\lambda-1)}$ in Sect. 10.3.

Before proceeding with the main family of this section, we consider a multiparameter family based on $m^\beta = 2^{-\beta}$ [i.e., $M(\beta) = 2^{(1-\beta)}$] and $R(\beta) = \exp[-\lambda(\log\beta)^\eta]$, with $\lambda > 0$ and $\eta > 0$. There is nothing new for the value $\eta = 1$, which brings us back to $r_\beta = \beta^{-\lambda} - (\beta+1)^{-\lambda}$. For all η, the criterion function (10.3) and (10.7) is $\alpha^*(\beta) \sim \beta/\lambda(\log\beta)^\eta$. Therefore, $\alpha^*(\beta) \to \infty$ as $\beta \to \infty$ and we deal with a left-sided multifractal. Solving $R(\beta(\epsilon)) = \epsilon$ yields $\beta(\epsilon) = \exp[(\log\epsilon^{-1/\lambda})^{1/\eta}]$. Substituting this in (10.6) shows that the minimal ϵ-coarse grained probability in this family behaves like $\mu_{\min} = 2^{1-\beta(\epsilon)}$. For $\eta < 1$, this decay is faster than any stretched exponential behavior, and this measure is thus even more "anomalous" than any measure obtained via $r_\beta \sim \beta^{(-\lambda-1)}$. For $\eta > 1$, the decay is slower than any stretched exponential, but does not reach down to the power-law decay typical of the skew binomial.

A far more interesting second multiparameter family is based on $m^\beta = 2^{-\beta}$ [i.e., $M(\beta) = 2^{(1-\beta)}$] and $R(\beta) = \exp[-\lambda(\beta-1)^\eta]$, where $\lambda > 0$ and $\eta > 0$. Again, the value $\eta = 1$ brings us back to $r_\beta = r^{\beta-1} - r^\beta$, with $r = e^{-\lambda}$, i.e., to the measure in Sect. 10.3. For very small η, β^η is "like" $\log\beta$ and r_β is "like" $\beta^{-(\lambda+1)}$. This suggests that this family can be said to include the range from the multifractal in Sect. 10.3 to the skew binomial of Sect. 10.2.

The criterion function (10.7) becomes $\alpha^*(\beta) = c(\beta-1)^{1-\eta}/\lambda$, and when $\eta < 1$, we find $\alpha^*(\infty) = \infty$ and thus that $\alpha_{\max} = \infty$. Thus we deal with a left-sided multifractal for $\eta < 1$. But the power $(\beta-1)^{1-\eta}$ increases far less rapidly than the ratio $\sim \beta/\log_2\beta$ in (10.8).

When $\eta > 1$ we find $\alpha^*(\infty) = 0$, which implies that $\alpha_{\min} = 0$. This possibility is discussed in Sect. 10.A.2, where we also show that $\alpha_{\min} = 0$ implies $q_{\text{top}} = 1$.

The measure $\mu([0,\epsilon])$ of the left-most interval of $[0,1]$ is easily found to be given by

$$\mu([0,\epsilon]) = \exp[-c(\log\epsilon^{-1/\lambda})^{1/\eta}],$$

using (10.6). Again, the constant c is nonintrinsic: here, it is $c = \log 2$, but taking $m_\beta = m^{\beta-1} - m^\beta$ with $m \neq 1/2$ would yield $c = -\log m$. The above results, that $\alpha_{\max} = \infty$ for $\eta < 1$ and $\alpha_{\min} = 0$ for $\eta > 1$, are direct consequences of $\mu([0,\epsilon])$ decaying faster than any power of ϵ in the former case, and slower than any power of ϵ in the latter case. There is power-law decay only for $\eta = 1$, which, as we will show after the following remark, is the only restricted multifractal in this family.

Reference [10.26] proposes a decay of $\mu_{\min}$ similar to the one obtained above for this family for $\eta = 1/2, \lambda = 1$, to describe the harmonic measure on DLA, in preference to the exponential decay discussed Sect. 10.3 (see also Sects. 1.8 and 4.7). This is not the proper place to discuss which of our several analytical expressions gives a better fit to the DLA data. But one reason for including the present example is to show that all the behaviors that have been proposed so far can fit very well in the framework of our multiplicative multifractals.

The shape of $\mathbf{f}(\alpha)$. Without being able to dwell on details, we remark that for all η and λ the graph of $f(\alpha)$ is made of two parts: for $\alpha < \alpha(0)$, one has a left side with a second-order maximum, and for $\alpha > \alpha(0)$, one has a horizontal right side. (Hence, we deal with a "phase transition" of order 2.) Let us write this $f(\alpha)$ as $f_{\text{true}}(\alpha)$.

A most interesting sharp discontinuity arises between $\eta = 1$ and η just below 1. Below $\eta = 1$, one must define two distinct $f(\alpha)$s. The left-sided "true" $f_{\text{true}}(\alpha)$ only concerns the far-out asymptotics. That is, $f_{\text{true}}(\alpha)$ only matters when ϵ is below a threshold that converges to 0 with $1 - \eta$. When ϵ is not small, the mass in an interval of length ϵ is for all practical purposes skew binomial, i.e., it is ruled by an $f(\alpha)$ that is $\cap$-shaped.

To describe what happens at intermediate and decreasing values of ϵ, it is best again to use our probabilistic definition of $f(\alpha)$ as a limit, a definition that will be touched upon in Sect. 10.5.2. It turns out that one should expect the histogram of α to be shaped like $\cap$, except for the added presence of an "anomalously" large number of "anomalously" large values that form a "tail" for large α. Using the statisticians' vocabulary, one would be tempted to consider these large values to be "outliers" generated by a "contaminating" mechanism unrelated to the mechanism that generates the $\cap$-shaped portion. For data in our spanning family, however, this would be a totally incorrect interpretation.

A symmetric sharp discontinuity arises between $\eta = 1$ and η just above 1, but now the apparent "outliers" are expected to be found to the left of the $\cap$ shape relative to $\eta = 1$.

10.5 Pre-asymptotics

10.5.1 Sampling of Multiplicatively Generated Measures by a Random Walk

The measures underlying physical processes are usually estimated by some sampling procedure. In dynamical systems [10.27], one can iterate the map in order to find the measure on the attractor. The harmonic measure on DLA clusters can be found by sending many random walkers and keeping track of how often the different growth sites are being visited [10.11]. It is important, therefore, to understand how the existence of a critical point and critical behavior can be established in the presence of a finite sampling. As a first step, we now describe a method to sample one of our left-sided measures up to a prescribed precision $\delta = 2^{-N}$.

The sampling is done by subdividing the interval $[0, 1]$ into 2^N bins of size δ. The measure μ_j in bin $[j\delta, (j + 1)\delta]$, $j = 1, \ldots, 1/\delta$, is the result of a particular multiplicative history $\{\beta_1, \ldots, \beta_n\}$ resulting in a box of size $\epsilon_n = \prod_{i=1}^{n} r_{\beta_i} \approx \delta$ and is given by $\mu_j = \prod_{i=1}^{n} m_{\beta_i} = \prod_{i=1}^{n} 2^{-\beta_i}$. Note that when a measure is constructed on a regular lattice of base r, one has $\epsilon_n = r^n$ independently of the position of the bin. In the present case, however, the r_β are not equal, hence n depends on the position of the bin. We can therefore sample the measure by generating random sequences $\{\beta_1, \ldots, \beta_n\}$, where β is chosen according to the probability distribution $\text{prob}(\beta) = m_\beta = 2^{-\beta}$ and n is the finite value such that $\epsilon_n \approx \delta$. The bin visited by this particular sequence is $[\Delta, \Delta + \epsilon_n]$, where $\Delta = \sum_{j=1}^{n} \epsilon_{j-1} R(\beta_j + 1)$, $\epsilon_0 \equiv 1$, and R was defined in (10.4). For the left-sided family introduced in Sect. 10.3, one thus has that $\Delta = \sum_{j=1}^{n} \epsilon_{j-1}(\beta_j + 1)^{-\lambda}$.

It is good to think of this process in terms of a random walk with $-\ln \mu$ along the x axis and $-\ln \epsilon$ along the y axis. At time 0, let our random walker start at the origin. Then the walker selects at each step from among an infinite number of possible next steps, indexed by $\beta = 1, 2, \ldots$, each with probability $2^{-\beta}$. When β has been chosen, the walker will jump by $\beta \ln 2$ in the x direction and by $-\ln r_\beta$ in the y direction, and so on. To reach the desired precision δ, the walker has to proceed until it crosses the line $y = -\ln \delta$. Since the jumps in the vertical direction are not equal, the required number of steps will be strongly dependent on the choice of jumps.

It is now possible to sample the measure μ, by probing it with a large number M of walks. Let $M\phi_j(\epsilon, M)$ be the number of times the bin j has been visited. Then $\phi_j(\epsilon, M)$ is an estimate of the measure $\mu_j(\epsilon)$ of bin j. With a finite number M of walks it is clearly impossible to sample adequately a bin whose actual measure is less than $1/M$. But for a restricted multifractal the estimates $\phi_j(\epsilon, M)$ will rapidly converge if M is made to increase faster than $\epsilon^{\alpha_{\max}}$. The frequencies $\phi_j(\epsilon, M)$ can then be combined into a partition function, which yields a sample estimate of $\tau(q)$. Finally, $f(\alpha)$ is *defined* as the Legendre transform of $\tau(q)$. In left-sided multifractals, in contrast, the situation is more complicated, as we will now show.

10.5.2 An "Effective" $f(\alpha)$

Let us consider the left-sided family of Sect. 10.3. From the behavior of the smallest ϵ-coarse grained probability given in (10.9), it follows that at least $M(\epsilon, \lambda) = \exp(\epsilon^{-1/\lambda} \log 2)$ walks are needed to adequately sample the measure to a spatial precision ϵ. For example, for $\lambda = 1, 0.5, 1.5$, one finds $M(2^{-8}, 1) \approx 10^{76}, M(2^{-5}, 1) \approx 10^9, M(2^{-4}, 1) \approx 35000, M(2^{-3}, 0.5) \approx 10^{19}$, and $M(2^{-8}, 1.5) \approx 10^{11}$. Figure 10.5 shows the estimated measure $(M^{-1}\phi_j)$ of a sample of the measure for $\lambda = 1$, obtained with $M = 4 \times 10^8$ walks and $\delta = 2^{-15}$. Similar samples were constructed for $\lambda = 0.5$ and 1.5.

The astronomical numbers of walks needed would not pose a serious problem if it were not for the fact that the properties of the measure change with scale in a nonconventional manner. This becomes immediately clear by noting that the largest value $\alpha_{\max}(\epsilon) \equiv \max_j[-\log \mu_j(\epsilon)/\log \epsilon]$, increases with $\epsilon \to 0$. Namely, from (10.8) it follows that $\alpha_{\max}(\epsilon) \approx -\epsilon^{1/\lambda}/\log \epsilon$. So for any finite M, however large, one will not be able to see the asymptotic $f(\alpha)$ discussed in Sects. 10.3 and 10.3.1.

One would like, however, to be able to define for each ϵ a notion of "effective" $f_\epsilon(\alpha)$. Unfortunately, the restricted theory has no room for such a notion. In Sect. 4.1 of [10.24] we briefly discussed how a blind application of the restricted formalism would perform in the above situation. But as will be shown below, the notion of effective $f(\alpha)$ which arose there does not have any meaning in the more fundamental, probabilistic theory of multifractals.

In this theory $f(\alpha)$ is not *defined* as the Legendre transform of $\tau(q)$. Instead, given that $D_0 = 1$ in the present case, the quantity $\rho(\alpha) = f(\alpha) - 1$ is introduced as a limit of the form $\rho(\alpha) = \lim_{\epsilon \to 0} \log \rho_\epsilon(\alpha)$, where $\rho_\epsilon(\alpha) = (1/\log \epsilon) \log p_\epsilon(\alpha)$. The quantities $p_\epsilon(\alpha)$ are the probability densities of the

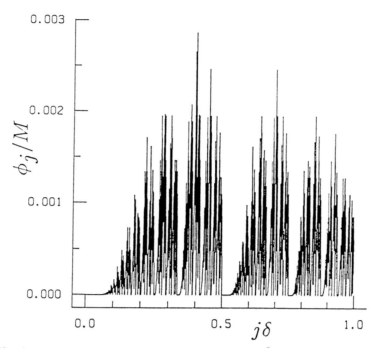

Fig. 10.5. The $\lambda = 1$ measure, sampled with $M = 4 \times 10^8$ walks. The size of the bins used to make this histogram is $\delta = 2^{-11}$

Hölder exponent α for given ϵ, defined as $\alpha = \log \mu(\epsilon)/\log \epsilon$. To form $\rho_\epsilon(\alpha)$ is to take the logarithm of a probability density, and to renormalize it by dividing by $\log \epsilon$. This is precisely how the Hölder exponent α is obtained from μ. To plot a function on doubly logarithmic coordinates is a cliché in the study of fractals. But here the plot is *not* a straight line! The fact that the sequence $\rho_\epsilon(\alpha)$ does indeed have a limit is a remarkable feature of the main tool of our theory, which is Harald Cramer's theory of large deviations [10.21]. As may have been expected, the Cramer theory does use the Legendre transform, but only to *evaluate* $f(\alpha)$, not to *define* it.

Now we come to an important distinction. For the restricted multifractals, $f(\alpha)$ is useful even for relatively large ϵ. The reason is that the convergence of $\rho_\epsilon(\alpha)$ to $\rho(\alpha)$ is acceptably rapid and uniform, and the "finite ϵ" corrections $\rho(\alpha) - \rho_\epsilon(\alpha)$ have not yet been found to be needed in physics. (They probably will !) For the special multipliers that yield left-sided multifractals, in contrast, the convergence of $\rho_\epsilon(\alpha)$ to $\rho(\alpha) = f(\alpha) - 1$ turns out to be extraordinarily slow. This becomes clear from the fact that $\rho(\alpha_{\max}) - \rho_\epsilon(\alpha_{\max}) = \infty$ for all finite values of ϵ. Also the shape of the approximant, even if ϵ is small, gives a totally misleading idea of the actual limit $f(\alpha)$. We have seen that, if $\lambda > 1$ (hence $\alpha(0) < \infty$), the limit of $\rho_\epsilon(\alpha)$ is identical to 0 for $\alpha > \alpha(0)$.

Since the limit $\rho(\alpha) = f(\alpha) - 1$ is not attained for any $\epsilon > 0$, this limit hardly matters at all when $\epsilon > 0$. Instead, what does matter greatly is the "preasymptotic" behavior of $\rho_\epsilon(\alpha)$. This paper gives us no room to go beyond asserting that, denoting $-\log \epsilon$ by k, one has

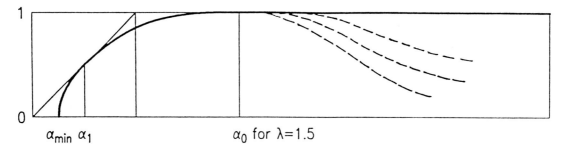

α_{min} α_1 α_0 for $\lambda = 1.5$

Fig. 10.6. This schematic view of the convergence of $\rho_k(\alpha) + 1$ to $f(\alpha)$ shows that the left-sidedness of the asymptotic $f(\alpha)$ fails to be reflected in the shape of finite sample approximants to $\rho_k(\alpha) + 1$. The value of k increases from bottom to top. Detailed studies of other classes of multifractals suggests that the form of $f(\alpha)$ to the left of α_0 is likely to differ markedly from $\rho_k(\alpha) + 1$ near its maximum. The difference or "bias" is likely to depend upon details of r_β for moderate β, and we prefer to discard it here

$$p_k(\alpha) \sim ck^{1-\lambda}\alpha^{-\lambda-1}$$

and

$$\rho_k(\alpha) \sim (1/k)\log[ck^{1-\lambda}\alpha^{-\lambda-1}]. \tag{10.15}$$

Figure 10.6 compares very schematically the shapes of $\rho_k(\alpha) + 1$ and $f(\alpha)$. It is clear form (10.5) and Fig. 10.6 that the expression $f_k(\alpha) = \rho_k(\alpha) + 1$ *is not the Legendre transform* of any function $\tau(q)$. One of many reasons is that all Legendre transforms are cap-convex, while our $\rho_k(\alpha) + 1$ is cup-convex for large α. In a rough way, $f_k(\alpha)$ is a modified box dimension that would only concern boxes of size e^{-k}. But there is no Hausdorff Besicovitch dimension behind this box dimension.

Mathematical digression. In the case of restricted multifractals, it is known that $f(\alpha)$ is a Hausdorff Besicovitch dimension. For the binomial measure, this property follows from difficult but standard theorems by Eggleston and also of Volkmann (see [10.21], p. 24). Turning to the present generalization and supposing that $\alpha(0) < \infty$ and $f(\alpha) = 1$ for $\alpha > \alpha(0)$, let us consider the Hausdorff Besicovitch dimension of the set of points where α satisfies $\alpha(0) < \alpha' < \alpha < \alpha''$, for given $\alpha' > \alpha(0)$ and $\alpha'' > \alpha'$. It is tempting to conjecture that this set's dimension is 1 for all α' and α''. Nevertheless, this set's linear measure (that is, its Lebesgue measure and its Hausdorff measure in the dimension 1) is necessarily 0, because the Hölder alpha equals $\alpha(0)$ almost surely. This conjecture is proven in the Appendix 10.B added in this second edition.

10.6 Miscellaneous Remarks

An issue of rigor. Sections 10.2 to 10.4 started with a formal equation (10.2) for $\tau(q)$ and continued by using diverse formal manipulations that are known to be valid for restricted multifractals. But the function $f(\alpha)$ obtained by these manipulations falls beyond the restricted class. This raises an issue of rigor which is tackled in the new Appendix 10.B. A genuine underlying complication is that $\tau(q)$ can be defined in either of two different ways: the partition function [10.5, 6] can be used when μ is a restricted multifractal, but in other cases one must use the earlier definition advanced in [10.1].

Randomness and its necessity in modeling. One point of the present chapter is that, even in the nonrandom case, the multiplicative multifractals introduced in [10.1] are more general than the multifractals introduced in [10.5,6]. Randomness is important for many physical applications of multifractal. A different advantage of the approach in [10.1], is that it applies immediately and rigorously to random multifractals. Of course, in the hands of many researchers, the theory in [10.5,6] has been informally extended to the random case by resorting to averaging. The reason we call these extensions *informal* is because they are not based on any theory. This is why they involve murky discussions about which of several methods of averaging is "the best". Our original theory of multiplicative multifractals tackles these issues in advance.

In the canonical random case [10.1,28], another anomaly is due to the presence of negative $f(\alpha)$s. When one starts with the usual formalism and then superposes diverse methods of averaging upon it, one finds that negative $f(\alpha)$s do occur with certain methods, but not with others, and their origin and meaning is totally obscure. In our approach, on the other hand, they are perfectly well understood and essential. In broad outline, the positive $f(\alpha)$s can serve to define a *typical* distribution of a random fractal measure, and the negative $f(\alpha)$s describe *fluctuations* one may expect in a finite-size sample.

Yet another cause of anomaly [10.28] may restrict q to lie below a different threshold, denoted by q_{crit} which satisfies $1 \leq q_{\text{crit}} \leq q_{\text{top}}$, and allow $\alpha < 0$. For discussions of these matters, and different anomalies, which are needed in the study of turbulence, we refer the reader to [10.22,28].

On left-sided multifractals that arise in dynamical systems. Let us mention that the "anomaly" of left-sided $f(\alpha)$s which is described in this chapter is not limited to multifractals in real space, like DLA, but does extend to certain dynamical systems. This fact broadens the impact of our criticism of the restricted multifractals, as being a tool of inadequate generality for the needs of physics. The new construction, described in [10.29] and in part in [10.30], is detailed in Appendix 10.C. The appendix follows closely [10.29].

Diffusion limited aggregation. The connections between the construction of Sects. 10.3 and 10.4 and the physics of DLA began to be understood in [10.15–

17]. For more recent results on the complexity of DLA we refer to [10.31–33]. Many of the most important facts do not appear until very large clusters are considered.

10.7 Summary

This chapter shows that a failure of the partition function of a measure to scale as a power law does not necessarily mean that this measure is not self-similar.

In Sect. 10.2 we defined a broad class of infinite base-deterministic multiplicative processes yielding exactly self-similar measures on the unit interval. A criterion was derived with which one can easily determine whether the partition function scales or fails to scale. In case the partition function scales for all q, we call the measure a "restricted multifractal".

In Sect. 10.3, we considered a specific family of measures, for which the minimum probability decayed like a stretched exponential as a function of the coarse-graining box size. This implies that the partition function fails to scale like a power law for $q < 0$. In such cases only the left side of the $f(\alpha)$ shows "normal" behavior. The right-hand side is straight. The $f(\alpha)$ of these measures are therefore called *left-sided*. An infinity of possible scenarios for the "phase-transition" at $q = 0$ are shown to exist. Namely, finite or infinite $\alpha(0)$ and all possible orders of transitions, ranging from 2 to ∞.

In Sect. 10.4 we discussed yet another family, whose minimal probability decays faster than any power law, but slower than any stretched exponential. This family therefore spans a bridge between the more familiar restricted multifractals and the family with stretched exponential decay. This family is "left-sided" for parameter values $\eta > 1$, and "restricted" for $\eta = 1$. For $\eta < 1$, a new "anomaly" arises from the restricted point of view, namely, the partition function fails to scale for $q > 1$. This is further discussed in Sect. 10.A.2.

In Sect. 10.5 we showed how a deterministic multiplicative process can be sampled, and discussed the difficulties associated with defining a notion of a pre-asymptotic $f(\alpha)$ as a Legendre transform of a pre-asymptotic $\tau(q)$. The use of a probabilistic approach was discussed and the importance of a knowledge of the pre-asymptotic behavior stressed.

Acknowledgements. Many discussions with A. Aharony have been invaluable. Aftereffects of an earlier collaboration of B.B.M. with M.C. Gutzwiller and inputs from R.Stinchcombe have been most helpful. We would like to thank D. Coppersmith for the proof presented in Sect. 10.A.1 and Y. Hayakawa [10.24]. One of us (B.B.M.) also acknowledges the hospitality of the Sackler Institute for Solid State Physics at Tel-Aviv University, where much of this chapter was written. This research has been financially supported in part by the Office of Naval Research, Grant N00014-88-K-0217.

Editorial note. The present chapter includes three appendices, 10.A, 10.B and 10.C. Appendix 10.A presents details of the solution of Eq. (10.2) as well as a discussion of the case $\alpha_{\min} = 0$. Two appendices 10.B (by R.H. Riedi and B.B. Mandelbrot) and 10.C (by B.B. Mandelbrot) are added in this second edition. Appendix 10.B refers to and sketches the full mathematical proof of a result that the body of the paper asserted without proof. Appendix 10.C describes a multifractal measure that is very close in spirit to those described in the paper. A complication is that it is not self-similar. A first attractive feature is that it enters though an ancient problem of analysis that can be interpreted in terms of a "dynamical system". A second attractive feature is that the definition itself goes back to H. Minkowski, yet this measure has not previously been studied as a multifractal, except in a heuristic preliminary paper by Gutzwiller and Mandelbrot [10.30].

10.A Details of Calculations and Further Discussions

10.A.1 Solution of (10.2)

We solve (10.2) for all λ in the limit of small positive values of q, i.e., $q \downarrow 0$. Let $\delta = q \ln 2$, $q > 0$, and write $\tau(q) = -1 + \varepsilon$. Since the nontrivial behavior of the measures studied here is caused by the behavior of the multipliers m_β for large values of β, we can use the approximation $r_\beta \simeq \lambda\beta^{-\lambda-1}$. Thus we can replace the reduction factors by $r_\beta = \beta^{-\lambda-1}/\zeta(\lambda+1)$, where the Riemann zeta function $\zeta(x) = \sum_{j=1}^{\infty} j^{-x}$ is introduced to ensure that the slightly changed lengths r_β add to 1. The original problem (10.2) then becomes

$$\sum_{\beta=1}^{\infty} e^{-\delta\beta} r_\beta^{1-\varepsilon} = \zeta(\lambda+1).$$

If we define $s_\beta = (1 - e^{-\delta\beta})r_\beta$ and $t_\beta = e^{-\delta\beta}(r_\beta^{1-\varepsilon} - r_\beta)$, the above equation becomes

$$\sum_{\beta=1}^{\infty} (s_\beta - t_\beta) = 0. \tag{10.16}$$

This sum is split into three parts, namely,

$$\sum(1) \equiv \sum_{\beta<1/\delta} s_\beta, \tag{10.17}$$

$$\sum(2) \equiv \sum_{\beta>1/\delta} s_\beta, \tag{10.18}$$

$$\sum(3) \equiv \sum_{\beta=1}^{\infty} e^{-\delta\beta}(\beta^{\varepsilon(\lambda+1)} - 1)/\beta^{\lambda+1}, \tag{10.19}$$

so that (10.16) becomes

$$\sum(1) + \sum(2) - \sum(3) = 0. \tag{10.20}$$

For $\sum(1)$ we find

$$\sum(1) = \sum_{\beta < 1/\delta} \frac{1}{\beta^{\lambda+1}} \left(\sum_{n=1}^{\infty} (-1)^{n+1} \frac{(\delta\beta)^n}{n!} \right).$$

Since this double sum is absolutely convergent we can rearrange terms to find for $\lambda \neq 1, 2, \ldots$, that

$$\sum(1) = \sum_{n=1}^{\infty} \{(-1)^{n+1}/n!\}\delta^n \sum_{\beta < 1/\delta} \beta^{n-1-\lambda} \tag{10.21}$$

$$= \sum_{n=1}^{\infty} \frac{(-1)^{n+1}}{n!} \delta^n \left\{ \left. \frac{\beta^{n-\lambda}}{n-\lambda} \right|_1^{1/\delta} \right\} \tag{10.22}$$

$$= \sum_{n=1}^{\infty} \frac{(-1)^{n+1}}{n!} \left\{ \frac{\delta^{\lambda-n}}{n-\lambda} + C_n + \text{lower-order terms} \right\} \tag{10.23}$$

$$= \delta^\lambda \left[\sum_{n=1}^{\infty} \frac{(-1)^{n+1}}{n!(n-\lambda)} \right] + \sum_{n=1}^{\infty} \frac{(-1)^{n+1}}{n!} \delta^n C_n + \text{lower-order.} \tag{10.24}$$

Note that for $1 < \lambda < 2$ the term in square brackets in (10.24) lies in the range $(-\infty, -2.8)$. It follows that

$$\sum(1) \approx \begin{cases} c\delta + c''\delta^2 + \ldots, & \lambda > 2, \\ c\delta + c'\delta^2 \log(1/\delta) + c'''\delta^2, & \lambda = 2, \\ c\delta + c_1\delta^\lambda + \ldots, & 1 < \lambda < 2, \\ c\delta \log(1/\delta) + c'\delta, & \lambda = 1, \\ c\delta^\lambda + c'\delta + \ldots, & 0 < \lambda < 1, \end{cases} \tag{10.25}$$

with $c_1 < 0$. The results for $\lambda = 1$ and $\lambda = 2$, follow from (10.21). Since $0.63 < c'_2 \equiv (1 - \exp(-\delta\beta)) < 1$ we can write $\sum(2)$ as

$$\sum(2) = c'_2 \sum_{\beta > 1/\delta} \beta^{-\lambda-1} = \frac{c'_2}{\lambda}\delta^\lambda + \text{lower-order terms} \sim c_3\delta^\lambda + \ldots,$$

where $c_2 = c'_2/\lambda$ and $|c_1| > |c_2|$ if $1 < \lambda < 2$. By Taylor's theorem with remainder

$$\sum(3) = \sum_{\beta=1}^{\infty} \frac{e^{-\delta\beta}}{\beta^{\lambda+1}} \left(\varepsilon(\lambda+1)\log\beta + O\left(\frac{\varepsilon^2}{2}(\lambda+1)^2 \log^2(\beta)\beta^{\varepsilon(\lambda+1)} \right) \right).$$

The first term has the upper bound

$$T(1) \equiv \sum_{\beta=1}^{\infty} \frac{e^{-\delta\beta}}{\beta^{\lambda+1}} \varepsilon(\lambda+1)\log\beta < \sum_{\beta=1}^{\infty} (\log\beta/\beta^{\lambda+1})\varepsilon(\lambda+1) \equiv c_3\varepsilon$$

and the lower bound

$$T(1) > \sum_{\beta<1/\delta} \frac{(1-\delta\beta)\log\beta}{\beta^{\lambda+1}} \varepsilon(\lambda+1) \tag{10.26}$$

$$= \sum_{\beta<1/\delta} \frac{\log\beta}{\beta^{\lambda+1}} \varepsilon(\lambda+1) - \delta \sum_{\beta<1/\delta} \frac{\log\beta}{\beta^{\lambda}} \varepsilon(\lambda+1) \tag{10.27}$$

$$= c_3\varepsilon - O\big(\delta^{\lambda}\log\left(1/\delta\right)\varepsilon\big) - \begin{cases} O\big(\delta\varepsilon\big), & \lambda > 1, \\ O\big(\delta\varepsilon\log^2\left(1/\delta\right)\big), & \lambda = 1, \\ O\big(\delta^{\lambda}\varepsilon\log\left(1/\delta\right)\big), & \lambda < 1. \end{cases} \tag{10.28}$$

The expression $\sum(3) - T(1)$ is of the order $O(\varepsilon^2)$, as long as $(1-\varepsilon)(1+\lambda) > 1$, so that

$$\sum(3) \simeq c_3\varepsilon + O\big(\varepsilon^2\big) + O\big(\varepsilon\delta^{\lambda}\log\left(1/\delta\right)\big) + \begin{cases} O\big(\delta\varepsilon\big), & \lambda > 1, \\ O\big(\delta\varepsilon\log^2\left(1/\delta\right)\big), & \lambda = 1, \\ O\big(\delta^{\lambda}\varepsilon\log\left(1/\delta\right)\big), & \lambda < 1. \end{cases}$$

Equation (10.20) then yields

$$\varepsilon = \begin{cases} c\delta + c''\delta^2 + \ldots, & \lambda > 2, \\ c\delta + c''\delta^2\log(1/\delta), & \lambda = 2, \\ c\delta + c''\delta^{\lambda} + \ldots, & 1 < \lambda < 2, \\ c\delta\log(1/\delta) + c''\delta, & \lambda = 1, \\ c\delta^{\lambda} + O(\delta) + O(\delta^{2\lambda}\log(1/\delta)), & 0 < \lambda < 1, \end{cases}$$

as announced in the body of this chapter.

10.A.2 The Case $\alpha_{\min} = 0$

First, a warning. In order that $\alpha_{\min} = 0$, a *sufficient* condition is that some point carries an "atom" of positive measure. But this condition is *not* necessary: $\alpha_{\min} = 0$ also holds when $\mu[t, t+\epsilon] \sim L(\epsilon)$, where $L(\epsilon)$ is any function (such as $1/|\log\epsilon|$) that $\to 0$ with ϵ, but more slowly than any positive power ϵ^{α}.

The argument proceeds roughly as for $\alpha_{\max} = \infty$. A sufficient condition for $\alpha_{\min} = 0$ is that one can identify at least one point where $\alpha = 0$. When $\alpha^*(\beta) \to 0$, this is indeed the case for the left-most point of $[0, 1]$. It is easy to see that, in that case, $0 < \alpha_{\max} < \infty$.

A consequence of $\alpha_{\min} = 0$ is that $q_{\text{top}} = 1$, meaning that for $q < 1$, $\chi(q, \epsilon)$ *fails* to scale like $\epsilon^{\tau(q)}$ and the function $\tau(q)$ *fails* to be defined.

Proof: observe that we now have $\max_j \mu_j(\epsilon) \sim \epsilon^{\alpha^*(\beta)}$. Hence, the existence of $\tau(q)$ would, again, require the inequality $\epsilon^{q\alpha^*(\beta)} \geq \epsilon^{\tau(q)}$. When $q > 0$ this becomes $\tau(q) \leq q\alpha^*(\beta)$, which would require $\tau(q) \leq 0$. Since $\tau(q) \geq 0$ for $q > 1$, we have proved that $\tau(q) = 0$ for $q > 1$, if $\tau(q)$ exists. But, if $q > 1$, it would follow from $\tau(q) = 0$ that $\chi(q, \epsilon) = $ const for all ϵ. This last relation only holds if the measure concentrates at $\alpha = 0$, which we shall see is not the case. Hence, $\tau(q)$ is not defined for $q > 1$.

Conclusion: when $\alpha^*(\beta) \to 0$, our infinite-base multiplicative multifractal satisfies $\alpha_{\min} = 0$, hence is not a restricted multifractal.

Reciprocity between the anomalies $\alpha_{\max} = \infty$ and $\alpha_{\min} = 0$. It happens that much about our new multiplicative multifractals does not depend on the functions m_β and r_β, i.e., on $M^*(\beta)$ and $R^*(\beta)$ taken separately. They depend solely on the behavior of the functions $M^*(R^*)$ and $R^*(M^*)$, obtained by eliminating β between the functions $M^*(\beta)$ and $R^*(\beta)$. The fact that two alternative functions are involved expresses that our input quantities m_β and r_β obey exactly the same constraints. It also follows that one can exchange their roles. This will exchange the roles of M^* and R^*, and replace the anomaly $\alpha_{\max} = \infty$ by the anomaly $\alpha_{\min} = 0$, or conversely.

The two measures obtained in this fashion, call them reciprocal and denote them by μ and $\tilde{\mu}$, must be closely related. Indeed, one can verify that the functions $\mu([0, \tilde{\mu}])$ and $\tilde{\mu}([0, \mu])$, which are monotone increasing, are the inverse of each other. Graphically, the relation between μ and $\tilde{\mu}$ is illustrated in the case of a very analogous measure by Figs. 1 and 6 of [10.30], where the graph of μ as function of μ and $\tilde{\mu}$ is called a *slippery staircase*. Its apparent horizontal steps are in fact "not quite" horizontal, which is why the measure μ and $\tilde{\mu}$ does not quite include atoms. We shall publish elsewhere the proof that the functions f and $\tilde{f}$ that characterize μ and $\tilde{\mu}$ are linked by $\tilde{f}(\alpha = \alpha f(1/\alpha)$.

Combining the anomalies $\alpha_{\max} = \infty$ and $\alpha_{\min} = 0$ through a bilateral generalization of our construction. Now we extend the range of the index β to also include the negative integers, i.e., $-\infty < \beta < \infty$. This generalization allows the relations $\lim_{\beta \to \infty} \alpha^*(\beta) = -\infty$ and $\lim_{\beta \to -\infty} \alpha^*(\beta) = 0$ to both be true. It is even easy to satisfy the identity $\tilde{f}(\alpha) = \alpha f(1/\alpha)$, which is a way to insure that the reciprocal measure $\tilde{\mu}$ is identical to μ. An example where this goal is fulfilled is when $m_{-\beta} = r_\beta$ and $r_{-\beta} = m_\beta$. This example shows that, even after the behavior of $\alpha^*(\beta)$ at $\beta \to \infty$ has been fixed, there exist an infinity of different "self-reciprocal" measures. When $\alpha_{\max} < \infty$ and $\alpha_{\min} = 1/\alpha_{\max} > 0$, these measures are restricted multifractals.

10.B Multifractal Formalism for Infinite Multinomial Measures, by R.H. Riedi and B.B. Mandelbrot

Chapter 10 introduced some interesting self-similar measures, each the sum of an *infinite* number of copies of itself, reduced in size by a factor r_β and in mass by a factor m_β ($\beta \in \mathbb{N}$). These measures provide examples of strictly multiplicative measures with left sided multifractal spectra. To compute the multifractal spectra, the preceding chapter gave strong heuristic reasons to believe that the multifractal formalism generalizes in the obvious way from the finite to the infinite case. That is,

$$d_{\mathrm{HD}}(K_\alpha) = f(\alpha) := \inf\{q\alpha - \tau(q) \,:\, q \in \mathbb{R}\} \qquad (10.29)$$

where K_α is the set of Hölder exponent α, $d_{\mathrm{HD}}(K_\alpha)$ denotes the Hausdorff dimension, and the function τ is defined (uniquely) through

$$\sum_{\beta=1}^{\infty} m_\beta{}^q r_\beta{}^{-\tau} = 1 \qquad (10.30)$$

if the equation has a solution and $\tau(q) = -\infty$ otherwise.

Reference [10.34] proves that the formula (10.29) holds for quite general sequences r_β and m_β. Here we give a short outline of the argument as it applies to the special self-similar measure μ studied in Chapter 10.

It is useful to recall its construction: Consider the uniform mass distribution on $I := [0,1]$. Divide I into subintervals I_β of length r_β and redistribute the mass in such a way that I_β carries a uniform distribution with total mass m_β. Do the same with I_β to obtain intervals $I_{\beta\beta'}$ such that the ratios of length and mass between I_β and $I_{\beta\beta'}$ are the same as between I and I_β. Continue inductively ad infinitum and obtain μ in the limit.

Considering covers by intervals $I_{\beta_1...\beta_n}$ with $\mu(I_{\beta_1...\beta_n}) \le |I_{\beta_1...\beta_n}|^\alpha$ gives the upper bound $d_{\mathrm{HD}}(K_\alpha) \le f(\alpha)$. In fact, the derivation of this bound is the same as in the finite case since Hausdorff dimension can be estimated using infinite covers.

To obtain a lower bound to $d_{\mathrm{HD}}(K_\alpha)$, [10.34] considers measures μ_M which are obtained the same way as μ but, except that mass is redistributed only among the first M intervals. This follows the idea of finite approximations which can reveal only details up to a certain degree. It is well known that the multifractal formalism holds for μ_M (a full proof is found in [10.35]: for $\alpha = \tau'_M(q)$

$$d_{\mathrm{HD}}(K_\alpha^M) = f_M(\alpha) = q\tau'_M(q) - \tau_M(q) \quad \text{and} \quad \sum_{\beta=1}^{M} \left(m_\beta / \sum_{i=1}^{M} m_i \right)^q r_\beta{}^{-\tau_M(q)} = 1.$$

To the Hölder exponent $\alpha = \tau'(q)$ of μ corresponds the Hölder exponent $\alpha_M = \tau'_M(q)$ of μ_M. Since μ has larger support than μ_M we have $K^M_{\alpha_M} \subset K_\alpha$ and hence $f_M(\alpha_M) \leq f(\alpha)$. What remains is a technical proof showing $\tau_M \to \tau$ and $\tau'_M \to \tau'$, hence $f_M(\alpha_M) \to f(\alpha)$.

Remark A: note that $f_M(\alpha_M)$ gives information about the Hölder exponents when computing μ up to a certain accuracy. In the case $r_\beta = \beta^{-\lambda} - (\beta + 1)^{-\lambda}$ the speed of convergence towards $f(\alpha)$ depends on λ in agreement with the results of Chapter 10.

Remark B: a straightforward adaption of the argument in [10.35,36] yields the (Lebesgue) almost sure Hölder exponent

$$\tau'(0+) = \sum_{\beta=1}^{\infty} \log(m_\beta) r_\beta / \sum_{\beta=1}^{\infty} \log(r_\beta) r_\beta$$

which can be infinite. On the other hand, the μ almost sure Hölder exponent is $\tau'(1) = \sum_{\beta=1}^{\infty} \log(m_\beta) m_\beta / \sum_{\beta=1}^{\infty} \log(r_\beta) m_\beta$.

10.C The Minkowski Measure and Its Left-sided $f(\alpha)$, by B.B. Mandelbrot

This appendix investigates the function $f(\alpha)$ for an important singular non-random measure $\mu(dt)$ defined in 1900 by Hermann Minkowski [10.37]. The value of α is shown to be almost surely infinite. Hence, $\alpha_{\max} = \infty$ and the graph of $f(\alpha)$ is left-sided. On the other hand, Mandelbrot's original approach to multifractals based on the distribution of the coarse Hölder α [10.1], also injects approximate measures $\mu_\epsilon(dt)$ that have been coarsened by replacing the continuous t by multiples of $\epsilon > 0$. This theory therefore involves a sequence of observable approximant functions $f_\epsilon(\alpha)$; their graphs are not left-sided.

As $\epsilon \to 0$, one has $f_\epsilon(\alpha) \to f(\alpha)$, which expresses that the theory behind the method of distribution is logically consistent. But the convergence is excruciatingly slow and extremely singular; very great care is needed to extrapolate the shape of $f(\alpha)$ from that of the $f_\epsilon(\alpha)$.

Those results illustrate a limitation of the thermodynamical theory behind $f(\alpha)$. In some cases $f(\alpha)$ is very poorly approximated by the pre-thermodynamical results one obtains when $\epsilon > 0$, however small ϵ may be. Contrary to the left-sided multifractals investigated in Chapter 10, the Minkowski measure is not multiplicative. Nevertheless, several properties of μ were first conjectured on the basis of an approximation of μ by a multiplicative multifractal and later proved to be correct.

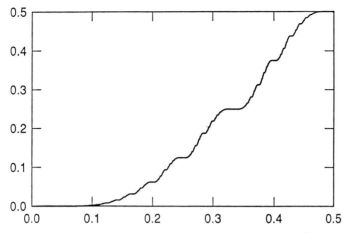

Fig. 10.7. Graph of the Minkowski function $M(x)$ for $0 < x < 1/2$. Contrary to the well-known Cantor devil staircase [10.3], plate 83), the graph of $M(x)$ has no actual steps, only *near steps* that led to its being called a *slippery staircase* in [10.30].

10.C.1 The Minkowski Measure on the Interval [0,1]

The Minkowski measure μ is simple to define and work with, but exhibits very interesting and totally unexpected peculiarities. This μ and the *inverse Minkowski measure* $\tilde{\mu}$ are, respectively, the differentials of two increasing singular functions: $M(x)$ and its inverse $X(m)$.

It is easier to start by defining the inverse Minkowski function $X(m)$, which is constructed step by step, as follows. The first step sets $X(0) = 0$ and $X(1/2) = 1/2$. The second step interpolates: $X(1/4)$ is taken to be the Fairey mean of $X(0)$ and $X(1/2)$, where the Fairey mean of two irreducible ratios (a/c) and (b/d) is defined as $(a + b)/(c + d)$. More generally, the k-th step begins with $X(m)$ defined for $m = p2^{-k}$, where p is an even integer, and uses Fairey means to interpolate to $m = p2^{-k}$, where p is an odd integer. Finally, $X(m)$ is extended to the interval $[1/2, 1]$ by writing $X(1 - m) = 1 - X(m)$. The resulting function $X(m)$ is continuous, it increases in every interval, and it is singular; that is, it has no finite derivative at any point. It has an inverse function $M(x)$ with the same properties, illustrated by the "slippery staircase" in Figure 10.7.

The differentials $\tilde{\mu}$ and μ of the functions $X(m)$ and $M(x)$ are singular measures. The two parts of Figure 10.8 illustrate the measure μ, as evaluated for intervals of length 10^{-5}.

10.C.2 The Functions $f(\alpha)$ and $f_\epsilon(\alpha)$ of the Minkowski Measure

The theoretical function f(α). For a derivation of the $f(\alpha)$ functions of the Minkowski measure μ and of the inverse Minkowski measure $\tilde{\mu}$, we must refer the reader elsewhere [10.38]. A first basic fact is that there is a theoretical $f(\alpha)$ for these measures: $f(\alpha)$ is the Hausdorff dimension of the set of points x such

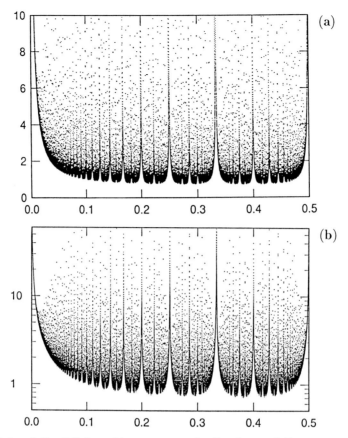

Fig. 10.8a,b. Plots of the Minkowski measure μ in the form of the coarse-grained Hölder exponent and of its logarithm. Taking $\varepsilon = 10^{-5}$, we evaluated the increments $\Delta M = M[(k+1)\varepsilon] - M[k\varepsilon]$. (a) plots the coarse-grained Hölder $\alpha = \log \Delta M / \log_\varepsilon$, and (b) plots $\log \alpha$. The theory described later in the paper shows that the fine-grained (local) Hölder almost prescribed α. In this figure, this property altogether fails to be reflected.

that the Hölder exponent $H(x)$ takes the value α. In the case of μ, the graph of $f(\alpha)$ has the following properties:

- $\alpha_{\min} = -1/\log_2 \gamma^2 \sim .7202\ldots$ where γ is the golden mean $\sim .6180\ldots$ (obtained in [10.39] and - independently - in [10.30])
- $\alpha_1 = [2 \int_0^1 \log_2(1 + x)dM(x)]^{-1} = .874\ldots$ [10.40]
- $f(\alpha) \to 1$ as $\alpha \to \infty$. This property has many restatements: $\alpha_{\max} = \infty$; almost surely, $H(\alpha) = 0$; the set of points x where $H(x) = \infty$ is of measure 1.

The problem of inferring the shape of f(α) from data. After $f(\alpha)$ has been specified analytically, one cannot rest. One must continue by asking whether or not this function can also be inferred when the mechanism of our dynamical system is unknown and only some empirical data are available. In this context, data may mean one of two things. It may denote the coarse (or coarse-grained or quantized) form of the function $M(x)$, as computed effectively for values of x restricted to be multiples of some quantum $\Delta x = \varepsilon$. Data may also denote a long orbit of the above dynamical system, that is, a long series of successive values of x; they too must be recorded in coarse-grained format. For many physical quantities in real space, coarse-graining is physically intrinsic; i.e., they are not defined on a continuous scale, but only for intervals whose length is a multiple of some $\Delta x = \varepsilon$ due to the existence of atoms or quanta; in other physical quantities, there are intrinsic limits to useful interpolation; for example, a turbulent fluid is locally smooth. In the present case, coarse-graining is the result of the necessary finiteness of actual computations and of observed orbits.

Given coarse data, there are at least two ways of seeking to extract or estimate $f(\alpha)$.

The method of moments as applied to the Minkowski distribution. The better-known way [10.5,6] towards $f(\alpha)$ deserves to be called the *method of moments*. It starts with the coarse-grained measures $\mu_\varepsilon(x)$ contained in successive intervals of length ε and proceeds as follows: a) evaluate the collection of moments embodied in the partition function defined by $\chi(\varepsilon, q) = \sum \mu_\varepsilon^q(x)$; b) estimate $\tau(q)$ by fitting a straight line to the data of $\log \chi(\varepsilon, q)$ versus $\log \varepsilon$, and c) obtain $f(\alpha)$ as the Legendre transform of $\tau(q)$.

When applied mechanically to the Minkowski μ, the method of moments either yields nothing or yields nonsense. More precisely, the more prudent mechanical implementations of the method do not fit a slope $\tau(q)$ without also testing that the data are straight (this can be done by eye). But the Minkowski data for $q < 0$ are not straight at all. Therefore, the prudent conclusion is that there is no $\tau(q)$. Since no such difficulty arises for $q > 0$, conclusions of this sort are is often accompanied by the assertion that the data are not quite multifractal. The less prudent mechanical implementations of the method of moments simply forge ahead to fit $\tau(q)$. Depending on a combination of the rule used to fit and of the details of how quantification is performed, those methods may yield an estimated $\tau(q)$ that is not convex. The resulting "Legendre transform" is not a single-valued function, and $f(\alpha)$ is a mystery. In other mechanical methods, the difficulty in estimating $\tau(q)$ is faced by first "stabilizing" the estimate in one way or another; such stabilization may yield some sort of $f(\alpha)$, but one can hardly say what it means and what purpose it serves.

A central feature of the method of moments should be mentioned at this point. The limit process $\varepsilon \to 0$ is invoked in estimating $\tau(q)$ from the data. But the preasymptotic data corresponding to $\varepsilon > 0$ do not define an approximate $f_\varepsilon(\alpha)$.

The method of distributions. A second way to estimate $f(\alpha)$ is the *method of distributions,* which is used in all my papers listed as references. I have been using it since 1974 and every new development motivates me to recommend it more strongly. The key is simple. While the method of moments rushes to compute the moments of $\mu_\varepsilon(x)$ embodied in the partition function $\chi(\varepsilon, q)$, the method of distributions considers, for every ε, the full frequency distribution of the $\mu_\varepsilon(x)$. These distributions are embodied in graphs statisticians call histograms.

First, the range of observed α's is subdivided into equal "bins" and one records the number of data in each bin. If the number of bins is too small, information is lost, but if there are too many bins, many are empty. In the case of the Minkowski μ, there are many α's a little above $\alpha_{\min}$ and few α's strung along up to very high values.

Denote by N_b the number of data in bin b. When N_b is large, $N_b/\Delta\alpha$ serves to estimate a probability density for α. When $N_b = 1$ and the neighboring bins are empty, one estimates probabilities by averaging over a suitably large number of neighboring bins; these probabilities are very small.

Having estimated the probability density $p_\varepsilon(x)$, one forms

$$f_\varepsilon(\alpha) = \frac{\log p_\varepsilon(\alpha)}{\log \varepsilon} + 1. \tag{10.31}$$

Thus, the method of distribution creates a sequence of functions $f_\varepsilon(\alpha)$. Because $f_\varepsilon(\alpha)$ is the normalized logarithm of a measure, each $f_\varepsilon(\alpha)$ is nothing but a histogram that was replotted in doubly logarithmic coordinates and was suitably weighted. These histograms should be evaluated for a series of values of ε. When the measure is multifractal, $f_\varepsilon(\alpha)$ converges to a limit $f(\alpha)$. That is, the function $f(\alpha)$ enters the theory as

$$f(\alpha) = \lim_{\varepsilon \to 0} f_\varepsilon(\alpha). \tag{10.32}$$

A physicist's typical reaction to histograms is, "Why bother? We all know that the information contained in the histograms is also contained in the moments; besides, moments organize information, and they are familiar and far easier to handle than histograms." Unfortunately, in the context of fractals and multifractals this typical reaction *won't* do.

In the study of fractals, the typical probability distributions are scaling (hyperbolic), and some of their population moments are infinite. The corresponding sample moments – sometimes even the sample average – behave in totally erratic fashion; they bring out no useful information and can be thoroughly misleading.

Now proceed to multifractals. When $f(\alpha)$ is truly ∩-shaped, with $f > 0$, moments raise no major issue, the method of moments works well, and the method of distributions is a less efficient way to obtain $f(\alpha)$. But in all delicate cases, the sample moments embodied in the partition function are treacherous. The method of distributions is the only way to go.

The method of distributions as applied to the Minkowski measure. In [10.30], Gutzwiller and I used histograms, and Figure 10.9 (which is copied from Figure 2 of [10.30]) reproduces the empirical $f_\varepsilon(\alpha)$ we obtained. To obtain this graph, we coarse-grained x, then (in effect) we coarse-grained M. The quantum of M was tiny, because it was simply the smallest $M(x + \varepsilon) - M(x)$ our

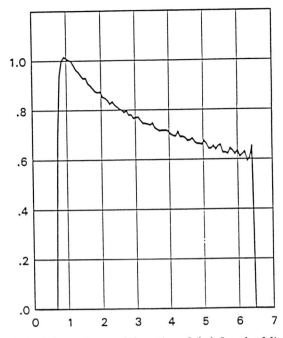

Fig. 10.9. An early plot of the estimated function $f_\varepsilon(\alpha)$ for the Minkowski measure. Reproduced from [10.30]

computer allowed in quadruple precision. Thus, the values of the computer could not distinguish from 0 (10% of the whole) were not used.

The resulting data-based curve is utterly different from the theoretical left-sided $f(\alpha)$. It begins with an unquestionably cap-convex left side — as usual. The middle part satisfies $f_\varepsilon(\alpha) > 1$, which cannot be true of $f(\alpha)$, but was expected; this is one of the inevitable biases of the method of distributions, and can be handled. Finally, there is cup-convex right side. This was totally unexpected, because a theoretical $f(\alpha)$ is necessarily cap-convex throughout.

We gave up seeking a better test of this cup-convexity. We did not come close to testing my further hunch, that the estimated $f_\varepsilon(\alpha)$ –if extended far enough– would become < 0 for large enough α. We showed that for $q < 0$ the moment $\chi(q, \varepsilon)$ was not a power law function of ε. But, to our disappointment, we did not succeed in evaluating $f(\alpha)$ analytically. We did conjecture the correct actual form of $f(\alpha)$ (but did not write it down) and were concerned by functions $f_\varepsilon(\alpha)$ and $f(\alpha)$ that differ to such extreme degree.

I have returned to this problem. Figure 10.10 was prepared using a method that does not compute $M(x)$ itself, but computes $M(x + dx) - M(x)$ directly. This can be done with arbitrary relative precision, therefore we can reach huge values of α. Figure 10.10 gives resounding confirmations of the earlier conjectures concerning the existence of a cup-convex right side in the empirical $f(\alpha)$ and of a negative tail.

This sharp mismatch between the theory and even the best experiments spurred me to a rigorous derivation of the theoretical $f(\alpha)$ and of the predicted

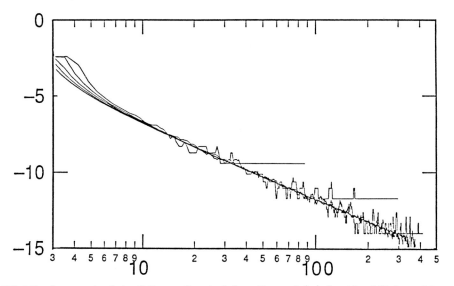

Fig. 10.10. A recent plot of the estimated functions $f_\varepsilon(\alpha)$ for the Minkowski measure restricted to the interval $[1/10, 1/9]$. We started with $f_\varepsilon(\alpha) - 1$ as the vertical coordinate and—in order to straighten $f_\varepsilon(\alpha)$—we chose $\log\alpha$ as the horizontal coordinate. Then we collapsed the five graphs that correspond to $\varepsilon = 10^{-5}, 10^{-6}, 10^{-7}, 10^{-8}$ and 10^{-9}. This figure strongly confirms the cup-convexity suspected in Fig. 10.9. It shows the occurrence of $f < 0$. Finally, the weighting rule shows that $1 - f_\varepsilon(\alpha)$ decreases as $\varepsilon \to 0$.

$f_\varepsilon(\alpha)$. The shape of $f(\alpha)$ has already been mentioned. For $f_\varepsilon(\alpha)$, it suffices to say that, for large ε,

$$f_\varepsilon(\alpha) \sim 1 - \text{ constant} \cdot \log\alpha/\log\varepsilon. \qquad (10.33)$$

Figure 10.10 verifies this dependence on the data.

Is f(α) a useful notion in the case of the Minkowski μ? Once again, our recent evaluations of $f_\varepsilon(\alpha)$ did not come close to reproducing the true shape of the graph of $f(\alpha)$, despite the fact that they involved precision that is totally beyond any conceivable physical measurement. Even the early evaluations [10.30] were well beyond the reach of physics.

Given the difficulties that have been described, should one conclude, in the case of the Minkowski measure, that $f(\alpha)$ is a worthless notion? Certainly this measure confirms that I have been arguing strenuously for a long time: that $f(\alpha)$ is a delicate tool. Its proper context is distributions, that is, probability theory. Moreover, it does not concern the best known and more "robust" parts of that theory, namely, those related to the law of large numbers and the central limit theorem. Instead, it concerns the probability theory of large deviations, which is a delicate topic.

10.C.3 Remark: On Continuous Models as Approximations, and on "Thermodynamics"

Why do physicists study limits that cannot be attained? Simply because it is often easier to describe a limit than to describe a finite structure that can be viewed as an approximation to this limit. In particular, this is why it is often taken for granted in the study of multifractals that a collection of "coarse-grained" approximations can be replaced by a continuous "fine" or "fine-grained" description. The latter involves Hausdorff dimensions and introduces the functions $\tau(q)$ and $f(\alpha)$ directly, not as limits.

For the Minkowski μ, however, the actual transition from coarse to fine-graining (as $\Delta x \to 0$ and $\Delta m \to 0$) is extraordinarily slow and many aspects of the limit differ qualitatively from the corresponding aspects of even close approximations. Therefore, the role of limits demands further thoughts, which I propose to describe elsewhere.

The continuous limit approximation has been described as ruled by a "thermodynamical" description. Thus, the fact that the convergence is slow and singular in the case of the Minkowski μ reveals a fundamental practical limitation of the thermodynamic description.

10.C.4 Remark on the Role of the Minkowski Measure in the Study of Dynamical Systems. Parabolic Versus Hyperbolic Systems

Minkowski ([10.37], Vol 2, p. 50-51), had called $M(x)$ the "?(x) function," which has few redeeming features. Little (if anything) was written about the measure μ until 1932, when Denjoy [10.41] observed that it has the following property. It is the restriction to $[.0, 1]$ of the attractor measure for the dynamical system on the line based on the maps

$$x \to \frac{1}{2} + \frac{1}{4(x - 1/2)}, \qquad x \to -x' \qquad \text{and} \qquad x \to 2 - x. \qquad (10.34)$$

To transform such a collection of functions into a dynamical system, the standard method is, of course, to choose the next operation at random. This method was used in Plates 198 and 199 of my book [10.3], and the current (and recent) term for it is IFS: *iterated function system.*

Thanks to this interpretation, $M(x)$ proves to have deep roots in number theory (modular functions) and in Fuchsian or Kleinian groups. From this paper's viewpoint, however, the main virtue of the above dynamical system lies in its extraordinary simplicity. I surmise that any complication or difficulty encountered in the study of its invariant measures will *a fortiori* appear in more complex systems grounded in physics. Moreover, one must keep in mind that the paper I wrote with Gutzwiller [10.30] had two motivations. I was concerned with the above maps, but he was concerned with an important Hamiltonian system in which x is the Liouville measure and m a second invariant measure yielding equally interesting information about individual trajectories.

To pinpoint the essential ingredient of the above special dynamical system, it is important to see how its properties change if the system itself is modified. If one wants $f(\alpha)$ to become intersect-shaped, it suffices to replace the first of our three maps by

$$x \to \frac{1}{2} + \frac{\rho}{x - 1/2}, \tag{10.35}$$

with $\rho < 1/4$. As $\rho \to 1/4$, the right side of $f(\alpha)$ lengthens and is pushed away to infinity, and the anomalies disappear asymptotically. Formally, the system changes from being hyperbolic to parabolic. Hence, the anomalies we have investigated are due to the system's parabolic. In terms of the limit $f(\alpha)$, the differences between parabolic and hyperbolic cases increase as $\rho \to 1/4$. But actual observations lie in a preasymptotic range; for a wide range of ε, $f_\varepsilon(\alpha)$ will be effectively the same for ρ close to $1/4$ as it is for $\rho = 1/4$.

10.C.5 In Lieu of Conclusion

The facts described in this appendix must not discourage the practically minded reader. Repeating once again a pattern that is typical of fractal geometry, it turns out that what had seemed strange should be welcomed and not viewed as strange at all.

Acknowledgements of Appendix 10.C. Since this work began in 1987, I had invaluable discussions with M. C. Gutzwiller, C. J. G. Evertsz, T. Bedford and Y. Peres. Figures 10.7, 8 and 10 were prepared by J. Klenk and D. Fracchia. Peres, who wrote his Ph.D. thesis on the Minkowski measure, informed me that, while both Gutzwiller and I had independently rediscovered μ before we joined forces to write about it in 1988, we had been anticipated by Minkowski and Denjoy. However, the $f(\alpha)$ function of μ was not discussed until our joint paper [10.30] .

References

10.1 B.B. Mandelbrot: J. Fluid Mech. **62**, 331 (1974)

10.2 B.B. Mandelbrot, in: *Statistical Physics 13*, ed. by D. Cabib, C.G. Kupper, I. Riess (Adam Hilger, Bristol 1977), p. 225

10.3 B.B. Mandelbrot: *The Fractal Geometry of Nature* (W.H. Freeman, New York 1982)

10.4 C.J.G. Evertsz and B.B. Mandelbrot, in: *Chaos and Fractals*, ed. by H.-O. Peitgen, H. Jürgens, D. Saupe (Springer, New York 1992); Reference [10.1] and other earlier papers are to be reprinted in B.B. Mandelbrot: *Multifractals and 1/f-Noise*, Selecta Vol. N (Springer, New York, forthcoming)

10.5 U. Frisch, G. Parisi, in: *Turbulence and Predictability of Geophysical Flows and Climate Dynamics*, ed. by M. Ghil, R. Benzi, G. Parisi (North-Holland, New York 1985), p. 84

10.6 T.C. Halsey, M.H. Jensen, L.P. Kadanoff, I. Procaccia, B.I. Shraiman: Phys. Rev. A **33**, 1141 (1986)

10.7 A. Aharony: Physica A **168**, 479 (1990)

10.8   T.A. Witten, L.M. Sander: Phys. Rev. Lett. **47**, 1400 (1981)

10.9   L. Niemeyer, L. Pietronero, H.J. Wiesmann: Phys. Rev. Lett. **52**, 1033 (1984)

10.10 C. Amitrano, A. Coniglio, F. di Liberto: Phys. Rev. Lett. **57**, 1016 (1986)

10.11 P. Meakin, in: *Phase Transitions and Critical Phenomena*, Vol. 12, ed. by C. Domb, J. Lebowitz (Academic Press, New York 1988), p. 335

10.12 J. Lee, H.E. Stanley: Phys. Rev. Lett. **61**, 2945 (1988)

10.13 C.J.G. Evertsz: *Laplacian Fractals* (Ph.D. thesis, Groningen 1989)

10.14 R. Blumenfeld, A. Aharony: Phys. Rev. Lett. **62**, 2977 (1989)

10.15 B.B. Mandelbrot, C.J.G. Evertsz: *Nature* **348**, 143 (1990)

10.16 B.B. Mandelbrot, C.J.G. Evertsz: Physica A **177**, 386 (1991)

10.17 C.J.G. Evertsz, P.W. Jones, B.B. Mandelbrot: J. Phys. A **24**, 1889 (1991)

10.18 R. Blumenfeld, Y. Meir, A. Aharony, A.B. Harris: Phys. Rev. B **35**, 3524 (1987)

10.19 T. Nagatani: J. Phys. A **20**, L381 (1987); Phys. Rev. A **36**, 5812 (1987); J. Phys. A **20**, L641 (1987); J. Phys. A **20**, 6135 (1987)

10.20 L. Pietronero, A. Erzan, C.J.G. Evertsz: Phys. Rev. Lett. **61**, 861 (1988) and Physica A **151**, 207 (1988)

10.21 B.B. Mandelbrot, in: *Fluctuations and Pattern Formation*, ed. by H.E. Stanley, N. Ostrowsky (Kluwer, Dordrecht 1988), p. 345; Pure and Appl. Geophys. **131**, 5 (1989)

10.22 B.B. Mandelbrot, in: *Fractals: Physical Origins and Properties*, ed. by L. Pietronero (Plenum, New York 1989)

10.23 B.B. Mandelbrot: Physica A **168**, 95 (1990)

10.24 B.B. Mandelbrot, C.J.G. Evertsz, Y. Hayakawa: Phys. Rev. A **42**, 4528 (1990)

10.25 H.G.E. Hentschel, I. Procaccia: Physica D **8**, 435 (1983)

10.26 S. Schwarzer, J. Lee, A. Bunde, S. Havlin, H.E. Roman, H.E. Stanley: Phys. Rev. Lett. **65**, 603 (1990)

10.27 M.H. Jensen, L.P. Kadanoff, A. Libchaber: I. Procaccia, J. Stavans: Phys. Rev. Lett. **55**, 2798 (1985)

10.28 B.B. Mandelbrot, in: *Frontiers of Physics*, ed. by E. Gotsman et al (Pergamon, New York 1989), p. 91

10.29 B.B. Mandelbrot, in: *Chaos in Australia*, ed. by G. Brown (World Scientific, Singapore 1993), p. 83

10.30 M.C. Gutzwiller, B.B. Mandelbrot: Phys. Rev. Lett. **60**, 673 (1988)

10.31 B.B. Mandelbrot: Physica A **191**, 95 (1992)

10.32 B.B. Mandelbrot, H. Kaufman, A. Vespignani, I. Yekutieli, C.-H. Lam: Europhys. Lett. **29**, 599 (1995)

10.33 B.B. Mandelbrot, A. Vespignani, H. Kaufman: forthcoming

10.34 R. H. Riedi, B. B. Mandelbrot: *Advances in Appl. Math.* **16**, xxx (1995)

10.35 M. Arbeiter, N. Patzschke: Self-Similar Random Multifractals, to appear in *Math. Nachr.* (1995)

10.36 R. Cawley, R. D. Mauldin: *Adv. Math.* **92**, 196 (1992)

10.37 H. Minkowski: *Gesammelte Abhandlungen* Vol 2, pp 50-51 (1911) (Chelsea, New York, reprint)

10.38 B. B. Mandelbrot: forthcoming

10.39 R. Salem: *Trans. Am. Math. Soc.* **53**, 427 (1943); reprinted in R. Salem: *Oeuvres Mathématiques,* pp. 282–294 (Hermann, Paris 1967)

10.40 J. R. Kinney: *Proc. Am. Math. Soc.* **11**, 788 (1960)

10.41 A. Denjoy: *Comptes Rendus* (Paris) **194**, 44 (1932); *Journal des Mathématiques Pures et Appliquées* **62**, 105 (1938); reprinted in A. Denjoy: *Articles et Mémoires,* **2** 1955, 925 (1955)

Subject Index

A. **Bunde, S.Havlin** (Eds.)

Fractals in Science

1st ed. 1994. 2nd printing 1995. XVI, 298 pp. 120 figs.,
10 color plates, Hardcover

Macintosh version **System requirements**: 3 1/2" Macintosh program diskette, System 6 or 7, b/w monitor 3MB RAM, 1 program needs a 256 color display. ISBN 3-540-56221-4

MS-DOS version **System requirements:** MS-DOS program diskette, EGA or VGA graphics card, 16 colors and 1 MB RAM, MS-DOS 3.30 or higher; a 386 DX or 486 DX processor with a numerical coprocessor is recommended but not necessary. ISBN 3-540-56220-6

The fractal concept has become an important tool for understanding irregular complex systems in various scientific disciplines. This book discusses in great detail fractals in biology, heterogenous chemistry, polymers, and the earth sciences. Beginning with a general introduction to fractal geometry it continues with eight chapters on self-organized criticality, rough surfaces and interfaces, random walks, chemical reactions, and fractals in chemistry, biology, and medicine. A special chapter entitled "Computer Exploration of Fractals, Chaos, and Cooperativity" presents 14 programs of fractal models.

Springer

Springer-Verlag, Postfach 31 13 40, D-10643 Berlin, Fax 0 30 / 82 07 - 3 01 / 4 48   e-mail: orders@springer.de

tm.BA95.11.06